AF404727

Nuclear Data

An independent-particle motion view

Online at: https://doi.org/10.1088/978-0-7503-5648-0

IOP Series in Nuclear Spectroscopy and Nuclear Structure

Series Editors

John Wood, Georgia Institute of Technology, USA
Kristiaan Heyde, Ghent University, Belgium
David Jenkins, University of York, UK

About the Series:
The IOP Series in Nuclear Spectroscopy and Nuclear Structure provides up-to-date coverage of theoretical and experimental topics in all areas of modern nuclear spectroscopy and structure. Books in the series range from student primers, graduate textbooks, research monographs and practical guides to meet the needs of students and scientists.

All aspects of nuclear spectroscopy and structure research are included, for example:
- Nuclear systematics and data
- Ab-initio models
- Shell model-based descriptions
- Nuclear collective models
- Nuclear symmetries and algebraic descriptions
- Many-body aspects of nuclear structure
- Quantum mechanics for nuclear structure study
- Spectroscopy with gamma-rays, charged particles and neutrons following radioactive decay and reactions
- Spectroscopy of rare isotopes
- Detectors employed for gamma-ray, neutron and charged-particle detection in nuclear spectroscopy and related societal applications.

Nuclear Data

An independent-particle motion view

David Jenkins

Department of Physics, University of York, Heslington, York YO10 5DD, United Kingdom

John L Wood

Department of Physics, Georgia Tech, Atlanta, GA, USA

IOP Publishing, Bristol, UK

ISBN   978-0-7503-5648-0 (ebook)
ISBN   978-0-7503-5646-6 (print)
ISBN   978-0-7503-5649-7 (myPrint)
ISBN   978-0-7503-5647-3 (mobi)

DOI   10.1088/978-0-7503-5648-0

Multimedia content is available for this book from https://doi.org/10.1088/978-0-7503-5648-0.

Version: 20241001

IOP ebooks

British Library Cataloguing-in-Publication Data: A catalogue record for this book is available from the British Library.

Published by IOP Publishing, wholly owned by The Institute of Physics, London

IOP Publishing, No.2 The Distillery, Glassfields, Avon Street, Bristol, BS2 0GR, UK

US Office: IOP Publishing, Inc., 190 North Independence Mall West, Suite 601, Philadelphia, PA 19106, USA

Contents

4 How are Nilsson configurations affected by rotations?   4-1

5 How are Nilsson states manifested in even-mass nuclei?   5-1

Preface

In seeking useful classifications of nuclear data by structural behaviour, the independent-particle motion view has an historical precedent. However, the achievements of descriptions of nuclear data strictly using independent-particle degrees of freedom have proven to be limited. In this volume we explore this limitation.

We begin the view with nuclei that possess zero deformation, usually termed spherical nuclei. These nuclei are centred on regions of doubly closed shells. For these nuclei the historic shell model plays a major role. Thus, the structure involved is described as independent-particle motion in a spherical mean field.

The existence of deformation in many nuclei naturally leads to consideration of independent-particle degrees of freedom in deformed potentials. This comes under the title of the Nilsson model. This is an extraordinarily successful model which, while not providing a quantitative description of data with high precision, organizes enormous amounts of data with only rare ambiguity. However, there is a proviso: deformed nuclei possess rotational degrees of freedom, and these cannot be completely separated from the independent-particle degrees of freedom of the Nilsson model. Details of this are handled herein, beyond some of the features already described in *Nuclear Data: A Collective Motion View*.

We retain the use of questions in our chapter titles. We will see that at a most basic level there are important questions that lack answers. Again, we encourage the reader to be skeptical of every interpretation; and to remain aware that the database is often deficient with respect to addressing basic questions.

A key issue in exploring independent-particle motion in nuclei is spectroscopic 'fingerprinting' of such a degree of freedom. The most direct spectroscopic probes are 'knockout' of protons by high-energy electron beams and one-nucleon transfer reactions using light-ion beams. Nearly all the useful data are from 30 to 60 years ago. Identification of collectivity associated with supposed independent-particle states in nuclei can be made by determining E2 strengths for excitation or decay of these states. There are some decay data (lifetime measurements) available; there are very few excitation data (Coulomb excitation) available. Therefore, our treatment of independent-particle motion in nuclei borders on the research frontier in many of the details.

The identification of independent-particle motion in deformed nuclei benefits enormously from the Nilsson model and clear patterns of rotational bands built on 'Nilsson' states. One-nucleon transfer reactions populate 'Nilsson' bands with characteristic patterns, so called transfer-reaction 'fingerprinting' of Nilsson bands. The details are more technical and can be avoided by the newcomer to this issue of nuclear structure because of the robust systematic maps that are available for Nilsson states and associated rotational bands. There are deviations from the Nilsson state maps: again, these lie at the research frontier.

An arena that manifestly resides at the research frontier is the question of the borderline between independent-particle motion in spherical nuclei versus in deformed nuclei. This issue devolves onto weakly deformed nuclei and the involvement of the rotational motion. Our treatment of the issue is to essentially isolate it as a topic for future treatment and, in our view, the need for substantial research.

Author biographies

David Jenkins

David Jenkins is Head of the Nuclear Physics Group at the University of York, UK. He is also a Fellow of the Institute of Advanced Study, University of Strasbourg (USIAS) and an Extraordinary Professor of the University of Western Cape in South Africa. His research in experimental nuclear physics focusses on several topics such as nuclear astrophysics, clustering in nuclei and the study of proton-rich nuclei. In recent years, he has developed a strong strand of applications-related research with extensive industrial collaboration. He has led the development of bespoke radiation detectors for homeland security, nuclear decommissioning, borehole logging and medical applications.

John Wood

John Wood is a Professor Emeritus in the School of Physics at Georgia Institute of Technology. He continues to collaborate on research projects in both experimental and theoretical nuclear physics. Special research interests include nuclear shapes and systematics of nuclear structure.

IOP Publishing

Nuclear Data
An independent-particle motion view
David Jenkins and John L Wood

Chapter 1

Where are the shell model states found in nuclei?

*Independent-particle motion in nuclei is introduced in the guise of shell model configurations; this is distinct from the shell model as a theory of nuclear structure where it is a quantum mechanical **basis** with model interactions. The focus is on nuclei at and adjacent to doubly closed shells. The examples pinpoint in which nuclei shell model states are approximately realized, and how the shell model breaks down.*

Concepts: shell model configurations, quantum numbers N, l, j, $\Omega(=m_j)$, shell model energy ordering, magic numbers, spin–orbit coupling, potential well flattening, particle states, hole states, pairing, quasi-particle, single-particle strength, spectroscopic factor, knockout reaction, one-nucleon transfer reaction, two-nucleon transfer reaction, particle–hole states, pair-correlated states, core-particle coupling, monopole migration, magnetic moment, Schmidt diagrams, effective charges, intruder states, fragmentation of j configurations.

Learning outcomes: The key data view from this chapter is the limitation of the shell model beyond a qualitative basis for interpreting states in nuclei adjacent to double-closed shells.

Asking in which nuclei shell model states are found appears to attack the foundations on which nuclear models are built. It is important to realize that 'the shell model' plays two basic roles in nuclear structure study. First, it is a model description of states in nuclei, which has its limitations. Second, it is a quantum mechanical basis for the description of protons and neutrons in nuclei. Specifically, it is an independent-particle basis: this is applied by taking A copies of the basis to describe a nucleus of mass A, ordered by the energies of the shell model, as a direct product subject to the applicable symmetries, e.g. identical fermions demand antisymmetric products.

A widely used depiction of the ordering of shell model energy levels is presented in figure 1.1. We emphasize, nuclei that might be considered to exhibit states manifesting shell model character must be handled with caution when making

doi:10.1088/978-0-7503-5648-0ch1　　　1-1　　　

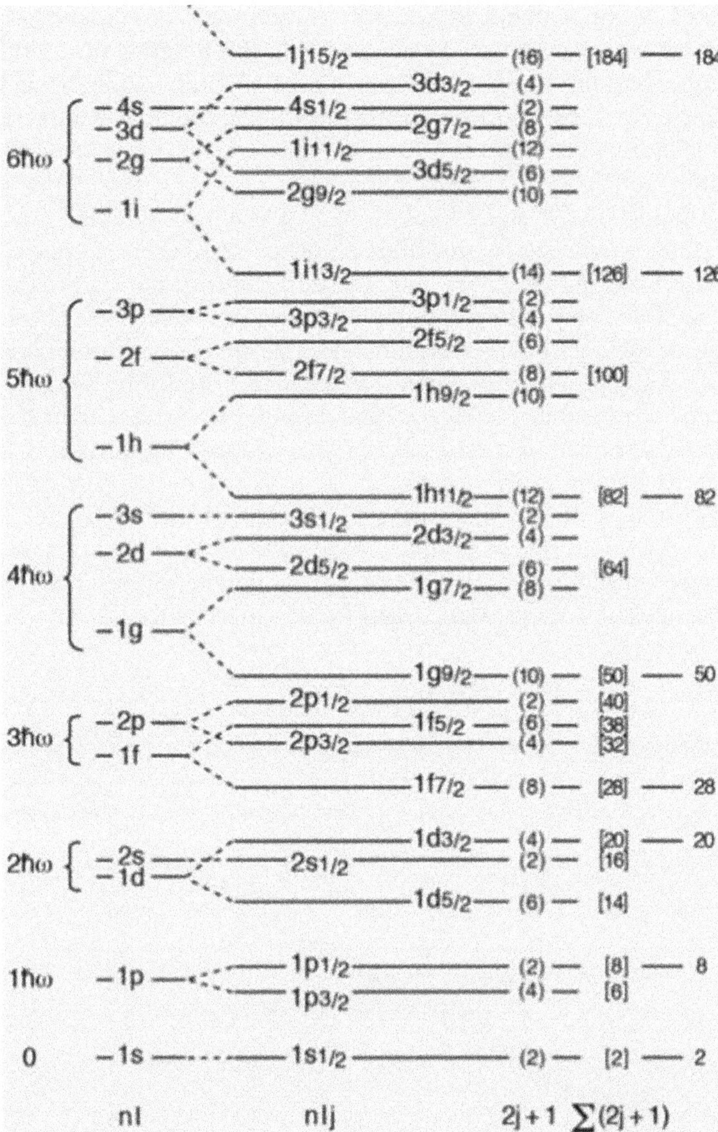

Figure 1.1. A shell model energy level diagram. The pattern of levels is only qualitative: it adopts an isotropic harmonic oscillator central potential, it recognizes the effect of 'well flattening', it incorporates a large empirically determined spin–orbit splitting. States are labelled with quantum numbers dictated by spin–orbit coupling, i.e. $j = l + s$, with n labelling the occurrence of each l value. Parity is determined as $\pi = (-1)^l$. Degeneracies of the j subshells, where $m_j = +j, +j - 1, \ldots, -j$, i.e. $(2j + 1)$ states per subshell, are indicated. The cumulative filling numbers of nucleons, separately for protons and neutrons, are given in square brackets. Energies are shown in units of oscillator frequency, ω multiplied by Planck's constant divided by 2π, $\hbar$. Reproduced from [1], copyright (2010) World Scientific Publishing Company.

interpretations because of the presence of correlations: this applies especially to where observed energies can only be equated with this diagram in carefully selected nuclei. Energies of states labelled with shell model quantum numbers, such as shown in the following figures, must be treated with considerable care when making inferences regarding energies of states that appear to match shell model expectations.

The shell model, as its name implies, had its origin in attempts to explain the observed (energy) shells or so-called 'magic numbers' in nuclei, viz. N, $Z = 2, 8, 20, 28, 50, 82$ and $N = 126$. The early history of this process is largely forgotten; but following the recognition that certain nucleon numbers appear to possess extra stability [2], some interesting ideas for shapes of central potentials were explored (see, e.g. [3]). The realization that what was needed lay entirely outside of the refinement of potential shapes, i.e. that strong spin–orbit coupling was needed [4, 5], could be described as a classical example of 'thinking outside of the box'.

The place to start in answering the question posed by the title of this chapter would appear to be a study of nuclei at and near closed shells. Figure 1.2 shows the nuclei established as exhibiting doubly closed shells. Figure 1.3 is a schematic view of doubly closed shell nuclei that are accessible to detailed study. The shaded boxes are the stable nuclei, which provide targets for one-nucleon transfer reaction study of

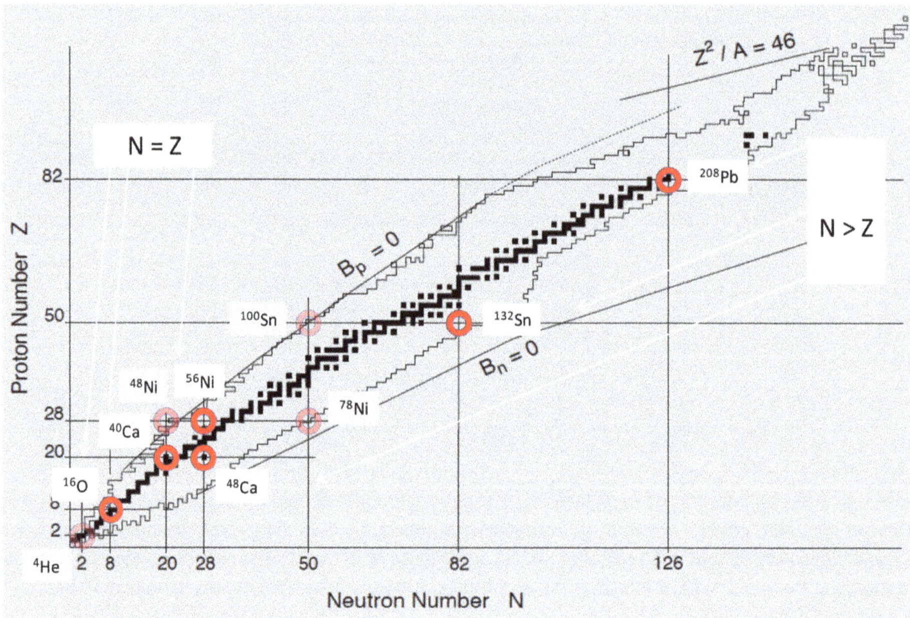

Figure 1.2. Chart of the nuclides showing the locations of nuclei with doubly closed shells. A lighter shading is used for ^{78}Ni and ^{100}Sn, two nuclei for which first data are now being obtained and where detailed structure can be anticipated soon. A lighter shading is also used for ^{4}He and ^{48}Ni: it is experimentally established that ^{4}He does not possess bound excited states; data are lacking for excited states in ^{48}Ni and it is likely that this nucleus does not possess bound excited states. Reproduced from [1], copyright (2010) World Scientific Publishing Company.

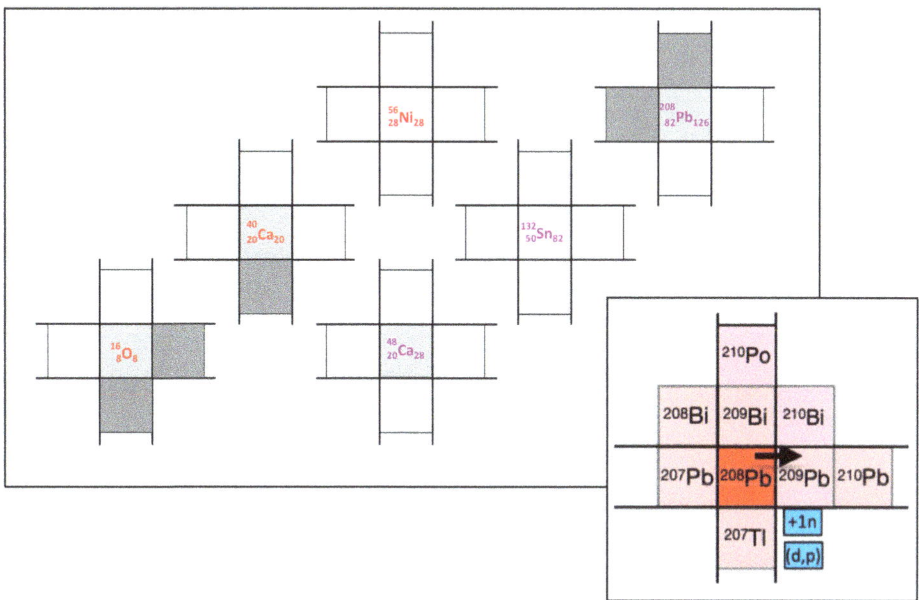

Figure 1.3. Schematic view of the doubly closed shell nuclei that are accessible to detailed study. The shaded isotope boxes indicate the nuclei that have stable isotopes and are available as target nuclei for one-nucleon transfer reactions. The lighter (heavier) shading indicates the stable isotopes available for one-nucleon transfer into odd-mass (even-mass) isotopes. The inset highlights nuclei that are discussed in this chapter in the ^{208}Pb region, with the example of a (d,p), one-neutron addition reaction indicated. Note that there are also some nucleon transfer reaction studies in the region of ^{208}Pb that use radioactive targets, specifically, targets of ^{210}Po [6], ^{210}Bi [7] and ^{210}Pb [8, 9]. See a video-based tutorial on the concept of one-nucleon transfer reactions in figure 1.39.

the isotopes at and adjacent to the double-closed shells: one-nucleon transfer reaction spectroscopy is a recurrent theme in the illustrations used in this chapter because it plays a critical role in identifying shell model configurations. A perusal of the spins and parities of states in odd-mass nuclei adjacent to double-closed shells compared to figure 1.1 suggests that the shell model is a successful description of nuclei adjacent to double-closed shells, cf figures 1.4–1.6. But already a surprise is evident. Shell model states from adjacent shells are observed at low energies. They appear to contradict the concept of an energy gap in association with the closed shells. The explanation of the unexpected low energies is given in figure 1.7. The presence of correlations must always be considered when assessing the evidence for independent-particle degrees of freedom in nuclei. Figure 1.7 also provides a schematic view of the various 'particle' and 'hole' excitations encountered in nuclei.

The simplest probe of independent-particle degrees of freedom in nuclei is to knock protons out of a nucleus using high-energy electrons. By using very high energy ($\sim$300 MeV) electrons, the process is simple: it is a pure electromagnetic interaction, and it is quasi-elastic, i.e. there is almost no change in the electron energy because the proton binding energies are relatively weak. Further, by using doubly closed shell target nuclei, correlations (whatever they may be) might be

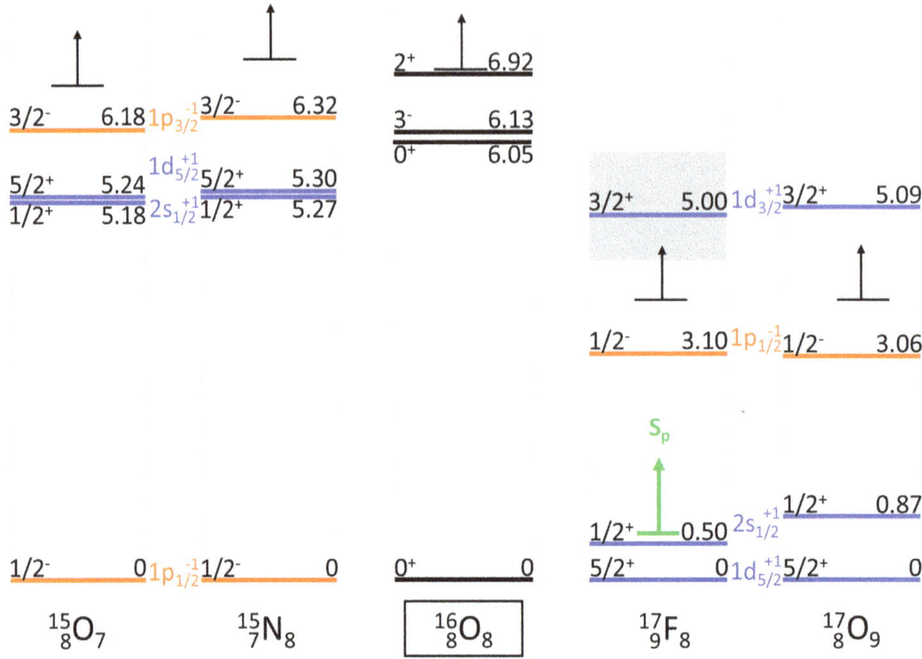

Figure 1.4. Interpretation of low-lying excited states in nuclei adjacent to the doubly closed shell with $N = Z = 8$, together with doubly closed shell, ^{16}O 'core' excitations. The shell model quantum number assignments N, l, j (as a subscript), and whether particle $(+1)$ or hole (-1) as superscripts, are shown (in blue —particles, in orange—holes). Note that these are naïve assignments, based on simple matching of spin-parity with shell model expectations, with no consideration of whether the states contain mostly independent-particle strength (see text). The energy threshold in ^{17}F for proton-unbound character of states is shown as a horizontal bar with vertical arrow and the label 'S_p' (in green), determined from the proton separation energy. Note that although these states are proton-unbound, the protons are confined by a Coulomb barrier; except the $3/2^+$, 5.00 MeV state, which has a large energy width (1.53 MeV), indicated by a grey shaded area. Energies above which states are omitted are indicated, also by horizontal bars with vertical arrows. The data are taken from ENSDF.

minimal. The result is shown in figure 1.8. The answer is that what we term an 'independent-particle' (proton) is never more than 70% of the story. This raises two questions. Where has the other 30+% of the independent-particle character gone? What has replaced the independent-particle character? The answer to both these questions is: we don't know. Indeed, quasi-elastic electron scattering knockout of protons from nuclei adds a further warning to the idea that the shell model is applicable to nuclei in its basic form: it reveals that proton numbers corresponding to doubly closed shells show significant occupancy of orbitals above the closed shell, i.e. the shells are not completely 'closed'. This is shown in figure 1.9.

Spectroscopic strengths or spectroscopic factors, S or S_ν, for a specified shell model configuration ν, are a much-debated quantity with respect to one-nucleon knockout and transfer reaction processes involving nuclei. Here, we note that it is a measure of the degree to which a state conforms to a naïve shell model

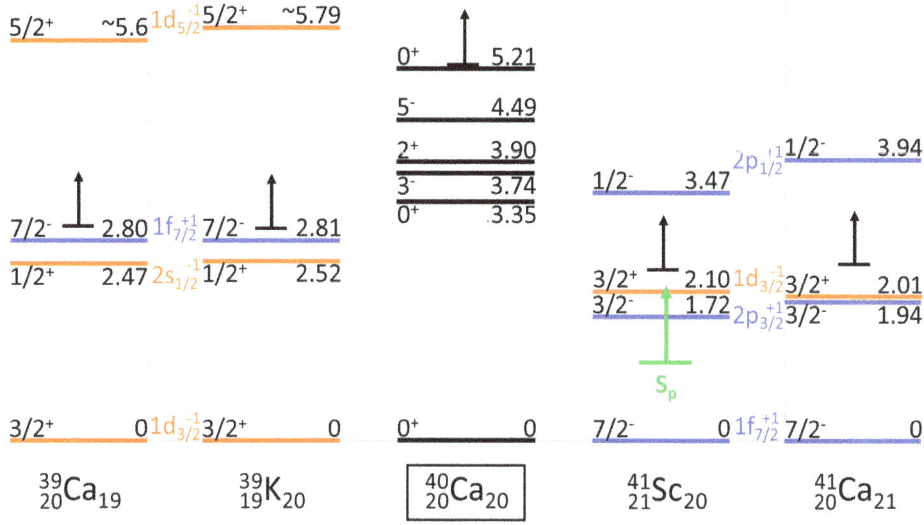

Figure 1.5. Interpretation of low-lying excited states in nuclei adjacent to the doubly closed shell with $N = Z = 20$, together with doubly closed shell, ^{40}Ca 'core' excitations. For an explanation of the labels, see the caption to figure 1.4. Energies of states with spin-parity 5/2⁺ in ^{39}Ca and ^{39}K are given with approximate energies because of the severe fragmentation of single-hole strength in these nuclei (discussed later). The data are taken from ENSDF.

configuration, j, l; $\{j, l\} = \nu$. In an independent-particle model view, such a state should exhibit an 'occupancy' ranging from zero to $2j + 1$. Where 'fragmentation' of strength occurs in contradiction to naïve expectations, energy-weighted sums can be constructed (see exercises).

The evidence that the shell model is never realized in nuclei in its most elementary form may not be as serious as it appears. The shell model can be viewed as an 'effective' theory, i.e. one can retain its simplicity by using effective two-body interactions between nucleons and effective charges for electromagnetic processes (and effective weak charges[1] for beta decay). But we underline that the use of effective charges for electric quadrupole properties of nuclei, quadrupole moments

---

[1] Just as nuclei couple to electromagnetic fields through the electrical charge of the protons, so they also couple to weak interaction fields through weak charges of the quarks in the protons and neutrons. Dynamically, leptons couple to nuclei through weak fields, although if the leptons are charged, electromagnetic interactions dominate. Coupling to the electromagnetic vacuum controls spontaneous radiative decay of excited nuclear states; coupling to the weak-field vacuum controls beta decay. (High-energy electron scattering is also beginning to probe weak interactions between electrons and neutrons in nuclei, see [14].)

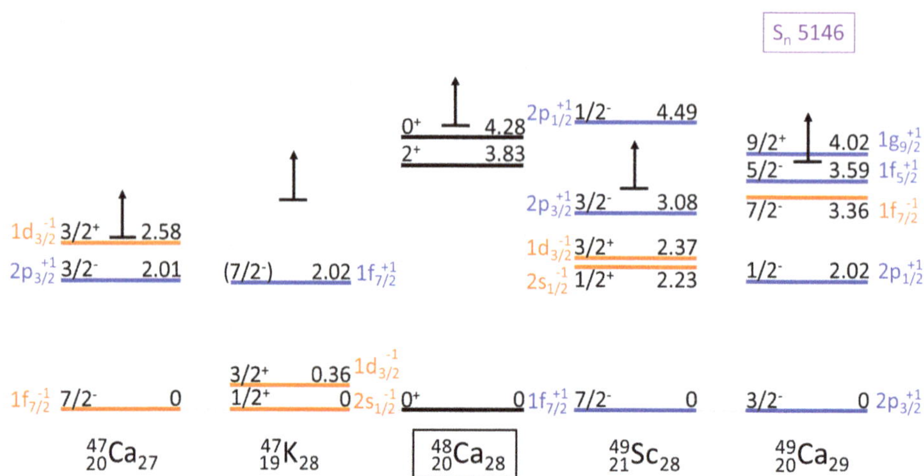

Figure 1.6. Interpretation of low-lying excited states in nuclei adjacent to the doubly closed shell with $N = 28$, $Z = 20$, together with doubly closed shell, ^{48}Ca 'core' excitations. For an explanation of the labels, see the caption to figure 1.4. The data are taken from ENSDF and from [10] for ^{47,49}Ca.

and E2 transition strengths, abandons any insight into the origin of quadrupole collectivity in nuclei[2].

Three primary types of data can be used for tests of the shell model as an effective theory: one-nucleon transfer reaction data, electromagnetic moment data, and electromagnetic decay data. We introduce these types of data as tests of the shell model. We state the proviso: a model is only as good as its failures. By this we mean that a model is only a 'shadow' of what is really happening in a many-body system, so it saves much time if one can find basic features of a model where failure is manifested.

---

[2] On this point there is a parallel between a shell model view of the nucleus and the so-called Standard Model view of particles and fields. The Standard Model has proven amazingly successful in the description of all observations involving particles (hadrons and leptons) and fields (electromagnetic, weak, strong); but it involves a plethora of parameters (particle masses, field coupling strengths, flavour mixing strengths) the origin of which is completely unknown. Lepton and quark masses presumably stem from the Higgs field, but the coupling strengths are not yet determined, and they will incur new parameters. We speculate that in the hadronization of quarks, it is possible that correlations involving virtual quark–antiquark pairs and deformation play a role, as particle–hole excitations and deformation play a role in nuclei. Hadron masses are >99% gluon fields; the Higgs field is not involved in the explanation of most of the 'mass' in the Universe. We especially note the hadronization of two up quarks and one down quark to form a proton is an extraordinarily complex process that is a current on-going topic of exploration that defies all preconceived notions of how quarks form a proton [15, 16].

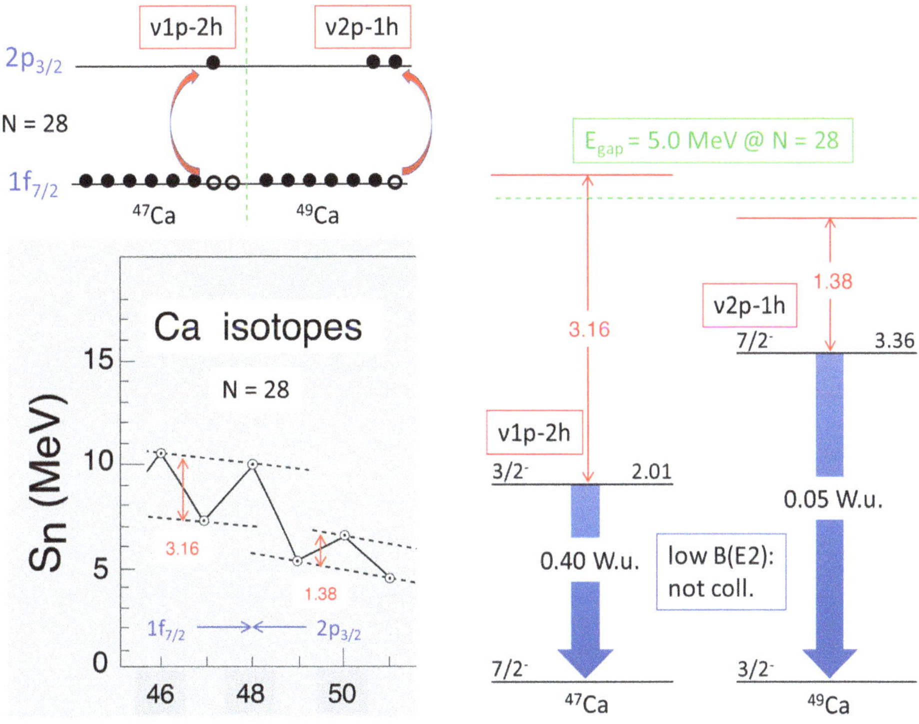

Figure 1.7. Intruder states in ^{47,49}Ca. The unexpectedly low energy of the $3/2^-$ state in ^{47}Ca and the $7/2^-$ state in ^{49}Ca, which are not collective excitations of the ground states in these nuclei (as indicated by the very small B(E2) values), can be understood as resulting from pairing correlations. The pairing correlation energy gain is explained using the information on the left-hand side of the figure. Thus, from a direct adoption of the odd–even staggering of S_n values, pairing correlation energies of 3.16 and 1.38 MeV, respectively, lower the energies of the 1p-2h state in ^{47}Ca and the 2p-1h state in ^{49}Ca by these amounts from an 'uncorrelated' energy of ∼5.0 MeV. The figure is reproduced, with updated information, from [11]. The data are taken from ENSDF and for ^{49}Ca are taken from [10]. See a video-based tutorial on the concept of intruder states in figure 1.42.

The approach of seeking how models fail was pursued for the rotor model [17], where the concept of an intrinsic quadrupole moment of a set of states (a rotational band) in a specified nucleus was found to be tenable at the level of current experimental precision, but the concept of a moment of inertia needs in-depth investigation. We recognize that adherents to a particular model view of nuclear structure may go beyond these bounds; that is a choice. The typical reader needs a more skeptical introduction, then they can make their own choice of which models to pursue. We point to the above data for quasi-elastic one-proton knockout reactions: already we are forced to adopt an effective shell model perspective.

In assessing the applicability of the shell model in its simplest terms to nuclei, we confine the narrative to the following: one-nucleon transfer reactions leading to excited states in odd-mass nuclei adjacent to doubly closed shells and magnetic dipole moments of nuclear ground states in odd-mass nuclei adjacent to doubly closed shells.

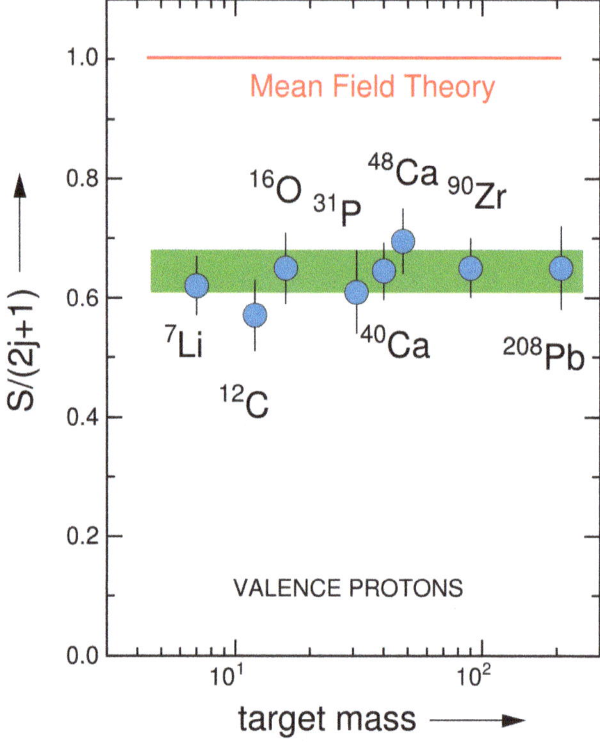

Figure 1.8. Spectroscopic strength, S for the valence orbitals extracted from exclusive A(e,e′p) reactions as a function of the mass number, A, of the target nucleus. Details are discussed in the text. The figure is taken with permission from [12] and is an adaptation of a figure from [13].

Shell model states in a doubly closed shell nucleus are considered for ^{208}Pb, and brief details are given for ^{48}Ca and ^{132}Sn. The doubly closed shell nuclei with $N = Z$, ^{16}O, ^{40}Ca, ^{56}Ni, possess low-energy shape coexistence as presented in [11], and so do not exhibit simple shell model structures (there is configuration mixing). Shell model states for odd–odd nuclei adjacent to doubly closed shells, e.g. ^{208}Bi are briefly detailed.

Nuclei with singly closed shells and nuclei adjacent to singly closed shells are considered in the next chapter.

1.1 Basic features of the model

The simplest model that describes the motion of independent particles in nuclei uses an isotropic harmonic oscillator potential and a spin–orbit coupling term. The oscillator potential must be modified and there is a simple prescription for this which involves including a term that depends on angular momentum. Thus, we consider a shell model Hamiltonian of the form

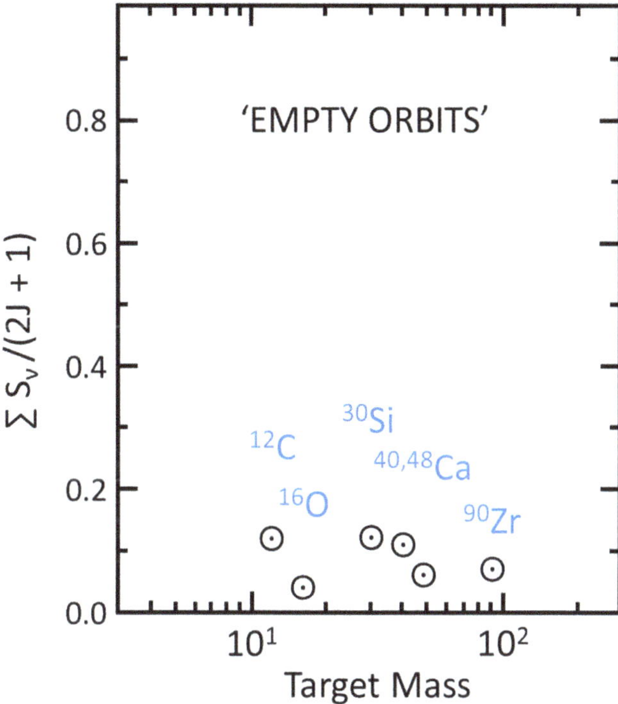

Figure 1.9. Quasi-particle strength, $\sum S_\nu/(2j+1)$ for states just above the Fermi surface, observed in the reaction (e,e′p) as a function of the mass of the target nucleus. All strengths were integrated to an excitation energy of about 20 MeV. Details are discussed in the text. The figure is an adaptation of one appearing in [13].

$$H = \frac{p^2}{2m} + \frac{m\omega^2 r^2}{2} + C\mathbf{l}\cdot\mathbf{s} + Dl^2, \tag{1.1}$$

where protons and neutrons are treated as particles of mass m, momentum p, radial position r, angular momentum l, spin s; ω, C, D are model parameters. The parameter ω is related to the size of the nucleus, which is fixed by the relationship $R = R_0 A^{1/3}$, $R_0 = 1.2$ fm, where A is the mass of the nucleus; it is called the oscillator frequency. The parameters C and D are determined to be negative by comparison with data.

The shell model Hamilton, equation (1.1) is exactly solvable in the basis $|Nlj\Omega\rangle$, $\Omega := m_j$

$$H|Nlj\Omega\rangle = E|Nlj\Omega\rangle, \tag{1.2}$$

whence

$$E = (N + 3/2)\hbar\omega + E_{ls} + Dl(l+1), \tag{1.3}$$

where

$$\begin{aligned}
E_{ls} &= +Cl\hbar^2/2, & j &= l + 1/2, \\
&= -C(l+1)\hbar^2/2, & j &= l - 1/2,
\end{aligned} \tag{1.4}$$

and standard spin-angular momentum quantum numbers are used. Because of the spin–orbit coupling term, m_l and m_s quantum numbers are mixed; but $m_l + m_s = m_j = \Omega$ is a good quantum number. The oscillator frequency is fixed by

$$\langle Nlj\Omega|\sum_{i=1}^{A}r_i^2|Nlj\Omega\rangle/A = R^2, \tag{1.5}$$

which yields the relationship

$$\hbar\omega = 41A^{-1/3} \text{ MeV}. \tag{1.6}$$

Typical values of C and D are -0.5 MeV and -0.15 MeV, respectively.

The literature provides variations[3] on the above model prescription; however, data reveal effects that cannot be accommodated by any of these model prescriptions and major examples are presented shortly. Further, the above basis is universally employed in all formulations of the shell model, and it possesses mathematical properties that are highly attractive when considering the symmetries manifested in the nuclear many-body problem. The focus herein is on a data view of independent-particle motion in nuclei and the above model prescription is well suited to this task.

1.2 What constitutes a shell model state?

In adopting a data-based view of nuclei, it is important to define what is meant by 'a shell model state' (as opposed to a configuration that constitutes an element of a shell model basis). A shell model state is an independent-particle state: this is a very restricted view of the degrees of freedom of protons and neutrons in the nucleus. The closest manifestation to such a state can be expected to appear at and near closed shells in nuclei.

The study of nuclei is limited to observable quantities. Energies, spin-parities, mean-square charge radii, magnetic dipole moments, electric quadrupole moments, electric hexadecapole moments, constitute what are termed diagonal moments, i.e. expectation values of operators. These are distinct from transition moments as deduced from Coulomb excitation and inelastic scattering; and as mediated by gamma decay, alpha decay, and beta decay. None of these observables are directly related to independent-particle behaviour of nucleons; but model-based estimates can be made. However, one-nucleon knockout reactions and one-nucleon transfer reactions play a key role in exploring independent-particle degrees of freedom in nuclei.

Knockout reactions such as (e,e'p) and transfer reactions such as (d,^{3}He), (^{3}He,d), (d,t), (d,p) (there are many more possibilities) provide a wide range of data for

---

[3] A widely used shell model prescription is a so-called Woods–Saxon potential, which obviates the need for the Dl^2 term. The Woods–Saxon potential also has a finite depth, unlike the oscillator potential which is infinitely deep. Other prescriptions replace the spin–orbit coupling term with a form that includes a factor which is a derivative of the potential. For the reader with more specialized interests, we refer to [18]. While a potential with finite depth is of concern for weakly bound states in nuclei, there is no evident difference relative to the oscillator potential for the data considered herein. The same qualification applies to prescriptions for the spin–orbit term and its possible dependence on the shape of the potential.

assessing the manifestation of independent-particle (and independent-hole) degrees of freedom in nuclei. However, there is a lack of consensus on whether such data constitute an observable related to particle and hole degrees of freedom in nuclei. Always, nucleon motion in the nucleus takes place in a modified environment. This environment is created by all the other nucleons, and notably by the correlations that exist in nuclear many-body systems. Sometimes, such nucleon degrees of freedom are termed 'dressed'. Further, these reactions incur so-called entrance channel and exit channel effects, e.g. Coulomb excitation of the residual nucleus by the exiting proton in the (e,e'p) reaction. Corrections can be applied for entrance and exit channel effects: these are termed coupled-channel descriptions of the reaction process. Indeed, models of the reaction process underlie all such spectroscopic probes of the nucleus. The present treatment focuses on what is *observed* in these reaction processes, without addressing details of the models describing the reaction processes. Models of the reactions used to characterize individual states in nuclei is a more advanced topic that lies beyond the present level of discussion. We underline that the present approach is a *view* of nuclear data.

1.3 Nuclei in doubly closed shell regions

Doubly closed shell nuclei can be expected to provide the 'focal point' where independent-particle (and independent-hole) degrees of freedom are most clearly manifested. Thus, the nuclei that differ by one particle or one hole from a doubly closed shell are of primary interest. The nuclei that differ from a doubly closed shell by two particles, two holes, or a particle and a hole must be viewed with the proviso that already two-body correlations will arise but may be amenable to descriptions using simple two-body interactions in a shell model basis.

Doubly closed shell nuclei with $N = Z$, notably ^{16}O and ^{40}Ca, must be treated with caution because the first excited states in these nuclei are strongly deformed. Thus, the neighbouring odd-mass nuclei will exhibit coexistence of spherical and deformed states. This brings the focus to nuclei in the regions of the doubly closed shell nuclei ^{48}Ca, ^{132}Sn, and ^{208}Pb.

1.3.1 Odd-mass nuclei

The nucleus ^{208}Pb has been the subject of one of the most detailed spectroscopic studies ever performed [19]. The adjacent odd-mass nuclei have all been studied sufficiently well for present purposes. Thus, we begin our look at data for these nuclei. Figure 1.10 shows the low-lying states in ^{208}Pb and its odd-mass neighbours, ^{207}Tl, ^{207}Pb, ^{209}Pb, and ^{209}Bi. Figures 1.11(a) and (b) compare the population of states in ^{207}Tl by the (e,e'p) and (t,α) reactions. Figure 1.12 shows the population of states in ^{207}Pb by the (p,d) reaction. Figure 1.13 shows the population of states in ^{209}Pb by the (d,p) reaction. Figure 1.14 shows the population of states in ^{209}Bi by the (^{3}He,d) reaction.

The effect of the spin–orbit coupling term can be assessed by subtracting E_{ls}, cf equation (1.4) with $C = -0.5$ MeV, from the experimental energies: this is depicted in figure 1.15. The underlying non-degeneracies due to angular momentum

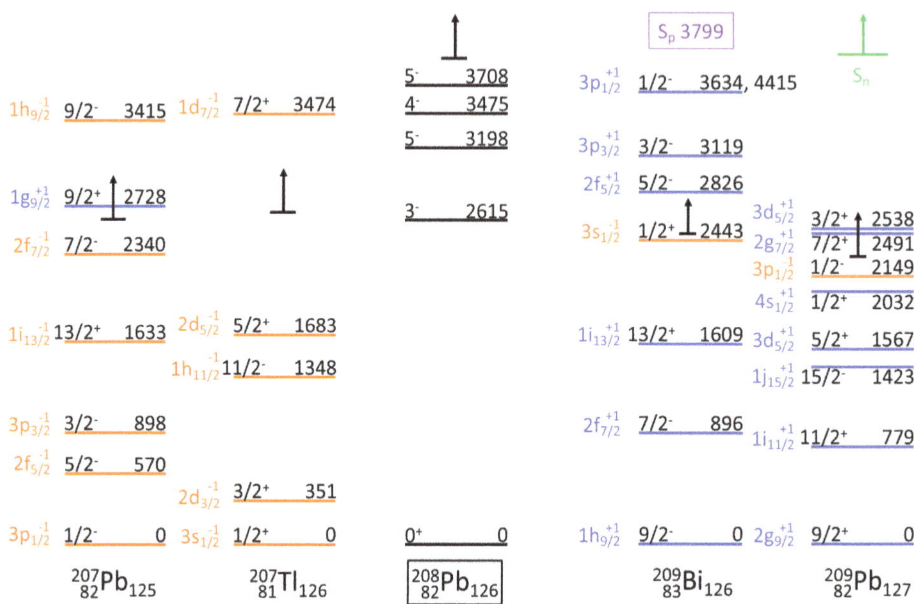

Figure 1.10. Interpretation of low-lying excited states in nuclei adjacent to the doubly closed shell, $Z = 82$, $N = 126$. The shell model quantum number assignments N, l, j (as a subscript), and whether particle $(+1)$ or hole $(−1)$ as superscripts, are shown (in blue—particles, in orange—holes). Note that these are naïve assignments, based on simple matching of spin-parity with shell model expectations, with no consideration of whether the states contain mostly independent-particle strength (see text). The energy threshold in ^{209}Pb for neutron unbound character of states is shown as a horizontal bar with a vertical arrow and the label S_n (in green), determined from the neutron separation energy.

eigenvalues are evident. Subtraction of a term $Dl(l + 1)$ with $D = −0.14$ MeV is shown. (Note that 'subtraction' of these terms, which involve parameters with negative values, translates into addition of the relevant energies.) The immediate question is: 'why are the adjusted energies non-degenerate?'. We look at this issue further below.

Figure 1.16 shows a comparison of neutron single-particle energies in the $N = 83$ nuclei ^{133}Sn and ^{147}Gd, with single-hole energies in the $N = 125$ nucleus ^{207}Pb. Evidently, states assigned as shell model configurations 'migrate' as the shell is filled, not only because of neutron filling from $N = 83$ to 125, but also by a concomitant filling of proton shell model configurations. Most notably, the neutron state labelled as $1h_{9/2}$ shifts by 3 MeV: this is a gain in 'binding energy' and can be attributed to the filling of the $1h_{11/2}$ subshell by 12 protons. This is a specific many-body effect due to the large overlap between the $1h_{9/2}$ and $1h_{11/2}$ orbitals and the attractive short-ranged nature of the nucleon–nucleon interaction. This effect lies entirely outside of the basic shell model description. It has been termed a monopole shift. There is also an increase in binding as a function of increasing angular momentum: this reflects the increasing mass and therefore increasing width of the independent-particle potential, which results in a lowering of higher angular momentum states.

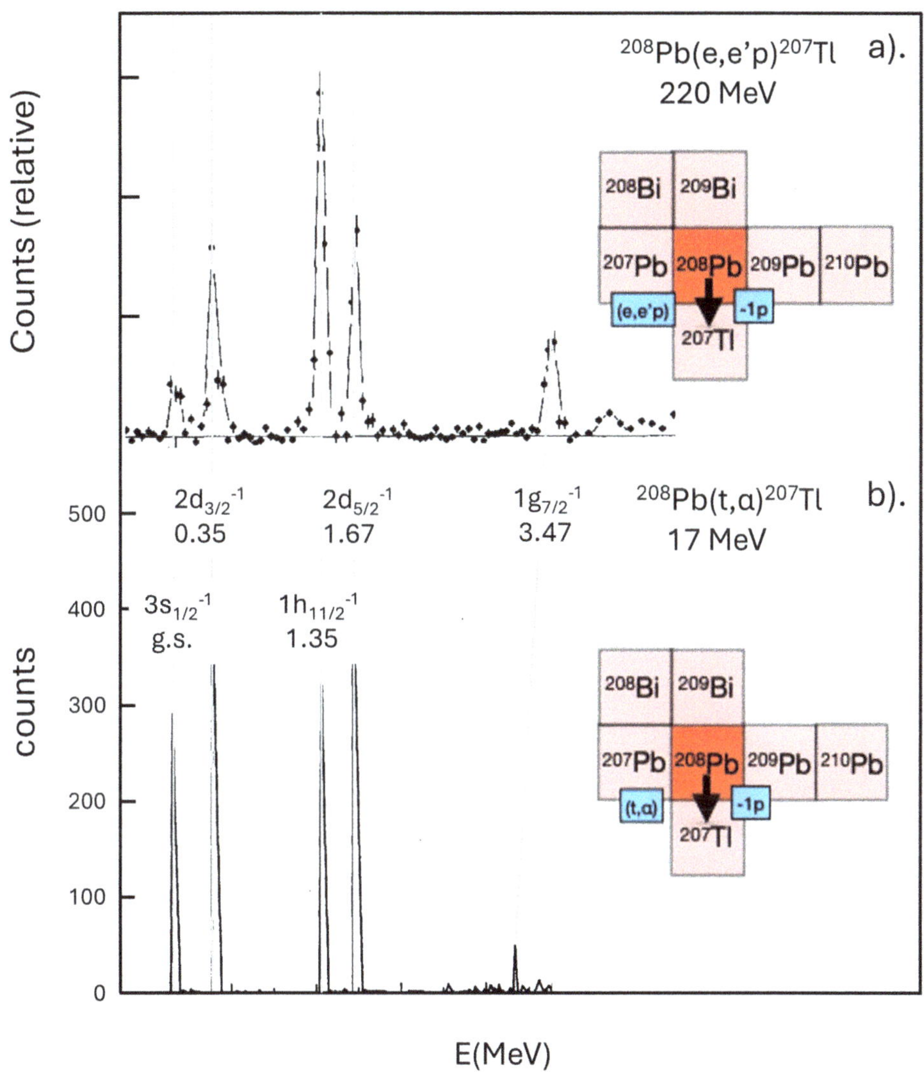

Figure 1.11. (a) Spectrum of protons knocked out of a ^{208}Pb target by inelastic scattering of high-energy electrons (quasi-elastic scattering). Reprinted from [20], copyright (1985) with the permission of Elsevier. (b) Spectrum of alpha particles following the (t,α) one-proton transfer reaction on a ^{208}Pb target using a polarized beam of 17MeV tritons. Reprinted from [21], copyright (1977) with the permission of Elsevier. The spectra show candidate states for expected shell model hole configurations in ^{207}Tl. Relative intensities are particular to the kinematics of the reactions and are beyond the present focus. (Note: there is an apparent non-linearity above 1.67 MeV in the triton spectrum taken from [21]. Thus, identification of the $1g_{7/2}$ state is not certain in [21].) The insets depict isotope boxes for orientation with respect to the doubly closed shell.

A first issue that arises with respect to treating nucleons as independent particles is: 'how well is the spin–orbit interaction characterized?' A survey of spin–orbit partner separation energies as deduced from naïve assignments of shell model configurations to states in odd-mass nuclei adjacent to closed shells suggests a value

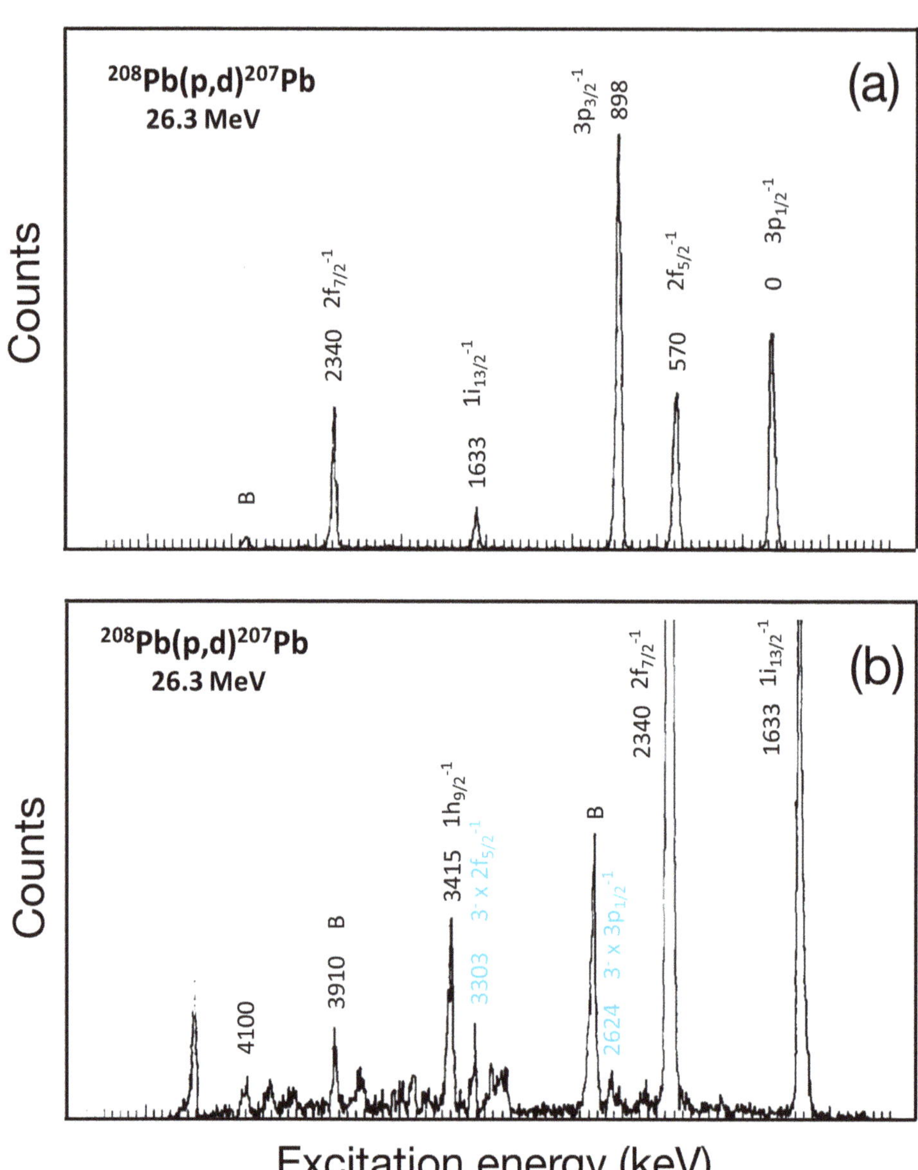

Figure 1.12. Spectrum of deuterons following the (p,d) one-neutron transfer reaction on a ^{208}Pb target using a beam of 26 MeV protons. Peaks are labelled by shell model quantum numbers. Peaks due to target impurities are labelled 'B'. The ground state is on the right-hand side of the spectrum in part a. Part b shows population of higher-energy states with an expanded counts scale (note that the peaks corresponding to the 1633 and 2340 keV excited states go off scale). Reprinted from [22], copyright (1982) with the permission of Elsevier.

of -0.5 MeV for the spin–orbit coupling strength, C for heavy ($A > 56$) nuclei. A more detailed view of spin–orbit coupling strength is presented in figure 1.17. (For further details, [18] should be consulted.) We term these assignments 'naïve' because

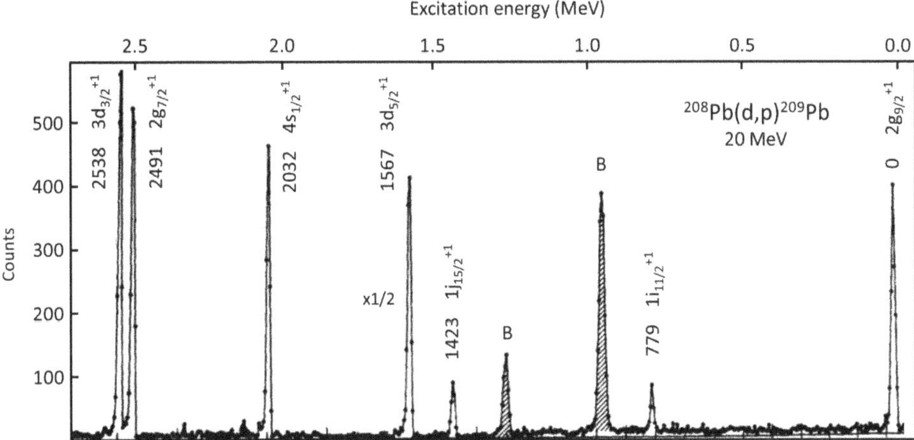

Figure 1.13. Spectrum of protons following the (d,p) one-neutron transfer reaction on a ^{208}Pb target using a beam of 20 MeV deuterons. Peaks are labelled by shell model quantum numbers and energies in keV. Peaks due to target impurities are labelled 'B'. The ground state is on the right-hand side of the spectrum. Note the scale reduction factor for the peak corresponding to the 1567 keV state in ^{209}Pb. Reprinted from [23], copyright (1974) with the permission of Elsevier.

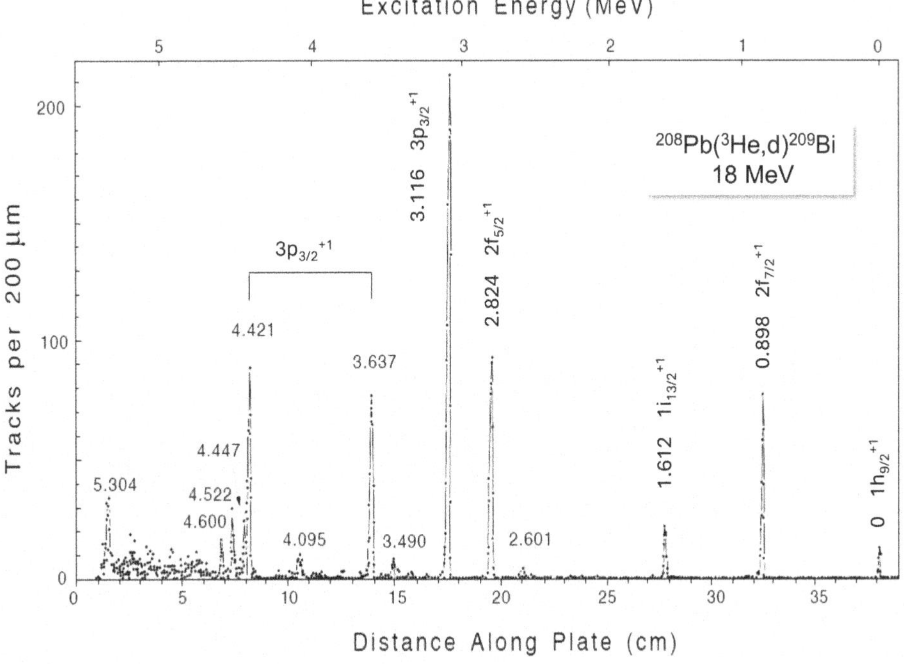

Figure 1.14. Spectrum of deuterons following the (^{3}He,d) one-proton transfer reaction on a ^{208}Pb target using a beam of 18 MeV ^{3}He ions. Peaks are labelled by shell model quantum numbers and energies (in MeV) of the states populated in ^{209}Bi. The $p_{1/2}$ strength is fragmented. The ground state is on the right-hand side of the spectrum. The figure is taken from [11], copyright IOP Publishing Ltd, all rights reserved, and is an adaptation of one found in [24].

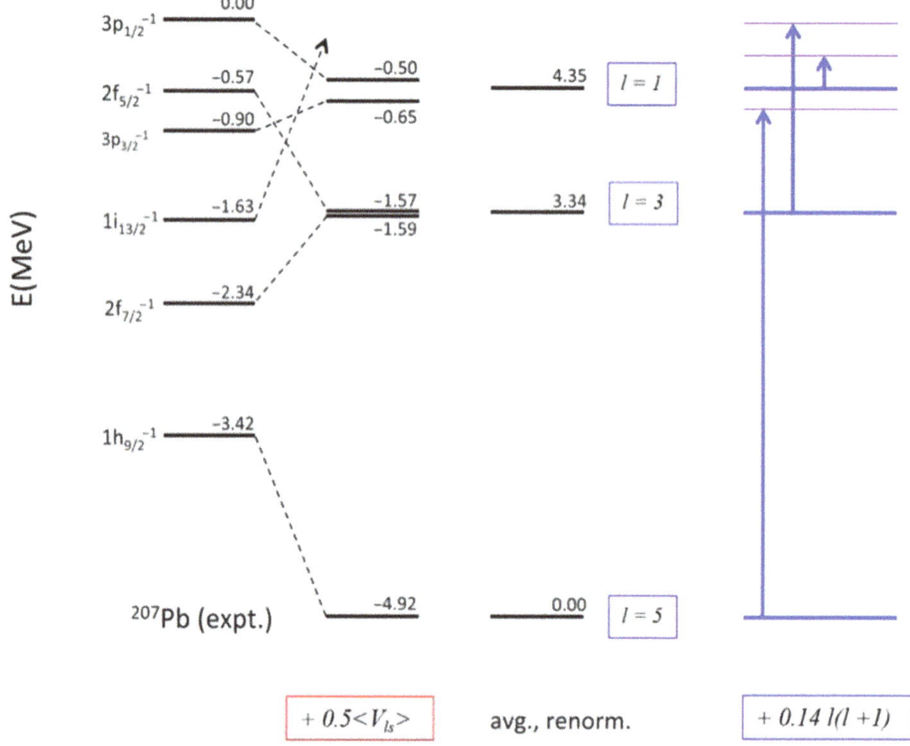

Figure 1.15. An illustration of the need to have a term in the mean-field Hamiltonian that separates shell model orbitals with different angular momenta in an ordered pattern. The strength of the spin–orbit potential is taken from global empirical data. The strength of the $l(l + 1)$ term is chosen to minimize the energy splitting shown here. The column labelled 'avg. renorm.' takes average energies of pairs of states after removal of the spin–orbit energy, i.e. $(-0.65 - 0.50)/2 = -0.58$, and renormalizes the ground state by adding 4.92 MeV, whence $4.92 - 0.57 = 4.35$ MeV for the excitation of the p ($l = 1$) configurations. The excitation pattern shown favours a 'flat-bottomed' potential. The figure is taken from [11], copyright IOP Publishing Ltd, all rights reserved. See a video-based tutorial on this concept in figure 1.40.

no consideration is given to whether these states are true independent-particle (and independent-hole) states. This will be addressed shortly.

A second issue is the degree to which a doubly closed shell nucleus can be treated as an 'inert core', i.e. simply as the origin of the mean-field that produces the shell model independent-particle potential, with no 'internal' degrees of freedom. A first look at this is depicted in figure 1.18. This shows a so-called 'weak coupling' of independent particles and holes to the first core-excited state in ^{208}Pb. We explore the issue of core-excited states in detail for ^{208}Pb below.

1.3.2 Excitations in ^{208}Pb

The simplest interpretation of excited states in ^{208}Pb is as 1p-1h excitations of protons and 1p-1h excitations of neutrons across the shell model energy gaps. This supposes that correlations are not important. Thus, the naïve expectation for

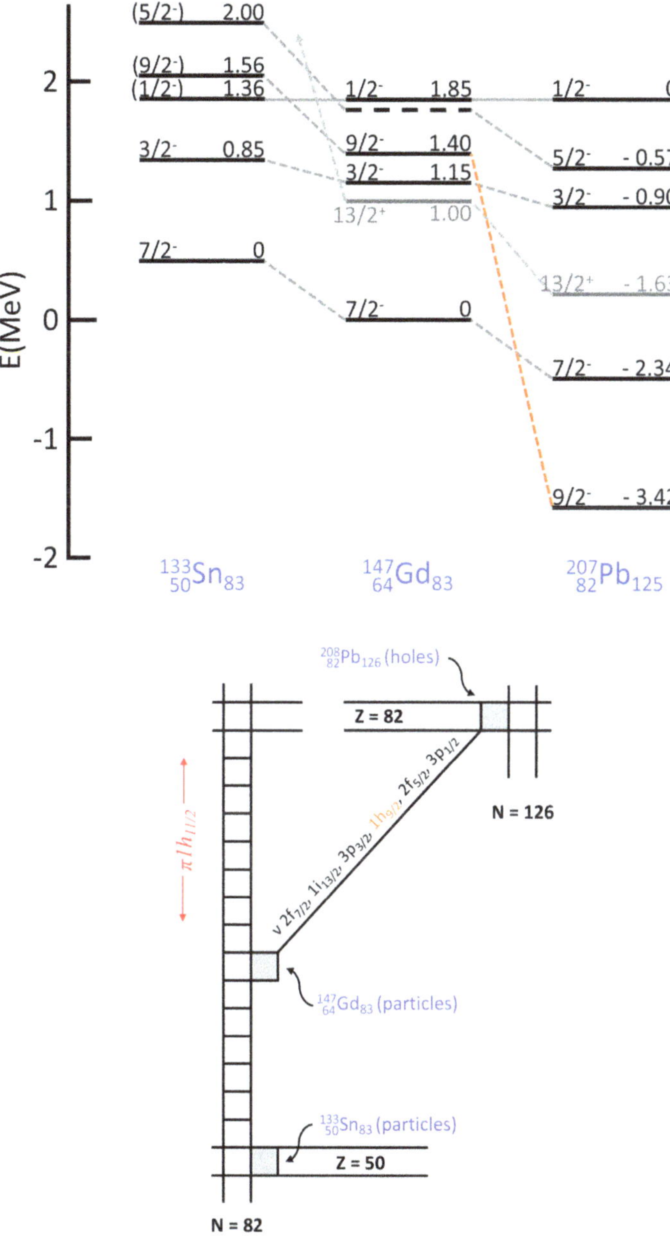

Figure 1.16. 'Migration' across the $82 < N < 126$ shell of states associated with independent-neutron configurations. Note that the general trend is for a lowering of the relative energies of higher angular momentum states. But the most dramatic shift in energy is for the 9/2⁻ state which is interpreted as resulting from the $\nu 1h_{9/2}$ configuration: this is due to a filling of the $\pi 1h_{11/2}$ orbital and an underlying interaction termed the proton–neutron monopole interaction. Thus, as protons fill the $1h_{11/2}$ orbital between $Z = 64$ and $Z = 82$ there is a progressive lowering of the $\nu 1h_{9/2}$ shell model state relative to the other neutron states. This is because of the large orbital overlap of the $\nu 1h_{9/2}$ and $\pi 1h_{11/2}$ orbitals and the short-ranged attractive nature of the proton–neutron interaction. The states in ^{207}Pb are hole states and are shown with negative energies. The data are taken from ENSDF. Further details are discussed in the text. See a video-based tutorial on this concept in figure 1.41.

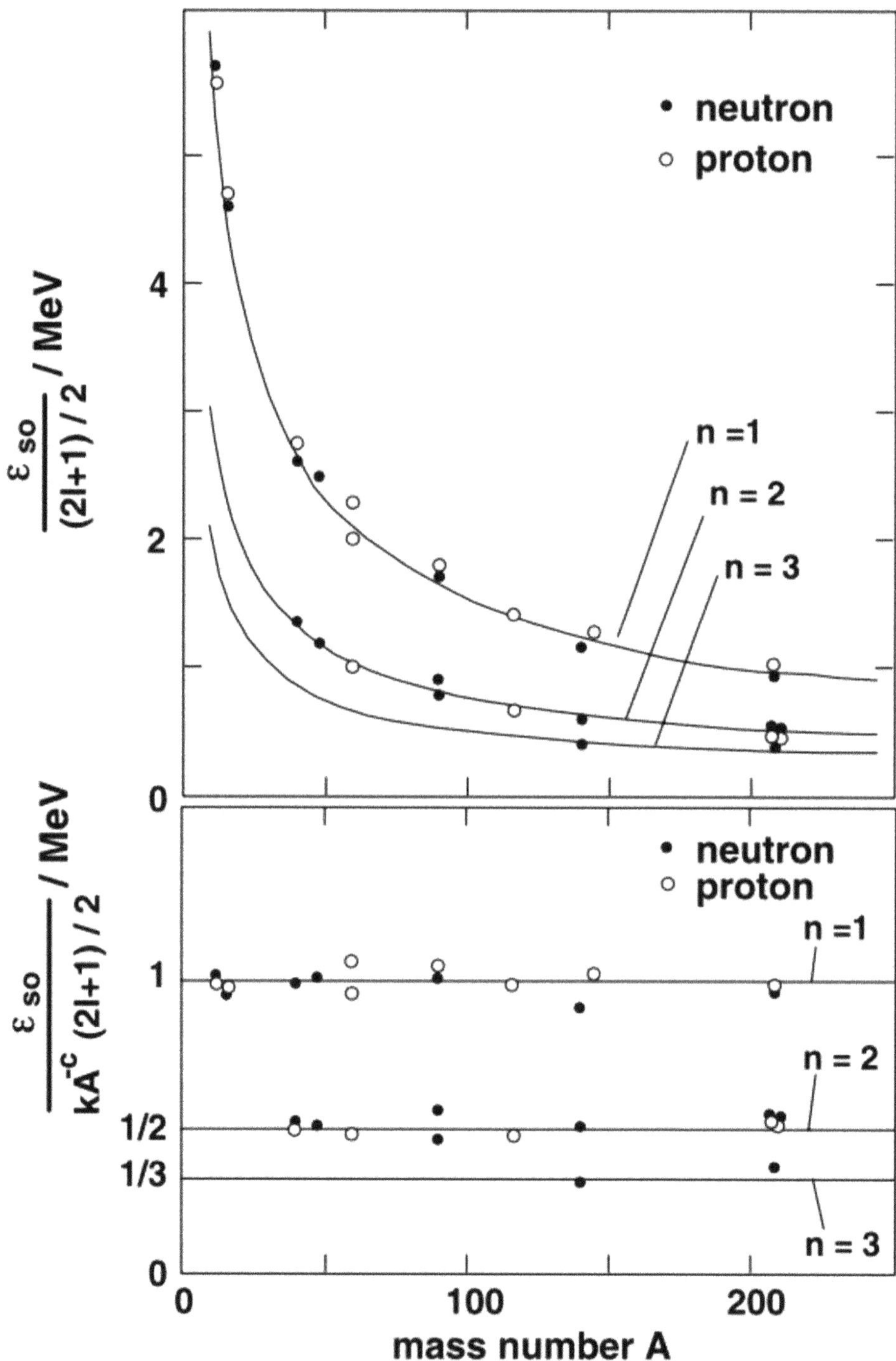

Figure 1.17. Top: experimental spin–orbit splitting for protons (open circles) and neutrons (solid circles) for $12 < A < 208$ with a scale factor $(2l + 1)/2$, cf equation (1.4). The systematic trends show a dependence on the oscillator shell quantum number, n and an A dependence of the form $f(A) = kA^{-c}$. Bottom: a further scaling by $f(A)$ reveals the dependence of the spin–orbit splitting energies on n. Reprinted from [18], copyright (2000) with the permission of Springer Nature.

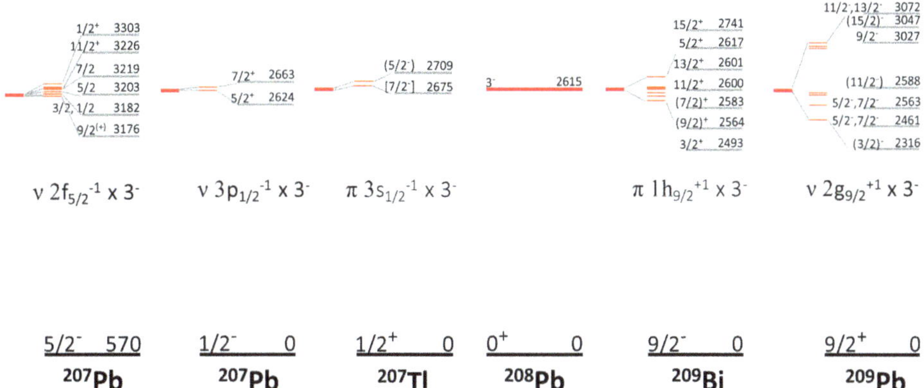

Figure 1.18. Weak coupling in odd-mass nuclei adjacent to ^{208}Pb. The 'parent' state for these structures is the 3^- state in ^{208}Pb at 2615 keV. The shell model configurations are indicated. Note that for ^{207}Pb the shell model configuration for the first excited state is also included. The 3^- state is a collective 'octupole' excitation and is discussed further in the text. The figure is an updated version of one appearing in [11], copyright IOP Publishing Ltd, all rights reserved. The data are taken from ENSDF.

the lowest excited states in ^{208}Pb is the configurations due to the neutron 1p-1h excitations $1g_{9/2}^{+1} \times 3p_{1/2}^{-1}$—$J^\pi = 4^-, 5^-$ and the proton 1p-1h excitations $1h_{9/2}^{+1} \times 4s_{1/2}^{-1}$—$J^\pi = 4^-, 5^-$. These combinations do not match the first excited state: this is consistent with the octupole collective character of the 3^- 2615 keV excited state. The collective character implies a mixture of many 3^- configurations, i.e. particle–hole correlations. But the next three excited states, cf figure 1.10, are $J^\pi(E_x \text{ keV}) = 5^-(3198), 3^-(3475), 5^-(3708)$. An immediate question is which 5^- state is the proton configuration and which one is the neutron configuration? This is directly answered by one-nucleon transfer reaction spectroscopy, as shown in figures 1.19(a) and (b): the $5^-(3198)$ state is dominated by the $\nu\nu$ $3p_{1/2}^{-1} \times 1g_{9/2}^{+1}$ configuration and the $5^-(3708)$ state is dominated by the $\pi\pi$ $4s_{1/2}^{-1} \times 1h_{9/2}^{+1}$ configuration.

The ^{208}Pb core excitations appear in ^{207}Pb, ^{209}Pb, ^{207}Tl, and ^{209}Bi as particle–core coupled states, not only for the $J^\pi(E_x \text{ keV}) = 3^-(2615)$ first excited state in the core, but also for the succession of core excitations, with a proviso. The core excitations involve proton and neutron particle and hole configurations, and these can be 'Pauli blocked' in the odd-mass nuclei. Specifically, the $J^\pi(E_x \text{ keV}) = 5^-(3198)$ core excitation is the configuration $\nu\nu$ $3p_{1/2}^{-1} \times 1g_{9/2}^{+1}$ and this cannot appear as a core excitation involving the coupling of $3p_{1/2}^{-1}$ in ^{207}Pb or of $1g_{9/2}^{+1}$ in ^{209}Pb. However, the $J^\pi(E_x \text{ keV}) = 5^-(3198)$ core excitation can appear for all proton particle and hole configurations in ^{209}Bi and ^{207}Tl, respectively.

A more subtle effect involves core excitations that can be described as 'pair' excitations. Figure 1.20 illustrates the spectroscopic evidence for such excitations in ^{208}Pb. Three states are of interest for present purposes, the $J^\pi(E_x \text{ keV}) = 0^+(4878)$, $2^+(5575)$ and $2^+(5828)$ states: these are usefully described as ^{210}Pb$(0_1^+) \times ^{206}$Pb(0_1^+) for the 4878 keV state and as a mixture of ^{210}Pb$(0_1^+) \times ^{206}$Pb(2_1^+) and ^{210}Pb$(2_1^+) \times ^{206}$Pb(0_1^+)

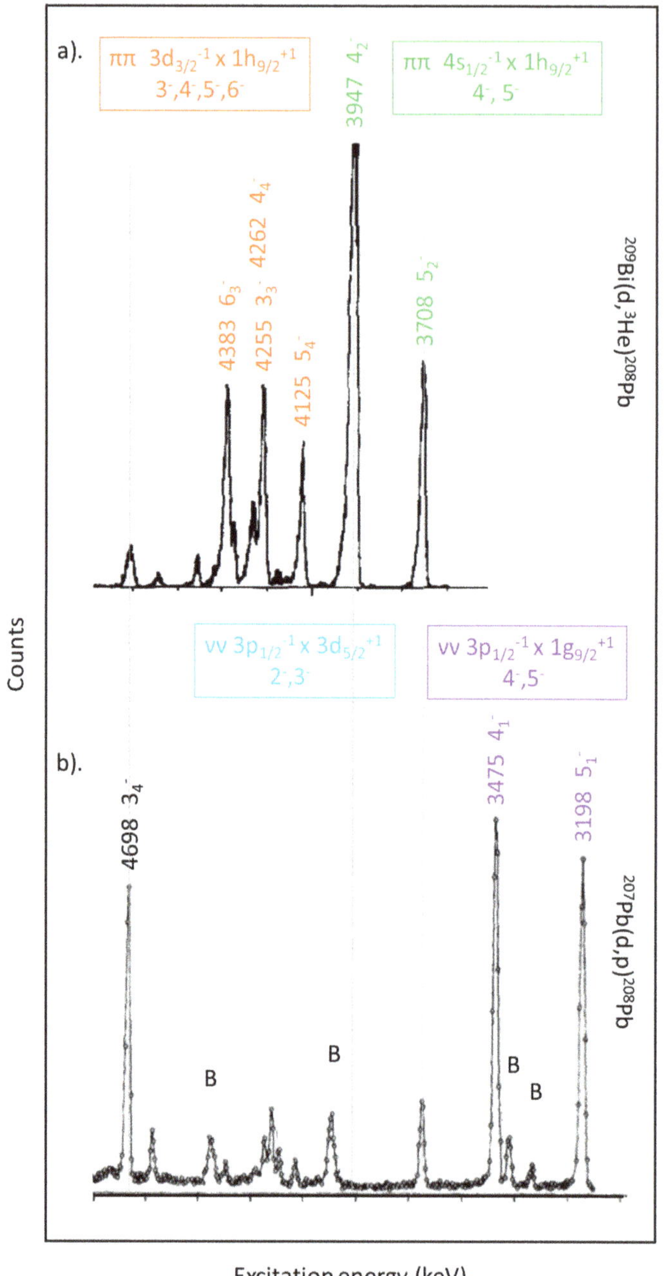

Figure 1.19. (a) Spectrum of ^{3}He ions following the (d,^{3}He) one-proton transfer reaction on a ^{209}Bi target using a beam of 45 MeV deuterons. Reprinted from [25], copyright (1987) with the permission of Elsevier. (b) Spectrum of protons following the (d,p) one-neutron transfer reaction on a ^{207}Pb target using a beam of 18 MeV deuterons. Reprinted from [26], copyright (1973) with the permission of Elsevier. (Note that $\Sigma = 0$ spin coupling is favoured energetically for like nucleons.) Energies are taken from ENSDF. Further details are given in the text.

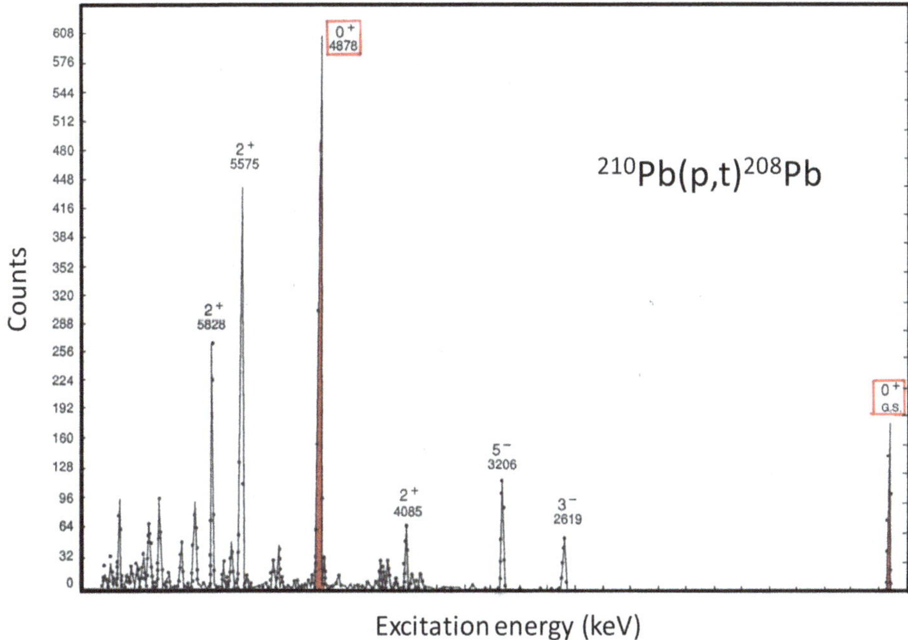

Figure 1.20. Spectrum of tritons following the (p,t) two-neutron transfer reaction on a ^{210}Pb target using a beam of 20 MeV protons. Energies below 5 MeV are taken from ENSDF. Details are discussed in the text. Reprinted from [9]. Copyright (1970) by the American Physical Society.

configurations for the 5575 and 5828 keV states. Note that the 2^+_1 state in ^{206}Pb has an excitation energy 803 keV and in ^{210}Pb it is 800 keV. Thus, a useful estimate of the unmixed 2^+ configurations in ^{208}Pb would be 5678 and 5681 keV.

The pair excitation degree of freedom is manifested in ^{207}Pb, ^{209}Pb, ^{207}Tl, and ^{209}Bi as 'two-particle-one-hole' (2p-1h) and 'one-particle-two-hole' (1p-2h) states. These states are observed in ^{209}Pb and ^{209}Bi via the ^{210}Pb(p,d)^{209}Pb and ^{210}Po(t,α)^{209}Bi reactions, shown in figures 1.21 and 1.22, respectively. Such a view of ^{207}Pb via ^{206}Pb(d,p)^{207}Pb is more complicated and is handled in exercise 1-9(b). The primary manifestations of such excitations are the $J^\pi(E_x \text{ keV}) = 9/2^+(2728)$ state in ^{207}Pb, the $J^\pi(E_x \text{ keV}) = 1/2^-(2149)$ state in ^{209}Pb and the $J^\pi(E_x \text{ keV}) = 1/2^+(2443)$ state in ^{209}Bi. Note the sum, $2728 + 2149 = 4877$ keV, cf the $J^\pi(E_x \text{ keV}) = 0^+(4878)$ state in ^{208}Pb. Note the fragmentation of the strength occurring due to mixing with some of the core-excited states shown in figure 1.18.

More complex excitations involving the particle pairs and hole pairs coupled to spin 2, 4, $\cdots$ can be considered and candidate states are observed in these nuclei. What is remarkable is that estimates of these excitations can be made based on a few prototype degrees of freedom and their observed energies. An immediate observation is that the pair excitations involve large correlation energies: for example, from the lowest $\nu(1p\text{-}1h)$ excitation in ^{208}Pb at 3198 keV, a naïve estimate of the $\nu(2p\text{-}2h)$ excitation in ^{208}Pb would be 6396 keV. Thus, one can infer that the $\nu(2p\text{-}2h)$ state in ^{208}Pb involves a correlation energy of 1518 keV. A refinement of this estimate would

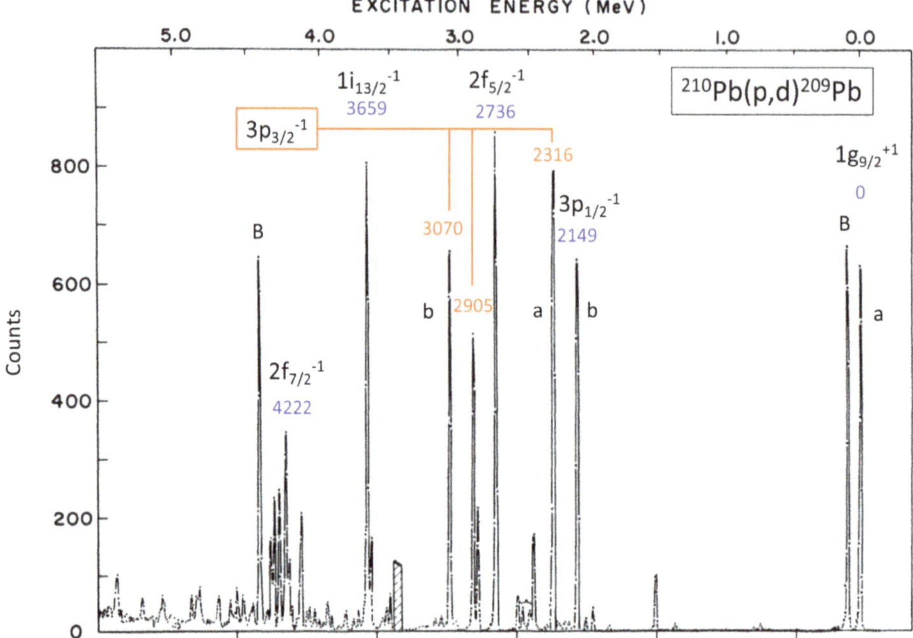

Figure 1.21. Spectrum of deuterons following the (p,d) one-neutron transfer reaction on a ^{210}Pb target using a beam of 20.6 MeV protons. Shell model configurations of strongly populated states are labelled, and they are hole states except for the ground state. Peaks marked 'a' are reduced in counts scale by $\times 1/2$, those marked 'b' by $\times 1/10$. Peaks marked 'B' are due to target impurities. Note the fragmentation of the $3p_{3/2}^{-1}$ configuration: the state at 2316 keV is a candidate for the $3^- \times 1g_{9/2}$ core excitation, cf figure 1.18; the main $3p_{3/2}^{-1}$ strength is at 3070 keV, cf energies of the hole states in ^{207}Pb as shown in figure 1.10. Energies in keV are taken from ENSDF. Further details are discussed in the text. Reprinted with permission from [8]. Copyright (1971) by the American Physical Society.

use the centroid of the doublet of states, $J^\pi(E_x \text{ keV}) = 5^-(3198), 4^-(3475)$, i.e. 3337 keV, yielding a ν(2p-2h) excitation energy estimate of 6673 keV and an implied correlation energy of 1795 keV for the observed ν(2p-2h) state in ^{208}Pb.

A further look at the core excitations in ^{208}Pb is presented in figure 1.23 which shows the spectrum for inelastic scattering of deuterons. This view can be matched, e.g. as shown in figure 1.24 for ^{209}Bi. Indeed, the dominance of the collective nature of the first-excited state in ^{208}Pb is remarkable, as already evident in figure 1.18. This pattern is termed 'weak coupling' in the terminology of particle–core coupling. There is a definitive study of excitations in ^{208}Pb [19] and we refer the more adventurous reader to this work.

1.3.3 Excitations in ^{208}Bi

The nucleus ^{208}Bi offers a unique example of shell model configurations that can be viewed from the perspective of a proton particle coupled to a neutron hole, as defined using states in ^{209}Bi and ^{207}Pb. This view is directly probed by the transfer

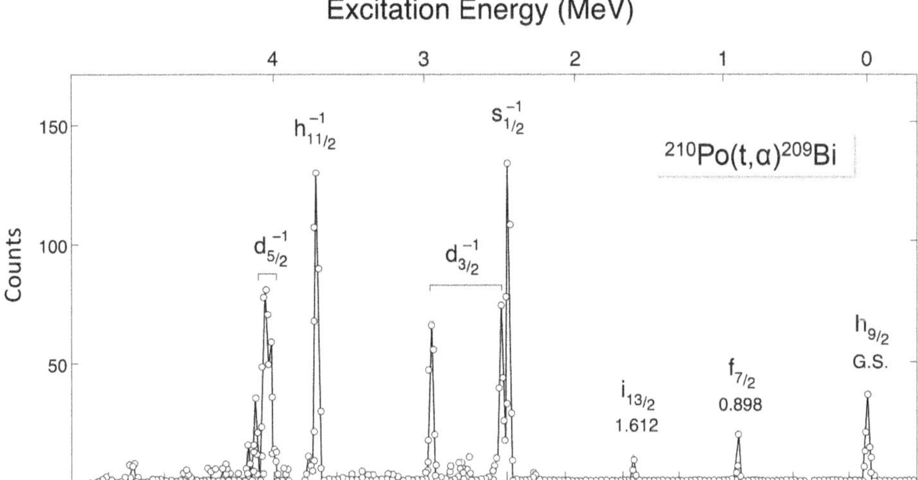

Figure 1.22. Spectrum of alpha particles following the (t,α) one-proton transfer reaction on a ^{210}Po target using a beam of 20 MeV tritons. Details are discussed in the text. Reproduced from [1], copyright (2010) World Scientific Publishing Company. The figure is based on original data found in [6].

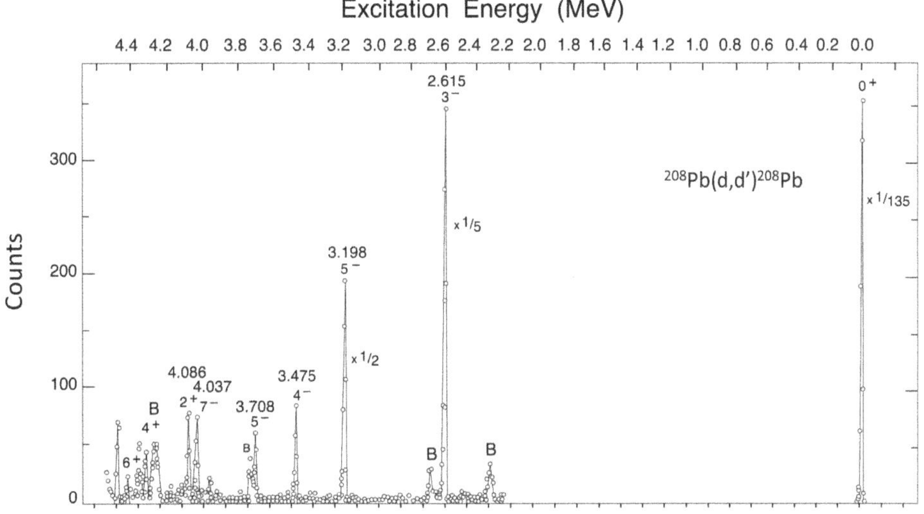

Figure 1.23. Spectrum of deuterons following the (d,d′) inelastic scattering reaction on a ^{208}Pb target using a beam of 13 MeV deuterons. Reproduced from [1], copyright (2010) World Scientific Publishing Company. The figure is based on original data found in [27].

reactions ^{209}Bi(p,d)^{208}Bi and ^{207}Pb(^{3}He,d)^{208}Bi. Figure 1.25 depicts the respective deuteron spectra from these two reactions.

The lowest two states in ^{208}Bi can be interpreted as due to the coupling $\pi 1h_{9/2}^{+1} \times \nu 3p_{1/2}^{-1}$, $J^\pi = 4^-$, 5^-. The ^{207}Pb(^{3}He,d)^{208}Bi reaction selectively populates configurations involving $\nu 3p_{1/2}^{-1}$ coupled to $\pi 1h_{9/2}^{+1}$, $\pi 2f_{7/2}^{+1}$, $\pi 1i_{13/2}^{+1}$, ...; the ^{209}Bi

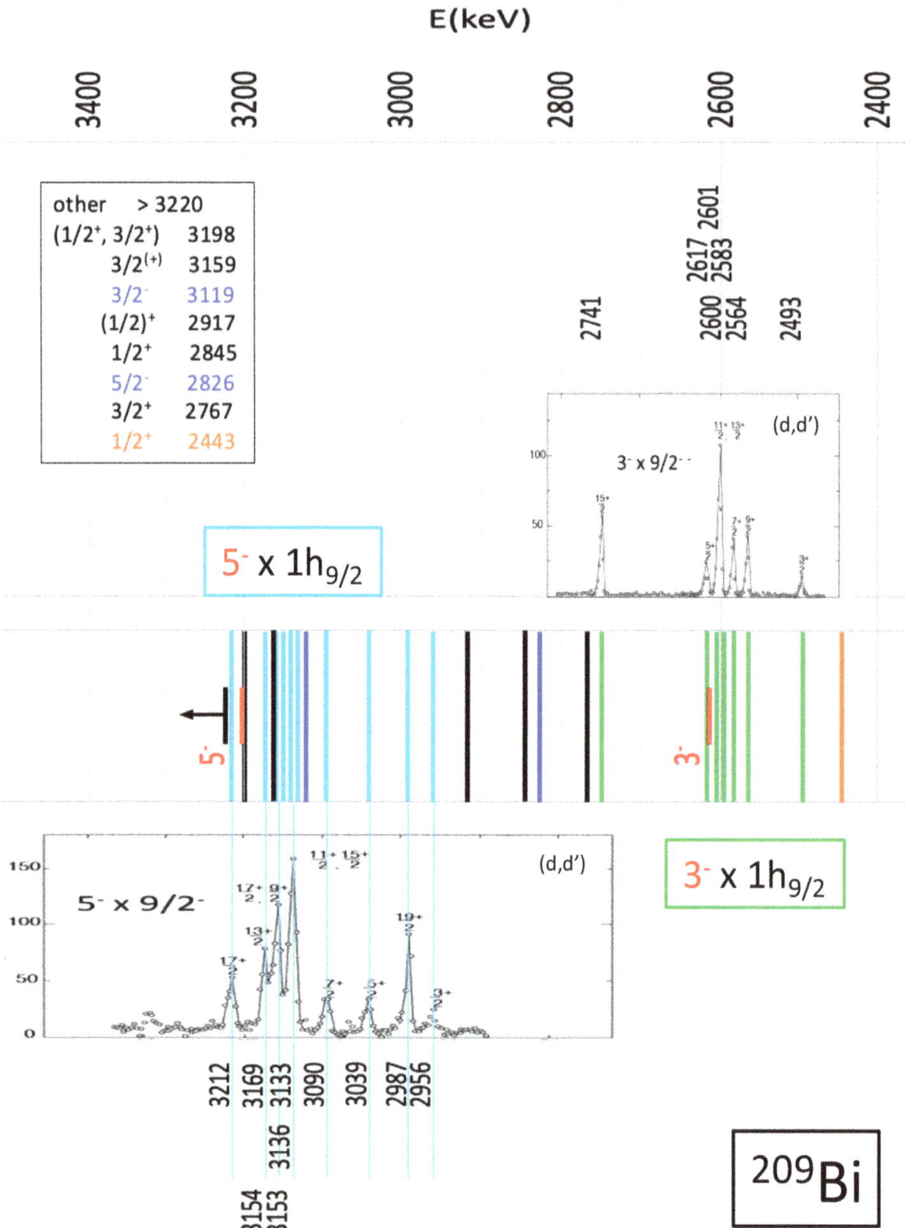

Figure 1.24. Spectrum of deuterons following the (d,d′) inelastic scattering reaction of 13 MeV deuterons on a target of ^{209}Bi. The states populated are interpreted as the $1h_{9/2}$ proton coupled to the 3^- and 5^- states in the ^{208}Pb core which are indicated in red. Other colour coding indicates: green—states assigned to the $3^- \times 1h_{9/2}$ septuplet; light blue—states assigned to the $5^- \times 1h_{9/2}$ 11-plet; orange—the $3s_{1/2}$ intruder state (q.v. figures 1.10 and 1.22); dark blue—the $2f_{5/2}$ and $3p_{3/2}$ configurations (q.v. Figures 1.10 and 1.14). The spins, parities, and excitation energies in keV are taken from the ENSDF. The figure is based on original data found in [27].

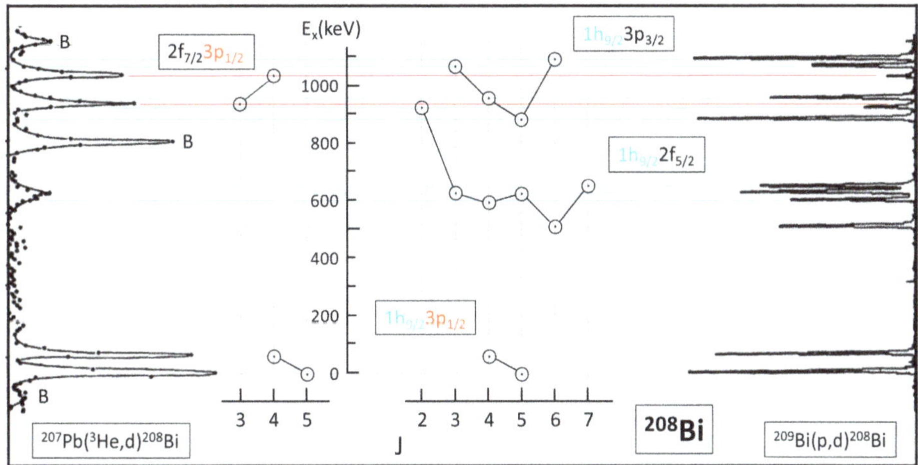

Figure 1.25. Low-lying states in ^{208}Bi organized into particle–hole multiplets by spin. (Left) States selectively populated in the ^{207}Pb(^{3}He,d)^{208}Bi transfer reaction. Reprinted with permission from [28]. Copyright (1971) by the American Physical Society. (Right) States populated in the ^{209}Bi(p,d)^{208}Bi single-nucleon transfer reaction. Reprinted with permission from [29]. Copyright (1973) by the American Physical Society. Other data are taken from ENSDF.

(p,d)^{208}Bi reaction selectively populates configurations involving $\pi 1h_{9/2}^{+1}$ coupled to $\nu 3p_{1/2}^{-1}$, $\nu 2f_{5/2}^{-1}$, $\nu 3p_{3/2}^{-1}$, $\cdots$. The resulting multiplets can be used to learn something about the way in which a proton particle and a neutron hole interact in a nuclear environment. Figure 1.25 organizes such data into a useful perspective.

1.3.4 Global view of nuclei in doubly closed shell regions

A broader look at nuclei in the region of doubly closed shells can be considered. The candidate nuclei can be identified using figure 1.2. Immediately, a division into two categories is needed: nuclei with $N \sim Z$ and nuclei with $N > Z$. The first category is illustrated in figure 1.26 and these are termed nuclei which involve low-energy shape coexistence. The second category is illustrated in figure 1.27. A key spectroscopic probe of shell model states is provided by nucleon transfer reactions, but these need stable or long-lived target nuclei. A 'map' of such nuclei is presented in figure 1.3. The overall view is that shell model states in nuclei are only accessible in a severely limited manner. In the following we explore what can and cannot be said about shell model states.

The foregoing presentation of nuclei centred on the doubly closed shell at $Z = 82$, $N = 126$ provides a baseline for discussion. First, it can be noted that the doubly closed shell nuclei ^{48}Ca and ^{132}Sn lack stable target nuclei for one-nucleon transfer reaction spectroscopy. Use of radioactive beams is beginning to solve this issue [30–35]; but spectroscopy is at present limited to the simplest details, which are already in part accessible by inference from decay spectroscopy. The nuclei adjacent to ^{132}Sn are shown in figure 1.28. The nuclei ^{133}Sn and ^{207}Pb are related to each other as

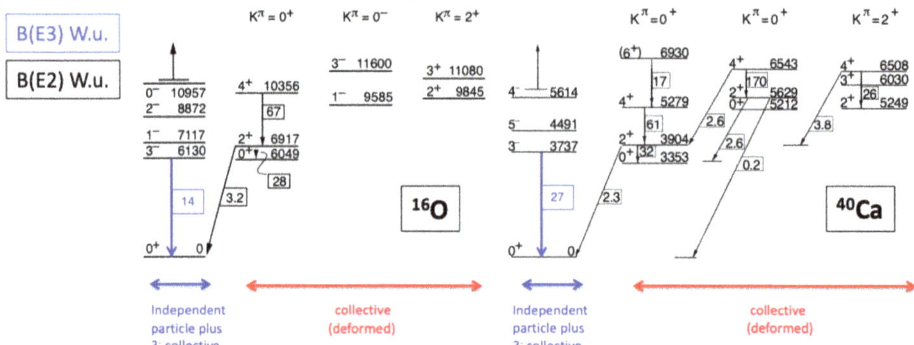

Figure 1.26. Excited states in the double-closed shell nuclei with $N = Z$. Some details of the collectivity of the 2^+ and 3^- states are given. The B(E2) and B(E3) values that support collectivity are shown. Band structure that is deduced from more detailed spectroscopy is indicated with inferred K quantum numbers. Thus, the 0^+ state at 6049 keV in ^{16}O is interpreted as a 4p-4h state. A similar interpretation is made for the 3353 keV state in ^{40}Ca. The 0^+ state in ^{40}Ca at 5212 keV is interpreted as an 8p-8h state: details are beyond the present scope for discussion. The figure is taken from [17], based on a figure from [1], copyright (2010) World Scientific Publishing Company.

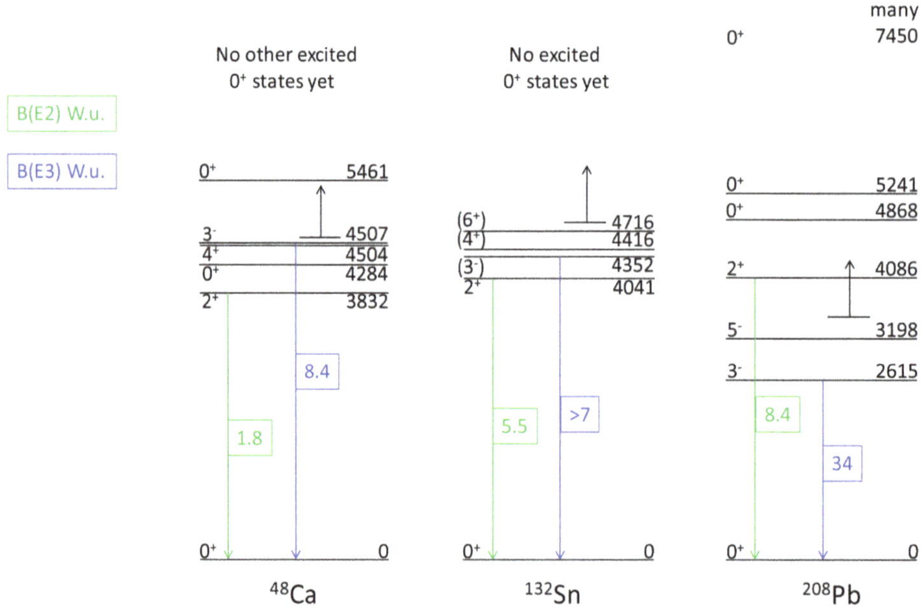

Figure 1.27. Excited states in the double-closed shell nuclei with $N > Z$. Some details of the collectivity of the 2^+ and 3^- states are given. The figure is taken from [11], copyright IOP Publishing Ltd, all rights reserved.

depicted in figure 1.16; a similar perspective is shown for ^{133}Sb and ^{207}Tl in figure 1.29. Removal of spin–orbit splitting energies, similar to that depicted in figure 1.15 for ^{207}Pb, is shown in figures 1.30(a)–(c) for ^{131}Sn, ^{209}Pb and ^{209}Bi, respectively.

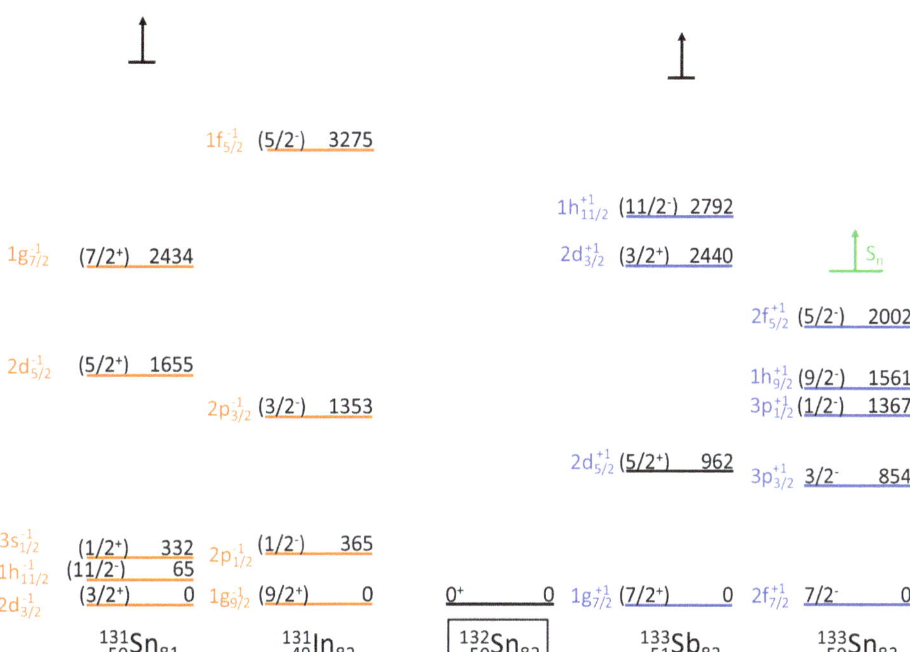

Figure 1.28. Interpretation of low-lying excited states in nuclei adjacent to the doubly closed shell, $Z = 50$, $N = 82$. For explanation of details, see caption to figure 1.4. The data are taken from ^{131}Sn—[34]; ^{131}In—[34]; ^{133}Sb—ENSDF; ^{133}Sn—[32]; and see [33].

An alternative look at shell model states in nuclei is provided by magnetic moment data. One starts with the magnetic moments of the bare proton and neutron noting that they have the values, $\mu_p = +2.793\mu_N$ and $\mu_n = -1.913\mu_N$ which are not the so-called Dirac moments that would be given by $\mu_N = e\hbar/m$ in SI units—the nuclear magneton. Manifestly, the neutron should have a magnetic moment of zero. The explanation lies in the composite character of the nucleon: it has a quark substructure with intrinsic quark spins and magnetic moments and their electrical charges and orbital angular momentum. There are also virtual meson (quark–antiquark pairs) that can contribute to the observed values for the proton and neutron magnetic moments. There are very recent developments that suggest the quark structure of the nucleon, which also involves gluons, is very complex. At the nucleon level, at the nuclear level and at the atomic level (inclusion of electron magnetic moments[4]), magnetic moments have vectorial character and so obey the rules for vector addition. Pairing of protons and separately of neutrons results in the paired nucleons contributing zero to nuclear magnetic moments.

The value of the above view is that nuclei of interest can be subjected to a variety of spectroscopic measurements which probe values of magnetic moments,

---

[4] The magnetic moment of the electron is (almost) equal to the Dirac moment. The deviation is due to quantum electrodynamic effects. This topic is beyond the scope of the present focus but provides a unique view of fundamental physics.

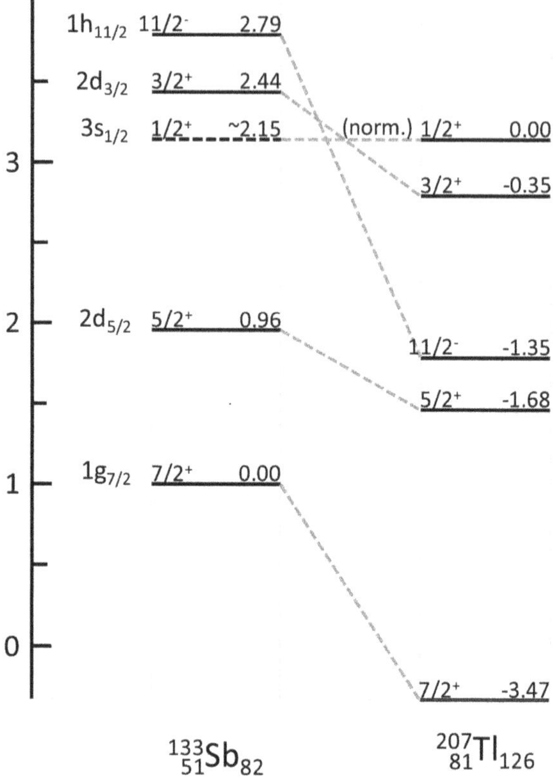

Figure 1.29. 'Migration' of states associated with independent-proton configurations across the $50 < N < 82$ shell. Note that the $2d_{5/2}$–$2d_{3/2}$ spin–orbit pair migrate in parallel. For an explanation of details, see caption to figure 1.16.

independently of whether they are near or far from stability. For example, laser hyperfine spectroscopy of atomic species, which is sensitive to nuclear magnetic moments (and quadrupole moments), can be studied in detail even when the half-life of the isotope is as short as milliseconds. The only requirement is that the isotopic species live long enough for transportation to the laser facility for study of its optical hyperfine spectrum. Note that a single atom or ion may be subjected to a very large number of excitation cycles (before it decays) using intense laser beams, thus providing an experimental probe with extraordinary sensitivity for an isotopic species of exceedingly low production rate.

The description of nuclear magnetic moments that is used adheres to the original work of Theodore Schmidt [36] who made estimates for nuclei characterized by extreme independent-particle behaviour, i.e. nuclei adjacent to doubly closed shells. Schmidt estimates for magnetic moments are given by the relationships, for an odd proton

$$\mu = j - 1/2 + \mu_p \qquad \text{for } j = l + 1/2, \tag{1.7}$$

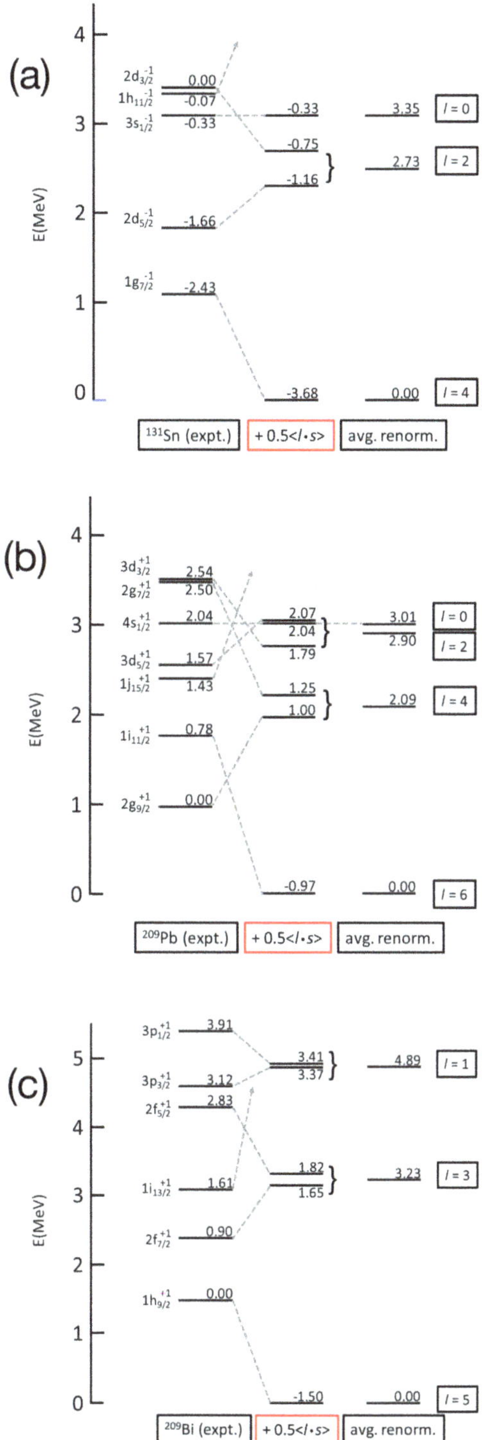

Figure 1.30. An illustration of the need to have a term in the mean-field Hamiltonian that separates shell model orbitals with different angular momenta in an ordered pattern, shown for (a) ^{131}Sn, (b) ^{209}Pb, (c) ^{209}Bi.

$$\mu = \frac{j}{j+1}(j + 3/2 - \mu_p) \text{ for } j = l - 1/2, \tag{1.8}$$

and for an odd neutron

$$\mu = \mu_n \qquad \text{for } j = l + 1/2, \; \mu = -\frac{j}{j+1}\mu_n \qquad \text{for } j = l - 1/2, \tag{1.9}$$

where $\mu_p = +2.793$ and $\mu_n = -1.913$ in nuclear magnetons. Such data are depicted using so-called Schmidt diagrams, cf figure 1.31 which presents magnetic moment values for states in odd-mass nuclei adjacent to doubly closed shells.

The agreement of magnetic moment data for nuclei adjacent to doubly closed shells with the Schmidt estimates is of limited value. This has led to considerable theoretical activity. This theory is complex, and the parameter fitting is intricate. Essentially there are two sources for corrections to the Schmidt values: configuration mixing; corrections to the magnetic moment operator at the sub-nucleon level (so-called 'mesonic effects'). The states which are viewed as ideal tests are those with $s_{1/2}$ character; this is because they have zero contributions from orbital angular

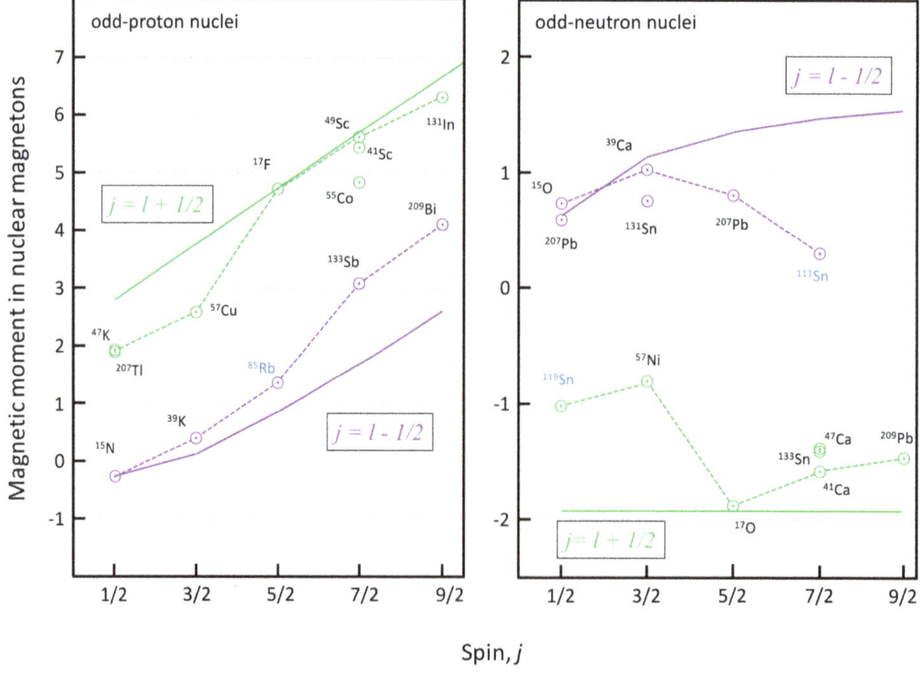

Figure 1.31. Schmidt diagrams showing magnetic moments for nuclei adjacent to doubly closed shells. Note the inclusion of values for ^{85}Rb, ^{111,119}Sn (shown in blue), which are adjacent to single-closed shells. These are selected because they manifest values for $j = 5/2$ $(j = l - 1/2)$, $j = 1/2$ $(j = l + 1/2)$, $j = 7/2$ $(j = l - 1/2)$, respectively; for these combinations of j and l there are no known values in nuclei adjacent to double-closed shells. The solid lines are the Schmidt estimates for the magnetic moments; the dashed lines are to guide the eye. Details are discussed in the text. The data are taken from ENSDF and [37].

momentum. The only known nuclei which provide such tests are ^{47}K and ^{207}Tl (see, e.g. [37, 38]).

Figure 1.32 depicts a comparison of states observed in the ^{40}Ca(d,p)^{41}Ca and ^{48}Ca(d,p)^{49}Ca one-neutron transfer reactions. The spectra are shown so that the

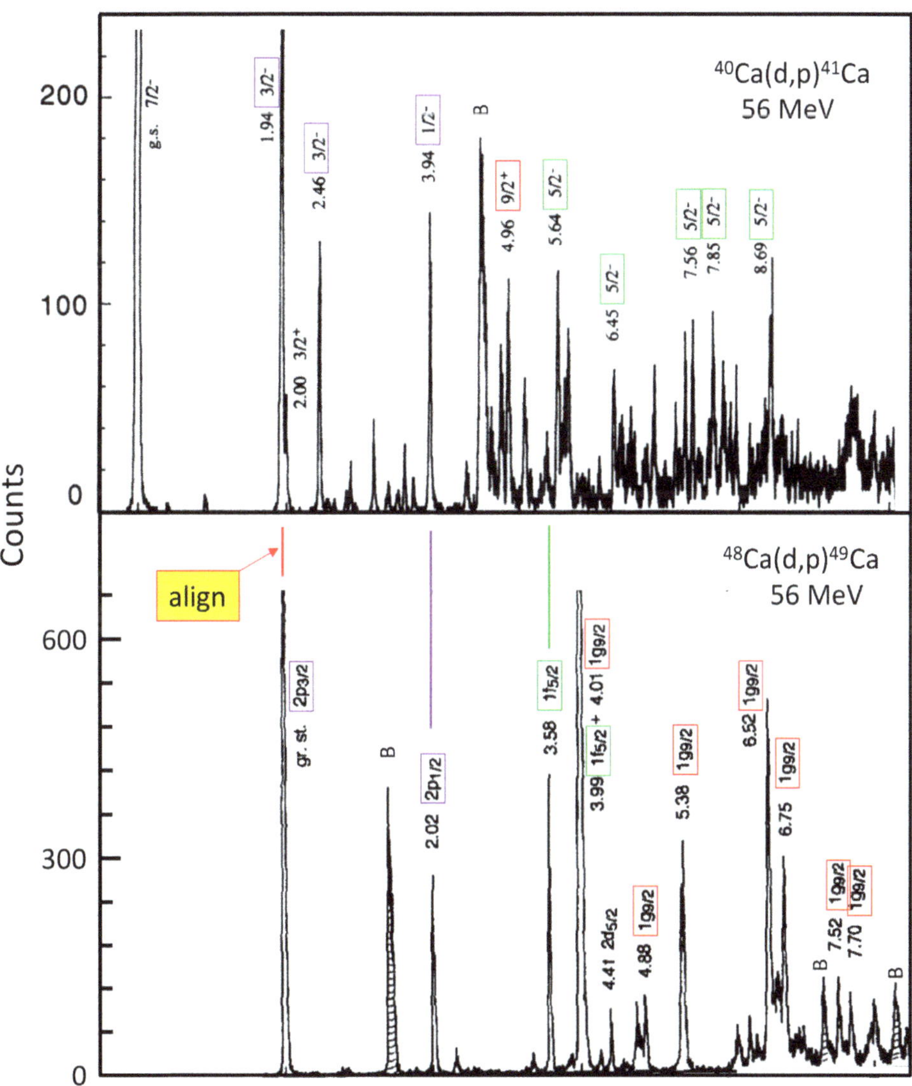

Figure 1.32. Spectra of protons following the (d,p) one-neutron transfer reaction on ^{40}Ca and ^{48}Ca targets using a beam of polarized 56 MeV deuterons. The spectra are shown so that a direct comparison can be made of the shell model states occurring in the shell above $N = 28$. The notation 'align' indicates that the proton peak corresponding to the 2p$_{3/2}$ ground-state of ^{49}Ca is aligned with the 2p$_{3/2}$ excited state in ^{41}Ca. Details are discussed in the text. Note the large fragmentation of 1f$_{5/2}$ strength in ^{41}Ca and of 1g$_{9/2}$ strength in ^{49}Ca. Reprinted from [39], copyright (1994) with the permission of Elsevier.

$2p_{3/2}$ ground state of ^{49}Ca is aligned with the $3/2^-$, 1943 keV state in ^{41}Ca. The following features should be noted. The $2p_{1/2}$ state in ^{49}Ca at 2023 keV and the $1/2^-$, 3945 keV state in ^{41}Ca 'align' closely ($1943 + 2023 = 3966$, cf 3945); the $1f_{5/2}$ state in ^{49}Ca at 3585 keV and the $5/2^-$, 5649 keV state in ^{41}Ca show a fair 'alignment' ($1942 + 3585 = 5517$, cf 5649). Thus, there is a persistence of shell model structure through the Ca isotopes. However, there is a $3/2^-$, 2462 keV state populated in ^{41}Ca which indicates that the configuration with $j = 3/2$, $l = 1$ is fragmented. This state can be identified as a member of a deformed band in ^{41}Ca, arising from the deformed structures identified in ^{40}Ca, cf figure 1.33. The spectrum shown for the ^{40}Ca(d,p)^{41}Ca one-neutron transfer reaction shows a sudden onset of complexity above 2.5 MeV and this is due to population of the large number of deformed states in ^{41}Ca above 2.5 MeV. Further discussion is pursued in chapter 3.

Figure 1.34 depicts states observed in the ^{40}Ca(p,d)^{39}Ca one-neutron transfer reaction. Besides the $J = 3/2^+$ ground state, interpreted as the $\nu 1d_{3/2}$ shell model configuration and the $J = 1/2^+$ first excited state, interpreted as the $\nu 2s_{1/2}$ shell model configuration, there are a large number of populated states, notably many

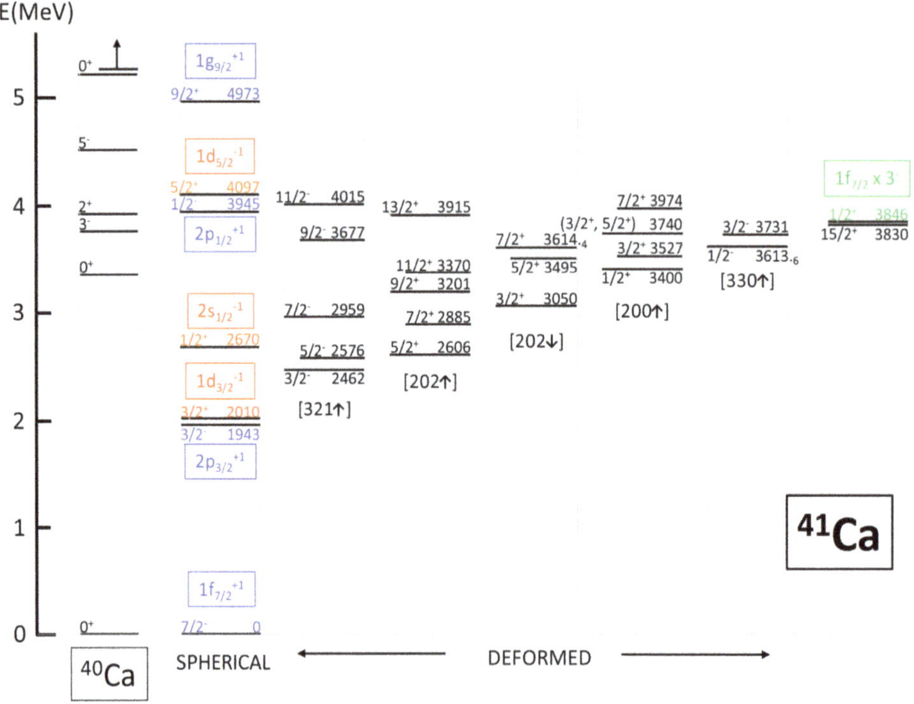

Figure 1.33. Excited states in ^{41}Ca organized into spherical and deformed structures following an interpretation suggested in [40]. The deformed states are organized into bands labelled by Nilsson model quantum numbers (see text for details). Particle (blue) and hole (orange) configurations are colour coded as per figure 1.5(b). The $1/2^+$ 3846 keV state, coloured coded in green, is assigned as the $3^- \times 1f_{7/2}$ configuration. The $15/2^+$ 3830 keV state is an isomer and the reader is referred to ENSDF for more details. There are other excited states above 3.7 MeV which are not shown. The data are taken from ENSDF.

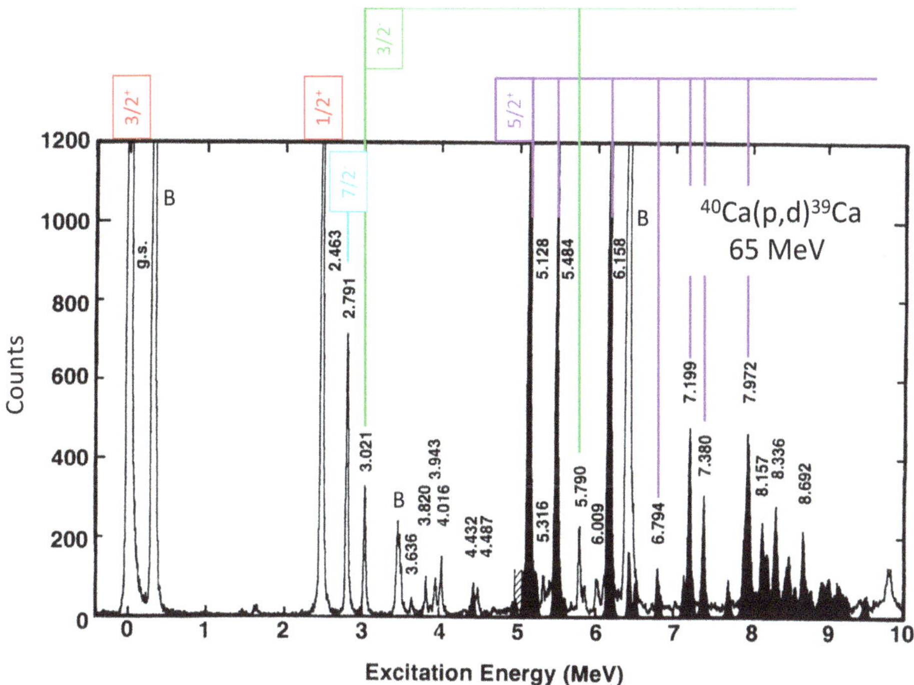

Figure 1.34. Spectrum of deuterons following the (p,d) one-neutron transfer reaction on a ^{40}Ca target using a beam of polarized 65 MeV protons. Key features to be noted are the severe fragmentation of population strength corresponding to the $1d_{5/2}$ shell model configuration and population of the $7/2^-$ state at 2796 keV which indicates that there is neutron occupancy above the shell closure in ^{40}Ca. Further details are discussed in the text. Reprinted with permission from [41]. Copyright (1993) by the American Physical Society.

states that contain fragments of $j = 5/2$, i.e. the shell model configuration $\nu 1d_{5/2}$ is severely fragmented. These states can be interpreted as deformed structures resulting from the ^{40}Ca core deformed and superdeformed structures. Such deformed structures give rise to Nilsson configurations $1/2^+$ [220], $3/2^+$ [211], $5/2^+$ [202], one set of three each for the deformed and for the superdeformed structures, a total of six fragments. This is explained in chapter 3.

Figure 1.35 depicts states observed in the ^{48}Ca(d,^{3}He)^{47}K one-proton transfer reaction. As with the ^{40}Ca(p,d)^{39}Ca one-neutron transfer reaction (see above), this is a spectroscopic view of the sd shell structure of these nuclei. A similar fragmentation of j strength is observed. While the ^{40}Ca core nucleus is known to possess multiple coexisting shapes, evidently a similar structure is implied for ^{48}Ca above an excitation of 5 MeV. This would be a new perspective on the structure of ^{48}Ca.

The foregoing view of the manifestation of shell model configurations as identifiable excitations in nuclei in the region of doubly closed shells leads to the following conclusions. Fragmentation of the population of states by one-nucleon transfer reactions, as viewed in terms of shell model configurations, is widespread. The fragmentation appears to increase with increasing excitation. We conjecture that this fragmentation is due to the presence at higher excitation of deformed (shape

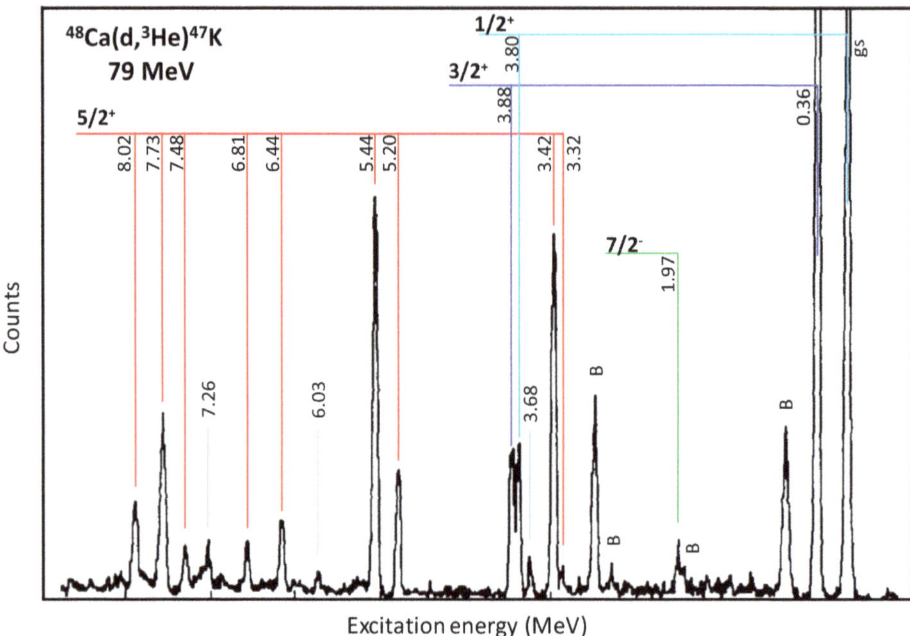

Figure 1.35. Spectrum of ^{3}He ions following the (d,^{3}He) one-proton transfer reaction on a ^{48}Ca target using a polarized beam of 79 MeV deuterons. Key features to be noted are the severe fragmentation of population strength corresponding to the $1d_{5/2}$ shell model configuration and population of the $7/2^-$ state at 1.97 MeV which indicates that there is proton occupancy above the shell closure in ^{48}Ca. Further details are discussed in the text. The ground state is on the right-hand side of the spectrum. Reprinted from [42], copyright (1985) with the permission of Elsevier.

coexisting) states: these states are populated in one-nucleon transfer reactions with known fragmentation of strength. This is discussed in chapter 4. The concept of a closed shell in nuclei appears not to be absolute, i.e. there is not always a sharp termination to the filling of shell model orbitals at the magic numbers.

1.4 Exercises

1-1 For a three-dimensional isotropic harmonic oscillator potential well the energy eigenstates can be labelled uniquely using the three quantum numbers n_x, n_y, n_z and spin-1/2 quantum numbers. Imposing the Pauli exclusion principle, show that the magic numbers (shell energy gaps) are 2, 8, 20, 40, 70, 112, 168.

1-2 Construct a view, similar to that shown in figure 1.7, for intruder states in nuclei adjacent to ^{208}Pb.

1-3 Construct a view, similar to that shown in figure 1.25, for low-lying states in ^{210}Bi.

1-4 Explore the energetics of spin couplings of broken-pair states in ^{208}Pb and ^{208}Bi using the observed energies of low-lying states in ^{207,209}Pb, ^{207}Tl, ^{209}Bi.

1-5 Explore Pauli blocking of ^{208}Pb core excitations in ^{207,209}Pb and ^{209}Bi.

1-6 Explore the excited states in ^{208}Pb and the assignment of proton and neutron 1p-1h configurations using the observed energies of low-lying states in ^{207,209}Pb, ^{207}Tl, ^{209}Bi.

1-7 Explore the excited states in ^{208}Bi and the assignment of proton and neutron configurations.

1-8 Explore the excited states in ^{210}Bi and the assignment of proton and neutron configurations.

1-9 Match relative energies of intruder states:

    (a) in ^{209}Pb to hole states in ^{207}Pb;

    (b) in ^{207}Pb to particle states in ^{209}Pb, cf figure 1.36;

    (c) in ^{209}Bi to hole states in ^{207}Tl.

Note the fragmentation of some of the transfer reaction strengths.

1-10 Using an 'inverse' of the procedure shown in figure 1.7, estimate the energy of the lowest intruder state in ^{131}Sn.

1-11 The magnetic moments of odd–odd nuclei adjacent to double-closed shells can be inspected for a simple additivity rule applied to the magnetic moments of the neighbouring odd-proton and odd-nuclei. For example, the magnetic moments for ^{133}Sb ($J^\pi = 7/2^+$, $\mu = +3.070$ μ_N) and ^{133}Sn ($J^\pi = 7/2^-$, $\mu = -1.410$ μ_N) add to $+1.660$ μ_N which can be compared with ^{134}Sb ($J^\pi = 7^-$, $\mu = +1.74519$ μ_N). Using data in ENSDF and XUNDL explore this simple additivity rule.

1-12 Figure 1.4 reveals a remarkable symmetry between ^{15}O and ^{15}N, and between ^{17}F and ^{17}O. These nuclei are described as 'mirror pairs' with respect to the proton and neutron numbers, i.e. $A = 15$, $\{N, Z\} = \{8, 7\}$ and $\{7, 8\}$ and similarly for $A = 17$. Using data in ENSDF, explore the

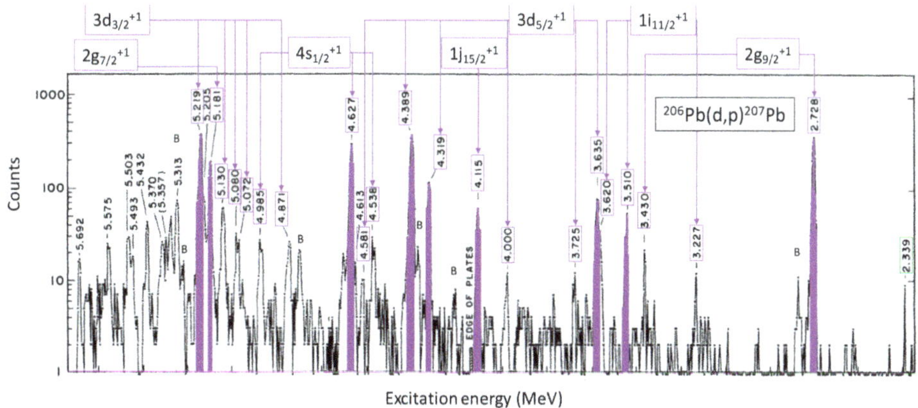

Figure 1.36. Spectrum of protons following the (d,p) one-neutron transfer reaction on ^{206}Pb target using a beam of 17 MeV deuterons. The peaks corresponding to 1p-2h configuration strengths are highlighted (in violet), note the fragmentation of shell model configurations. Peaks marked 'B' are due to target impurities. Further details are discussed in the text. Reprinted with permission from [43]. Copyright (1970) by the American Physical Society.

survival of the similarity as the excitation energy is increased in these pairs of nuclei. Consider also applying this view to the mirror pairs in figure 1.5.

1-13 The matching of states in mirror-pair nuclei can sometimes be used to make spin-parity assignments in one nucleus (usually the one further from the beta stability line), where the assignment is uncertain, by matching to the other nucleus where the assignment is made with some confidence. Apply this idea to ^{41}Sc/^{41}Ca, using data in ENSDF and figure 1.33 as a guide.

1-14 Just as one finds mirror-pair nuclei, so mirror triplets occur, e.g. $A = 42$, $\{N, Z\} = \{22, 20\}$—^{42}Ti, $\{21, 21\}$—^{42}Sc, $\{20, 22\}$—^{42}Ca. Using data in ENSDF match the states with similar energies in these three nuclei. Note there are additional states in ^{42}Sc. Use the 'm-scheme' (see section 4.2 of [11]) to explain this. No knowledge of the interactions behind these patterns is needed: the data reveal that the p–p, p–n, and n–n interactions must be essentially identical.

1-15 While the doublets in exercises 1-12 and 1-13 and the triplet in exercise 1-14 are nearly identical with respect to relative energies, their absolute energies are very different. Using nuclear mass data, what are the energy differences?

1-16 The energy differences between members of isobaric multiplets are dominated by Coulomb repulsion of the protons. Using a simple model of a uniformly charged sphere of radius, $R = 1.2A^{1/3}$ fm, estimate these Coulomb energy differences. Suggestion: make the simplistic approximation of a proton in surface contact with the $A - 1$ nucleus, taking the proton to be a point charged object.

1.5 Video-based tutorials

1-17 For those reading the e-book, explore preliminary views of nuclei and concepts around the energy scale of nuclear structure in the video-based tutorial found in figure 1.37.

1-18 For those reading the e-book, explore the concept of magic numbers and spherical nuclei in the video-based tutorial found in figure 1.38.

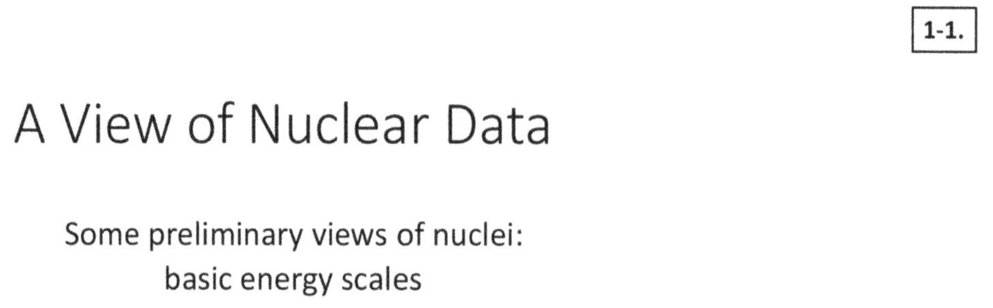

1-1.

A View of Nuclear Data

Some preliminary views of nuclei:
basic energy scales

Figure 1.37. Video tutorial: Some preliminary views of nuclei. Video available at https://doi.org/10.1088/978-0-7503-5648-0.

1-19 For those reading the e-book, explore more about the relevance of one-nucleon transfer reactions as an aid to nuclear structure in the video-based tutorial found in figure 1.39.

1-20 For those reading the e-book, explore more about the need for spin–orbit coupling and a flat-bottomed potential in the video-based tutorial found in figure 1.40.

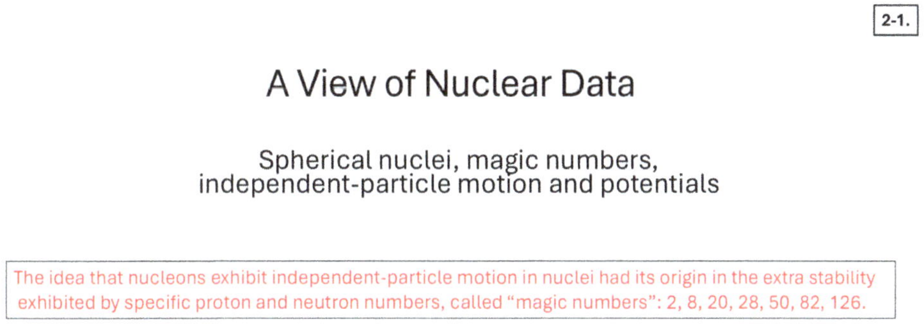

2-1.

A View of Nuclear Data

Spherical nuclei, magic numbers,
independent-particle motion and potentials

The idea that nucleons exhibit independent-particle motion in nuclei had its origin in the extra stability exhibited by specific proton and neutron numbers, called "magic numbers": 2, 8, 20, 28, 50, 82, 126.

Figure 1.38. Video tutorial: Magic numbers. Video available at https://doi.org/10.1088/978-0-7503-5648-0.

3-1.

A View of Nuclear Data

One-nucleon transfer reaction spectroscopic view of
independent-particle states in nuclei

Figure 1.39. Video tutorial: One-nucleon transfer reactions. Video available at https://doi.org/10.1088/978-0-7503-5648-0.

4-1.

A View of Nuclear Data

The need for the spin-orbit coupling term and a
"flat-bottomed" potential

Spin-orbit coupling is the key insight that launched the shell model of the nucleus and, indeed it resulted in the first quantum mechanical view of nuclear structure.

Here we show what "removal" of an empirically determined spin-orbit coupling energy does to the view of states in an odd-mass nucleus adjacent to a doubly closed shell.

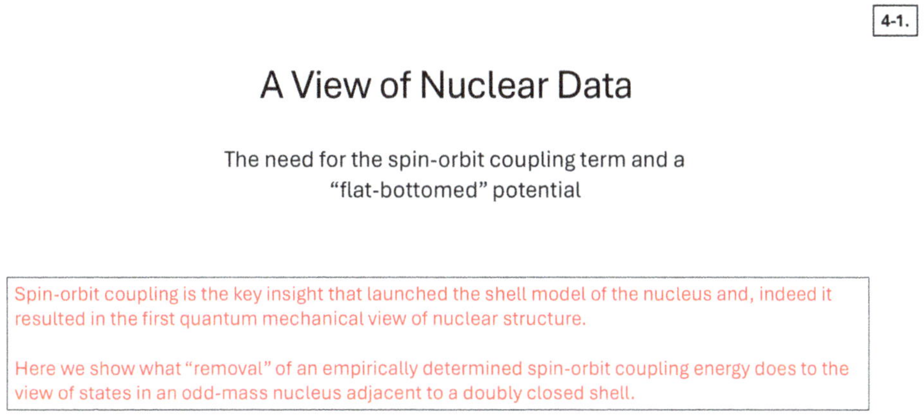

Figure 1.40. Video tutorial: Spin–orbit coupling. Video available at https://doi.org/10.1088/978-0-7503-5648-0.

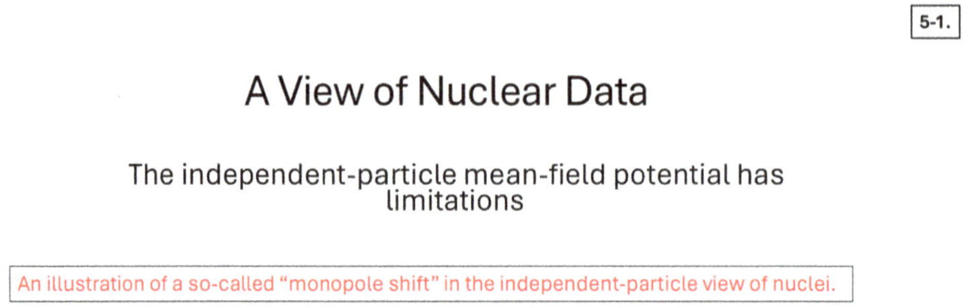
5-1.
A View of Nuclear Data
The independent-particle mean-field potential has limitations
An illustration of a so-called “monopole shift” in the independent-particle view of nuclei.

Figure 1.41. Video tutorial: Monopole shift. Video available at https://doi.org/10.1088/978-0-7503-5648-0.

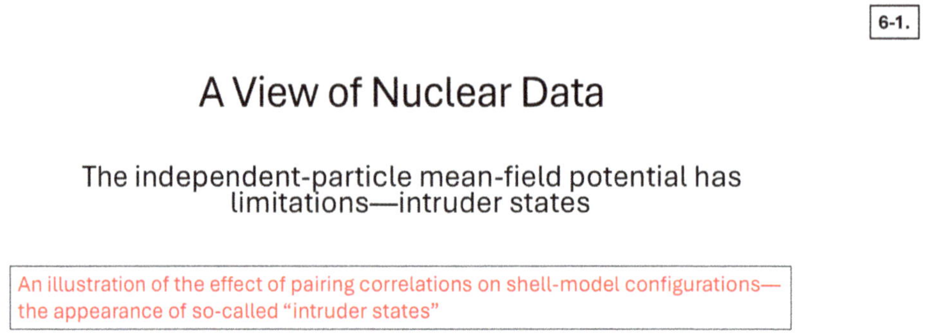
6-1.
A View of Nuclear Data
The independent-particle mean-field potential has limitations—intruder states
An illustration of the effect of pairing correlations on shell-model configurations—the appearance of so-called “intruder states”

Figure 1.42. Video tutorial: Intruder states. Video available at https://doi.org/10.1088/978-0-7503-5648-0.

1-21 For those reading the e-book, explore more about the relevance of the monopole shift in the video-based tutorial found in figure 1.41.

1-22 For those reading the e-book, explore more about the origin of intruder states in the video-based tutorial found in figure 1.42.

References

[1] Rowe D J and Wood J L 2010 *Fundamentals of Nuclear Models: Foundational Models* (Singapore: World Scientific)

[2] Mayer M G 1948 On closed shells in nuclei *Phys. Rev.* **74** 235

[3] Feenberg E and Hammack K C 1949 Nuclear shell structure *Phys. Rev.* **75** 1877

[4] Goeppert-Mayer M 1949 On closed shells in nuclei II *Phys. Rev.* **75** 1969

[5] Haxel O, Hans J, Jensens D and Suess H E 1949 On the ‘Magic numbers’ in nuclear structure *Phys. Rev.* **75** 1766

[6] Barnes P D, Romberg E, Ellegaard C, Casten R F, Hansen O, Mulligan T J, Broglia R A and Liotta R 1972 Proton-hole states in ^{209}Bi from the ^{210}Po(t,α) reaction *Nucl. Phys.* A **195** 146

[7] Cleary T P, Callender W D and Sheline R K 1980 ^{210}Bi(d,t) reaction; ^{210}Bi^{m}(9^-) parentage of states in ^{209}Bi *Phys. Rev.* C **21** 2244

[8] Igo G, Flynn E R, Dropesky B J and Barnes P D 1971 ^{210}Pb(p,d)^{209}Pb reaction at 20.6 MeV *Phys. Rev.* C **3** 349

[9] Igo G J, Barnes P D and Flynn E R 1970 Mixing of two-particle, two-hole states in ^{208}Pb *Phys. Rev. Lett.* **24** 470

[10] Crawford H L *et al* 2017 Unexpected distribution of $\nu 1f_{7/2}$ strength in ^{49}Ca *Phys. Rev. C* **95** 064317

[11] Jenkins D G and Wood J L 2021 *Nuclear Data: A primer* (Bristol: IOP Publishing)

[12] Vanhalst M 2014 Quantifying short-range correlations in nuclei *PhD Thesis* Ghent University, Ghent

[13] Lapikás L 1993 Quasi-elastic electron scattering off nuclei *Nucl. Phys.* A **553** 297–308

[14] Adhikari D *et al* 2021 Accurate determination of the neutron skin thickness of ^{208}Pb through parity-violation in electron scattering *Phys. Rev. Lett.* **126** 172502

[15] Ji X, Yuan F and Zhao Y 2021 What we know and what we don't know about the proton spin after 30 years *Nat. Rev. Phys.* **3** 27–38

[16] Riedl C 2022 Probing nucleon spin structure in deep-inelastic scattering, proton-proton collisions, and Drell-Yan processes *Acta Phys. Pol.* **B53** 5–A2

[17] Jenkins D G and Wood J L 2023 *Nuclear Data: A Collective Motion View* (Bristol: IOP Publishing)

[18] Mairle G and Grabmayr P 2000 A phenomenological spin-orbit and nuclear potential for ^{208}Pb *Eur. Phys. J.* A **9** 313–26

[19] Heusler A *et al* 2016 Complete identifiction of states in ^{208}Pb below $E_x = 6.2$ MeV *Phys. Rev. C* **93** 054321

[20] Lápikas L 1985 Recent results of high-resolution (e,e'p) reactions at Nikhef-K *Nucl. Phys.* A **434** 85c–96c

[21] Flynn E R, Hardekopf R A, Sherman J D, Sunier J W and Coffin J P 1977 Spectroscopy of proton hole states in the thallium nuclei with the (t,α) reaction *Nucl. Phys.* A **279** 394–412

[22] Dickey S A, Kraushaar J J and Rumore M A 1982 The energy dependence of the spectroscopic factors for the ^{208}Pb(p,d)^{207}Pb reaction *Nucl. Phys.* A **391** 413–31

[23] Kovar D G, Stein N and Bockelman C K 1974 Excitation of two-particle-one-hole states by the ^{208}Pb(d,p)^{209}Pb reaction *Nucl. Phys.* A **231** 266–300

[24] Ellegaard C, Patnaik B and Barnes P D 1970 Fragmentation of the single-particle strength in ^{209}Bi *Phys. Rev. C* **2** 2450

[25] Grabmayr P, Mairle G, Schmidt-Rohr U, Berg G P A, Meissburger J, von Rossen P and Tain J L 1987 Proton particle-hole states in ^{208}Pb *Nucl. Phys.* A **469** 285–312

[26] Vold P B, Andreassen J O, Lien J R, Graue A, Cosman E R, Dünnweber W, Schmitt D and Nüsslin F 1973 The ^{207}Pb(d,p)^{208}Pb reaction with high resolution *Nucl. Phys.* A **215** 61–78

[27] Ungrin J, Diamond R M, Tjøm P O and Elbek B 1971 Inelastic deuteron scattering in the lead region *Mat. Fys. Medd. Dan. Vid. Selsk.* **38** 34

[28] Alford W P, Schiffer J P and Schwartz J J 1971 Shell-model structure of ^{208}Bi *Phys. Rev. C* **3** 860

[29] Crawley G M, Kashy E, Lanford W and Blosser H G 1973 High-resolution study of the particle-hole multiplets in ^{208}Bi *Phys. Rev. C* **8** 2477

[30] Jones K L *et al* 2010 The magic nature of ^{132}Sn explored through the single-particle states of ^{133}Sn *Nature* **465** 454–7

[31] Kozub R L *et al* 2012 Neutron single particle structure in ^{131}Sn and direct neutron capture cross sections *Phys. Rev. Lett.* **109** 172501

[32] Allmond J M *et al* 2014 Double-magic nature of ^{132}Sn and ^{208}Pb through lifetime and cross-section measurements *Phys. Rev. Lett.* **112** 172701

[33] Vaquero V *et al* 2017 Gamma decay of unbound neutron-hole states in ^{133}Sn *Phys. Rev. Lett.* **118** 202502

[34] Vaquero V *et al* 2020 Fragmentation of single-particle strength around the doubly magic nucleus ^{132}Sn and the position of the 0f $_{5/2}$ proton-hole state in ^{131}In *Phys. Rev. Lett.* **124** 022501

[35] Jones K L *et al* 2022 Neutron transfer reactions on the ground state and isomeric state of a ^{130}Sn beam *Phys. Rev. C* **105** 024602

[36] Schmidt T H 1937 Über die magnetischen Momente der Atomkerne *Z. Phys.* **106** 358

[37] Neugart R, Stroke H H, Ahmad S A, Duong H T, Ravn H L and Wendt K 1985 Nuclear magnetic moment of ^{207}Tl *Phys. Rev. Lett.* **55** 1559

[38] Papuga J *et al* 2013 Spins and magnetic moments of ^{49}K and ^{51}K: establishing the $1/2^+$ and $3/2^+$ level ordering beyond N = 28 *Phys. Rev. Lett.* **110** 172503

[39] Uozumi Y *et al* 1994 Single-particle strengths measured with ^{48}Ca(d,p)^{49}Ca reaction at 56 MeV *Nucl. Phys. A* **576** 123–37

[40] Röpke H 2004 Systematics of deformed states around doubly-magic ^{40}Ca *Eur. Phys. J. A* **22** 213–30

[41] Matoba M *et al* 1993 ^{40}Ca(p,d)^{39}Ca reaction at 65 MeV *Phys. Rev. C* **48** 95

[42] Banks S M *et al* 1985 The ^{48}Ca(d,^{3}He)^{47}K reaction at 80 MeV *Nucl. Phys. A* **437** 381–96

[43] Moyer R A, Cohen B L and Diehl R C 1970 Excitation of states in Pb^{207} by Pb^{208}(d,t), Pb^{206}(d,p), and Pb^{207}(d,d′) reactions *Phys. Rev. C* **2** 1898

IOP Publishing

Nuclear Data
An independent-particle motion view
David Jenkins and John L Wood

Chapter 2

How widely are shell model states seen in nuclei?

The manifestation of shell model configurations away from doubly closed shell regions is surveyed. Most notably, singly closed shell regions are considered. The In and Sb isotopes are selected as case studies. Some structural features of singly closed shell nuclei are briefly sketched.

Concepts: shell-model state systematics, particle states, hole states, intruder states and their deformation, particle–core coupling, weak coupling, Coulomb excitation of collective states, enhanced and retarded electromagnetic transitions, one-nucleon transfer reactions spectroscopy, transfer reaction spectroscopic strengths, energy-weighted centroids of spectroscopic strengths, spin–orbit energy splitting, magnetic moments, isotope shifts, quadrupole moments, deformation parameters.

Learning outcomes: The limitations of the shell model for interpreting low-energy states in nuclei adjacent to closed shells is illustrated.

A much wider view of shell model descriptions of nuclei is available if nuclei adjacent to singly closed shells are considered. However, this comes at the price of greater complexity. Most notably, intruder states appear at medium-low excitation energy: such states are strongly deformed. Odd-mass closed shell nuclei beyond those adjacent to double-closed shells require a thorough treatment of pairing correlations. This is a more advanced topic which is beyond the present level of discussion.

Because of the structural complexity of nuclei adjacent to closed shells, it is necessary to establish a baseline of behaviour where detailed spectroscopic infor-mation is available. We choose the In ($Z = 49$) and Sb ($Z = 51$) isotopes to introduce such structure. These two isotopic chains are chosen because spectroscopic data extend across the entire $50 < N < 82$ shell and there are many different types of spectroscopic information.

The In and Sb isotopic chains reveal a variety of basic structural features besides independent-particle and -hole degrees of freedom, notably intruder states and particle- and hole–core coupling (where 'core' refers to the neighbouring tin, $Z = 50$

doi:10.1088/978-0-7503-5648-0ch2 2-1

isotopes): these structures are presented in some detail. Indeed, these degrees of freedom must be carefully characterized in order to extract the basic spectroscopic information pertinent to the shell model.

2.1 A case study view: the In ($Z = 49$) isotopes

Figure 2.1 provides an overview of the odd In isotopes for which there are useful spectroscopic data. The range of information spans the complete 50–82 shell of neutrons. The naïve shell model view would predict $J^\pi = 9/2^+$ for the ground states, resulting from a proton hole in the $1g_{9/2}$ configuration expected for $Z = 49$, and this is observed. The first excited state with $J^\pi = 1/2^-$ conforms to the shell model expectation of a proton hole in the $2p_{1/2}$ configuration. However, the second excited state with $J^\pi = 3/2^-$ exhibits an energy separation with respect to the $1/2^-$ state that does not conform to expectations of a $2p_{3/2}$–$2p_{1/2}$ spin–orbit pair; albeit in ^{131}In this energy separation (988 keV) is consistent with the expectation for spin–orbit coupling—the value deduced for the parameter C (cf equation (1.4)) is -0.66 MeV. In the mid-shell region intruder states are observed; and high-spin positive-parity states are observed above an excitation $\sim$1 MeV and can be associated with the $1g_{9/2}$ proton hole coupled to the $^{A+1}$Sn 2^+ core excitation. We address each of these structural features below.

Figure 2.2 shows the lowest energy negative-parity states observed in the odd In isotopes. The systematics show a compression in energy which is greatest near mid-shell. The departure of the $3/2^-$–$1/2^-$ energy difference from that expected for

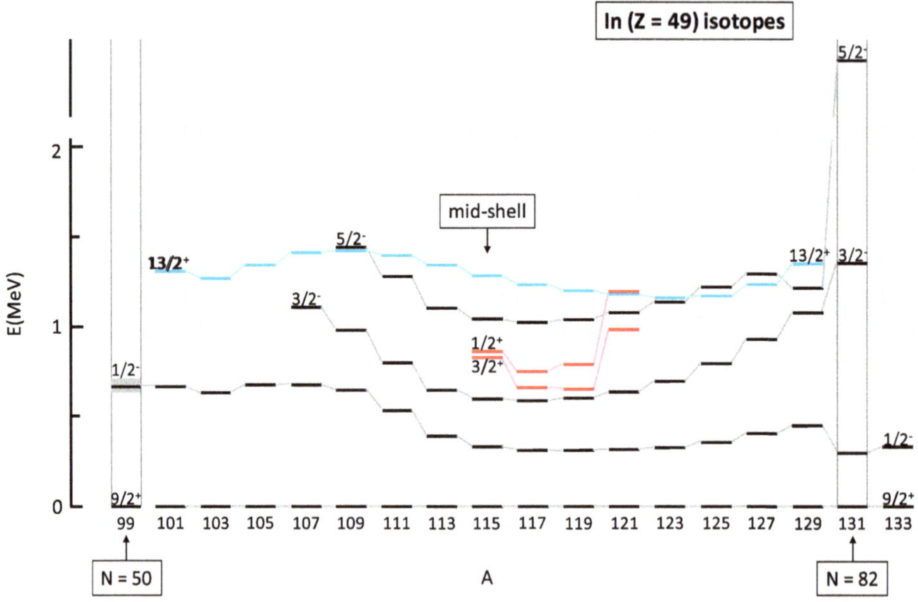

Figure 2.1. Systematics of selected states in the odd-mass indium isotopes. The states highlighted in red are intruder states. The states highlighted in blue are due to excitations in the $^{A+1}$Sn cores. The details for these and associated states are presented in following figures. The data are taken from [1–7] and ENSDF.

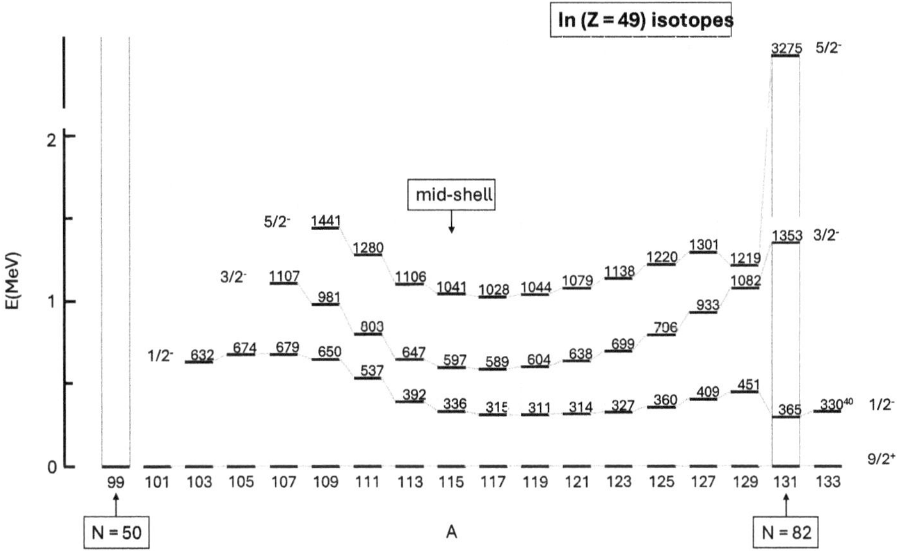

Figure 2.2. Systematics of the negative-parity states in the odd-mass indium isotopes. Naively, the negative-parity states could be interpreted as the shell model single proton–hole configurations $2p_{1/2}$, $2p_{3/2}$, and $1f_{5/2}$; but E2 transition strengths would be desirable before such an interpretation is made. The 'parabolic' energy trend suggests interactions with neutrons throughout the shell with a characteristic energy minimum near the mid-shell point ($N = 66$), as indicated. Note that there is a severe deficiency of data for electromagnetic decay strengths: there is one half life, for the 589 keV state in [117]In, and the E2/M1 mixing ratio for the decay of this state is ambiguous. Note that the energy separation of the $1/2^-$ and $3/2^-$ states, while consistent with spin–orbit splitting for $2p_{1/2}$–$2p_{3/2}$ in [131]In, is seriously at odds with a spin–orbit splitting strength progressively as the mid-shell point is approached. Data for [107]In are from [8] and for [131]In are from [9]; other data are taken from ENSDF.

spin–orbit splitting is noted above. The $1/2^-$, $3/2^-$, $5/2^-$ states possibly form a weakly deformed band, but spectroscopic data needed to assess this are not available. To investigate this, lifetimes, and δ(E2/M1) mixing ratios for the $\Delta I = 1$ transitions are needed. There is only a lifetime reported for the $3/2^-$ state in [117]In, as shown. Mixing ratios are limited to [115,117]In, with ambiguity for the value in [117]In.

Figure 2.3 shows the evidence for intruder configurations in [115,117]In. The highly retarded and enhanced electromagnetic transition strengths are the signature of such a structure; this was an historical 'first' for the identification of intruder states in heavy nuclei [10]. Figure 2.4 confirms the intruder structure in [117]In from the observed differences in population of states via one-proton addition and one-proton removal transfer reactions. Indeed, the difference reveals that the closed shell at $Z = 50$ must be well fulfilled: the (d,^{3}He) proton-removal reaction does not observably populate the intruder states.

Figure 2.5 provides a detailed view of all of the low-energy excited states in [115,117,119]In, organized into candidates for shell model proton–hole states, intruder states, and $1g_{9/2}^{-1} \times {}^{A+1}\mathrm{Sn}(2_1^+)$ states. The spectroscopic evidence for the $1g_{9/2}^{-1} \times {}^{A+1}\mathrm{Sn}(2_1^+)$ states in [115]In is shown in figure 2.6 which summarizes Coulomb excitation data for this nucleus. The candidate states forming the

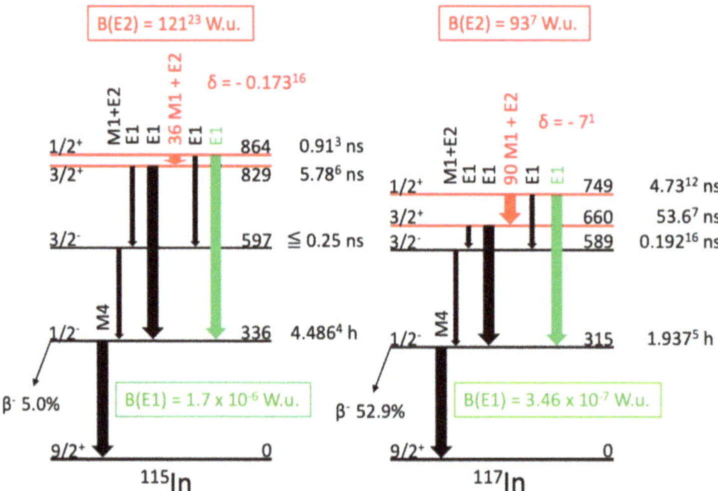

Figure 2.3. Intruder states in ^{115,117}In (levels highlighted in red) and the electromagnetic transition strengths that establish the shape coexistence associated with them. The 'intruder' nature reflects the impossibility of such states arising from a simple 2^+ core excited state coupling to the $9/2^+$ ground state, i.e. the $1/2^+$, $3/2^+$ spins must arise from a structure 'from somewhere else'. The figure resembles that in the original work [10] with only minor changes resulting from the most up-to-date information. The widths of the arrows qualitatively reflect the intensities of the transitions. Further details are discussed in the text. The juxtaposition of highly collective and highly retarded electromagnetic transitions, involving a common state, is a strong spectrscopic fingerprint of shape coexistence. The data are taken from ENSDF.

$5/2^+$–$13/2^+$ quintuplet resulting from the $1g_{9/2}^{-1} \times {}^{A+1}Sn(2_1^+)$ coupling are shown for all the In isotopes where observed in figure 2.7. Note that there is mixing implied in ^{115}In for the $5/2^+$ and $9/2^+$ configurations stemming from the intruder band and the core-hole coupling, as revealed in the Coulomb excitation study. A similar mixing in ^{117}In is also implied from the transfer data, cf figure 2.4. Figure 2.8 shows the systematics of the intruder band members. These conform to the expectation of a $1/2^+[431]$ Nilsson configuration. The decoupling parameter would be $a = -3$ in ^{117}In, cf figure 4.8.

There is a notably complete view of moment data for the odd In isotopes. Figure 2.9 shows the systematics of the ground-state magnetic moments values, i.e. associated with the $1g_{9/2}$ proton–hole configuration. Figure 2.10 shows a similar view for the $1/2^-$ first excited state systematics (note that this state is a long-lived isomer and so is easily isolated and transported to a laser hyperfine spectroscopy experimental setup). Figure 2.11 shows the systematics of the ground-state spectroscopic, Q_s quadrupole moment values. Figure 2.12 shows the In isotope shift, IS data.

The moment data show smooth systematic trends as a function of the changing neutron number. We make the following observations:

1. The ground-state $1g_{9/2}$ configurations show associated collective excitations that can be interpreted as weak coupling to the $^{A+1}$Sn(2_1^+) core configurations q.v. Figure 2.7; but deformation is implied from the spectroscopic quadrupole moment data q.v. Figure 2.11.

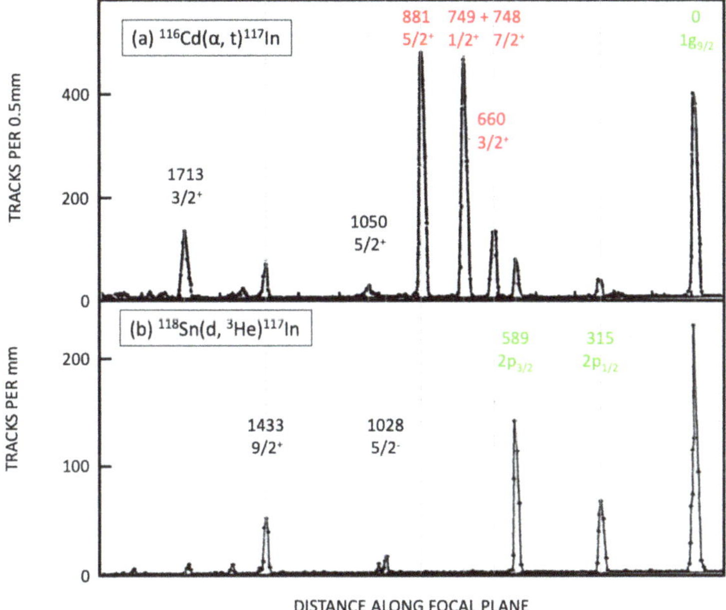

Figure 2.4. Spectroscopic proof of the 'particle' and 'hole' nature of low-lying states in ^{117}In. (a) Spectra for magnetically analysed tritons from the ^{116}Cd(α,t)^{117}In reaction. Reprinted from [11], copyright (1972) with the permission of Elsevier. (b) Spectra for magnetically analysed ^{3}He from the ^{118}Sn(d,^{3}He)^{117}In reaction. Reprinted with permission from [12]. Copyright (1971) by the American Physical Society. The peaks, corresponding to the ground state and excited states in ^{117}In, are labelled by the energies (in keV) of the states and by shell model configurations for spherical states and by spin-parities for deformed intruder states (cf figure 2.5). Further details are discussed in the text.

2. The magnetic moment data for both the $1g_{9/2}$ and the $2p_{1/2}$ configurations imply important collective contributions based on the smooth systematic trends q.v. Figures 2.9 and 2.10. Note that while the $2p_{1/2}$ magnetic moments pass through the extreme independent-particle (Schmidt) value, this is part of a broad and smooth systematic trend and so is likely fortuitous.

3. One-proton transfer reaction data imply that $Z = 50$ is a good shell closure, at least in ^{118}Sn q.v. Figure 2.4.

4. With modern accessibility to radioactive beams, Coulomb excitation of the $1/2^-$ isomeric states would provide an excellent test of the collectivity associated with this configuration.

5. We note that radioactive beams are now widely available, especially for atomic hyperfine structure studies using lasers and more recently for Coulomb excitation studies and nucleon knockout reaction studies (of the beam species). Such studies benefit from high beam intensities and availability across long chains of isotopes of some elemental species (such as the In isotopes [3, 6, 7]). Ultimately there is the limitation of short half lives; current laser hyperfine spectroscopy of atoms has reached, e.g. ^{11}Li ($T_{1/2} = 8.75$ ms). Manifestly, excited states which are isomeric progressively

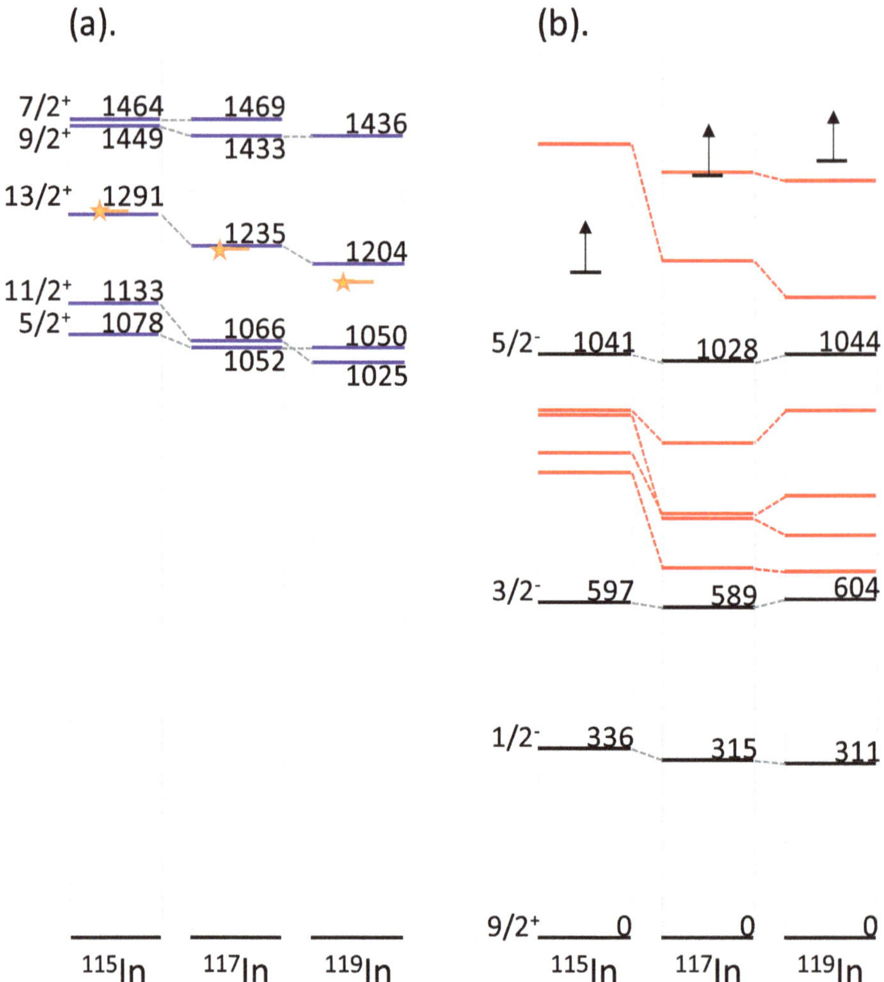

Figure 2.5. Systematics of the low-energy states in ^{115,117,119}In up to $\sim$1.2 MeV. (a) States that are candidates for the $^{A+1}\mathrm{Sn}(2_1^+) \times 1\mathrm{g}_{9/2}^{-1}$ core-particle coupling. (b) States that are candidates for the $1\mathrm{g}_{9/2}^{-1}$, $2\mathrm{p}_{1/2}^{-1}$, $2\mathrm{p}_{3/2}^{-1}$, $1\mathrm{f}_{5/2}^{-1}$ shell model configurations and the intruder band (Nilsson $1/2^+[431]$ configuration highlighted in red). Horizontal bars with vertical upwards-pointing arrows indicate excitation energies above which states are omitted. The data are taken from ENSDF.

become accessible to study on an equal footing with ground states. We underline, the study of nuclear structure is preeminently the study of many-body quantum mechanics as realized in finite systems: as such, *systematic* study of long chains of isotopes and isotones is crucial for any hope of a comprehensive view.

The overall view of the shell model provided by the odd-mass In isotopes suggests that collectivity is emerging in the negative-parity shell-model state candidates, but the ground state (positive-parity, spin 9/2 state) may be dominated by the shell

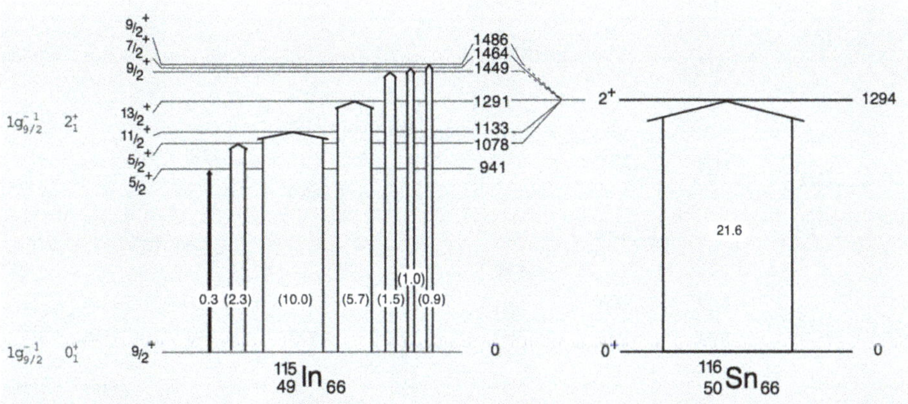

Figure 2.6. Coulomb excitation of ^{115}In. The states populated include the $1g_{9/2} \times{}^{116}$Sn(2_1^+) quintuplet, cf figure 2.5(a), and the intruder states, $J^\pi(E_x$ keV) $= 5/2^+(941)$, $9/2^+(1486)$, cf figure 2.5(c). The E2 excitation strengths are in units of e^{2}fm$^4 \times 100$. Note that the summed excitation strength in ^{115}In is 21.7 units, cf 21.6 units in ^{116}Sn. Note also that there are spin factors underlying the observed intensities for Coulomb excitation: these factors take the form $(2I_f + 1)/(2I_i + 1)$, reflecting the fact that the final-state spin, I_f provides accessibility to a multiplicity of final spin substates m_{I_f} and the initial state, with spin, I_i is limited to a single component of the spin substates m_{I_i} per excitation from the ensemble of nuclei. The E2 data are taken from [13]; other data are taken from ENSDF.

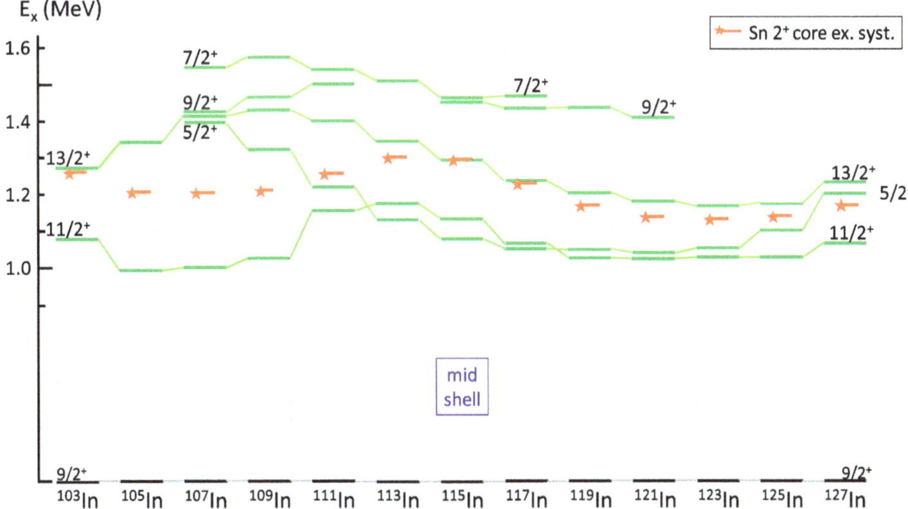

Figure 2.7. Systematics of the candidate states for $^{A+1}$Sn(2_1^+) $\times 1g\,\bar{9}/2$ core-particle coupling across the majority of the In isotope sequence. The energies of the $^{A+1}$Sn 2_1^+ states are shown in orange with stars. The data are taken from ENSDF.

model $1g_{9/2}$ configuration. These observations are based on the lowest $1/2^-$, $3/2^-$, $5/2^-$ states suggesting a band structure and the non-intruder positive-parity states suggesting weak-coupling, $j = 9/2 \times {}^{A+1}$Sn(2^+) quintuplets. A subtle and profound point is the evidence that j mixing is possible for the proton–hole, shell-model

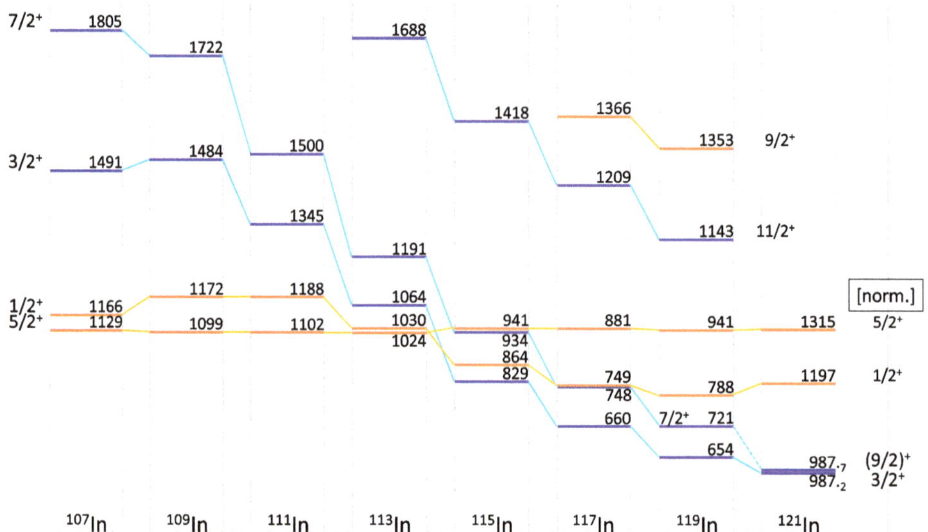

Figure 2.8. Systematics of the candidate states for the Nilsson $1/2^+$[431] intruder band across the In isotopes for which there is confidence in its identification, shown relative to the $5/2^+$ band member (arbitrarily chosen). The $1/2^+$, $5/2^+$, $9/2^+$, $\cdots$ and $3/2^+$, $7/2^+$, $11/2^+$, $\cdots$ spin sequences are distinguished by orange and blue, respectively, colour coding. The data are taken from ENSDF.

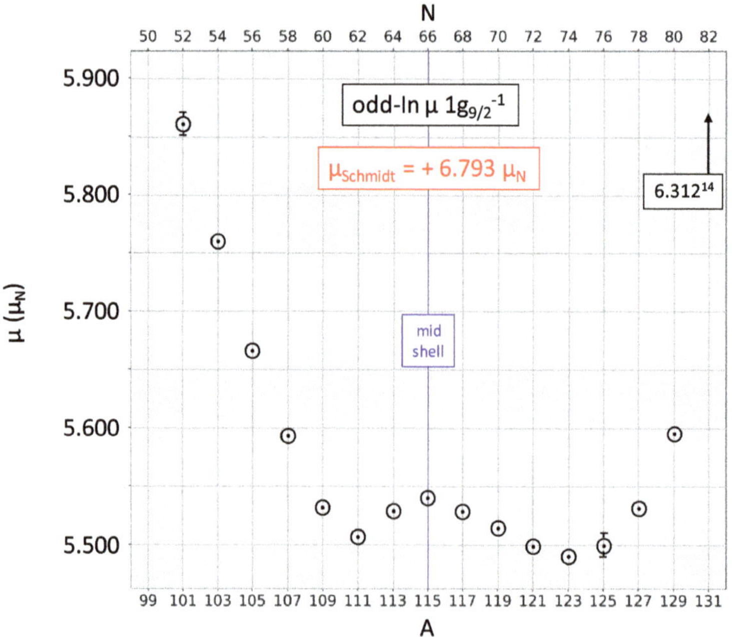

Figure 2.9. Systematics of the magnetic moments for the odd-In ground states, the $1g_{9/2}^{-1}$ configuration. The value for ^{131}In is off scale. Note the scale used is such that the variation in the magnetic moments shown is only ~4%; and the average deviation from the Schmidt value is only ~15%. The value for the singly closed shell isotope ^{131}In, $6.312^{14}\mu_N$ differs from the Schmidt value by 7%. The data are taken from [7, 14].

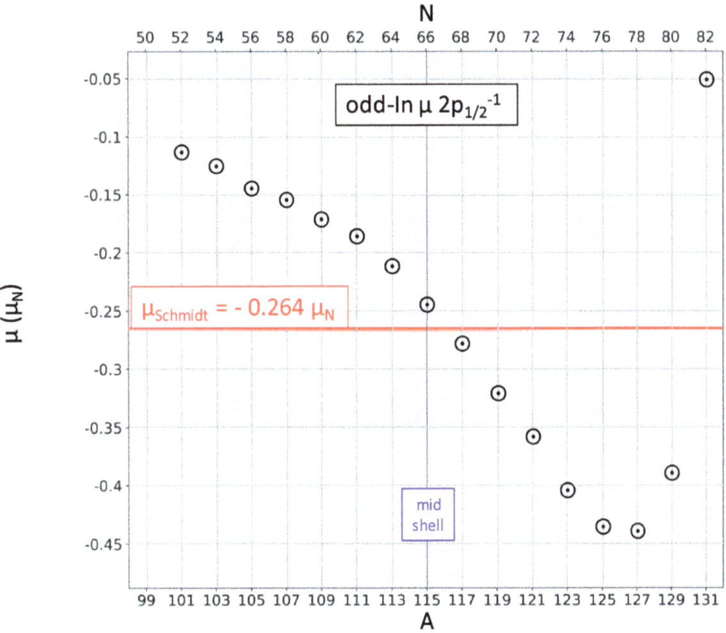

Figure 2.10. Systematics of the magnetic moments for the odd-In long-lived isomeric states, the $2p_{1/2}^{-1}$ configuration. Note that while the Schmidt value is intercepted, the variation in some of these magnetic moments relative to the Schmidt value is large, 50%–200%. The value for the isomer in the singly closed shell isotope ^{131}In, $-0.0513\ \mu_N$ differs in a major way from the Schmidt value. Some discussion of this is given in [6]. The data are taken from [7, 14].

configurations with negative parity but there is only the lone positive-parity configuration ($1g_{9/2}$). The mixing between the proton–hole positive-parity configuration and the intruder positive-parity (one-particle-two-hole) configuration is weak. We note that, fundamentally, j is a good quantum number in a spherical mean field but not in a deformed mean field; conversely, j mixing produces deformation. We discuss these observations further in the next section on the Sb isotopes.

2.2 A case study view: the Sb ($Z = 51$) isotopes

Figure 2.13 provides an overview of the odd Sb isotopes for which there are useful spectroscopic data. The range of information spans almost the complete 50–82 shell of neutrons. The naïve shell model view would predict $J^{\pi} = 7/2^{+}$ for the ground states, resulting from a proton particle in the $1g_{7/2}$ configuration. But there is a surprise: the systematics reveal a dramatic shift in the energy of the lowest $J^{\pi} = 7/2^{+}$ state in the Sb isotopes. This is not expected based on the naïve shell model, and it has received widespread discussion in the literature spanning many decades. We discuss this feature below, based on recent data. We also note that there is no sign of such a dramatic shift in the In isotopes, cf above, which possess the same closed shell Sn 'cores'.

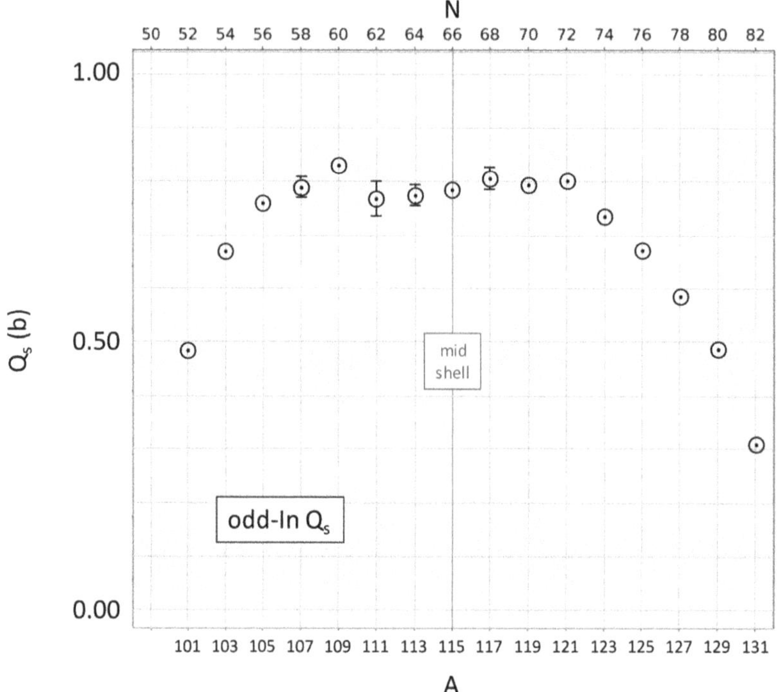

Figure 2.11. Systematics of the spectroscopic quadrupole moments for the odd-In ground states, the $1g_{9/2}^{-1}$ configuration. Note that all spectroscopic quadrupole moments for the low-spin In isomers are zero because they have $J = 1/2$ (see text and exercises). Uncertainties, where larger than the data points, are indicated by error bars. The data are taken from [7, 14].

Figures 2.14(a)–(c) show a summary of spectroscopic strengths for one-proton addition reactions on tin targets. This is a unique view, which covers all the stable tin targets and has been obtained with attention to uniform conditions for data acquisition and reduction to spectroscopic strengths. The distribution of strength as a function of excitation energy is complex and is beyond the present level of discussion, but the energy-weighted strengths can be summed to produce the so-called 'centroids' of single-particle strength which is the focus of figures 2.14(a)–(c). The centroids are not shown for the shell model configurations $2d_{5/2}$ and $2d_{3/2}$ because the data are unable to distinguish between them due to the identical angular distribution patterns for population of states with $J^\pi = 5/2^+$ and $J^\pi = 3/2^+$. Further details are discussed below.

The most important feature to note in the spectroscopic strengths for the odd-mass Sb isotopes is that the summed strength for the $J^\pi = 7/2^+$ states is essentially constant in energy with changing neutron number, unlike the energy of the lowest $J^\pi = 7/2^+$ state. This feature of the Sb isotopes has received considerable attention under the title of 'monopole energy shifts'. This is a more advanced topic which is not discussed here. Generally, the summed strengths match the shell model expectation that they should be near unity, i.e. the unpaired proton in the Sb isotopes is

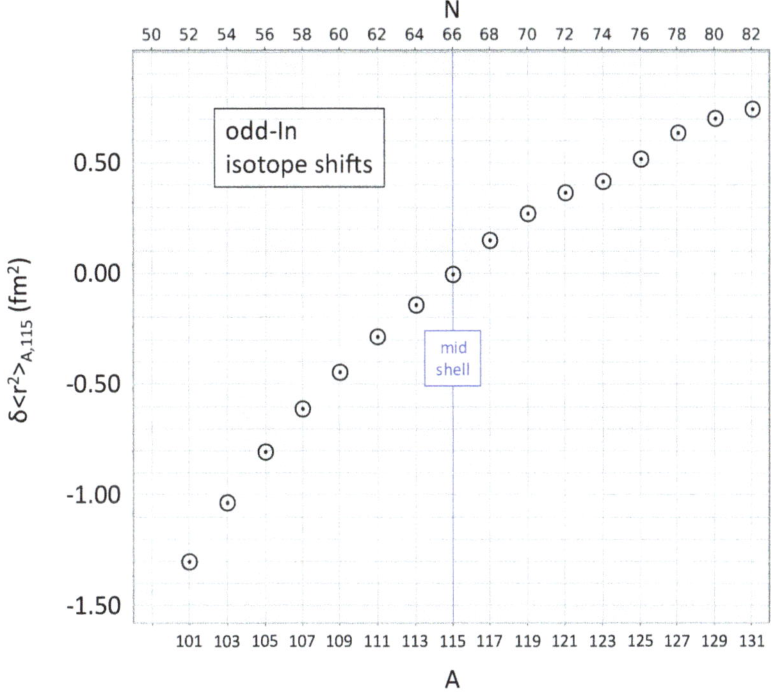

Figure 2.12. Systematics of the isotope shifts for the odd-In ground states, the $1g_{9/2}^{-1}$ configuration. The reference mass is $A = 115$. Values are given in femtometres squared. Uncertainties are generally smaller than the size of the data points. We note a discontinuity in the isotope shift values for $A > 121$; this may be correlated with the decreasing values of Q_s as seen in figure 2.11 for $A > 121$. The data are taken from [7, 14].

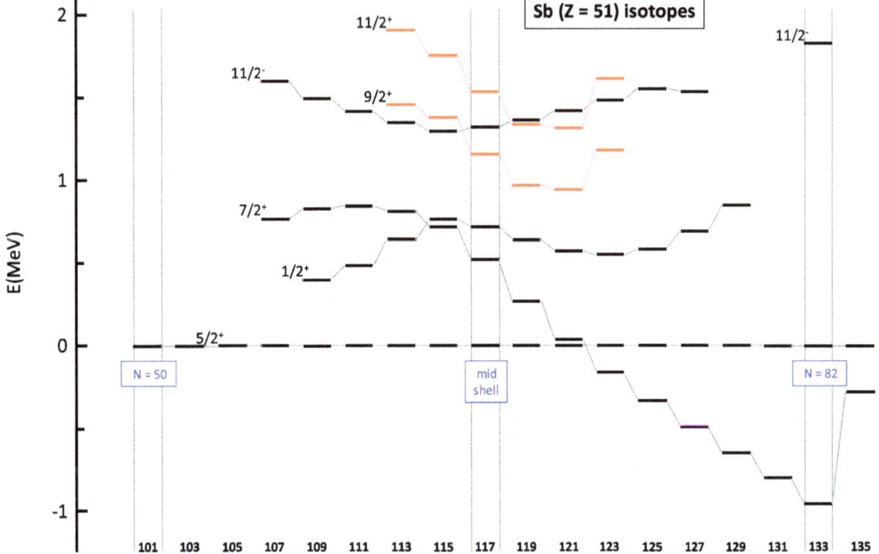

Figure 2.13. Systematics of selected states in the odd-mass antimony isotopes. The states highlighted in orange are intruder states. The details for these and associated states are presented in following figures. The data are taken from ENSDF.

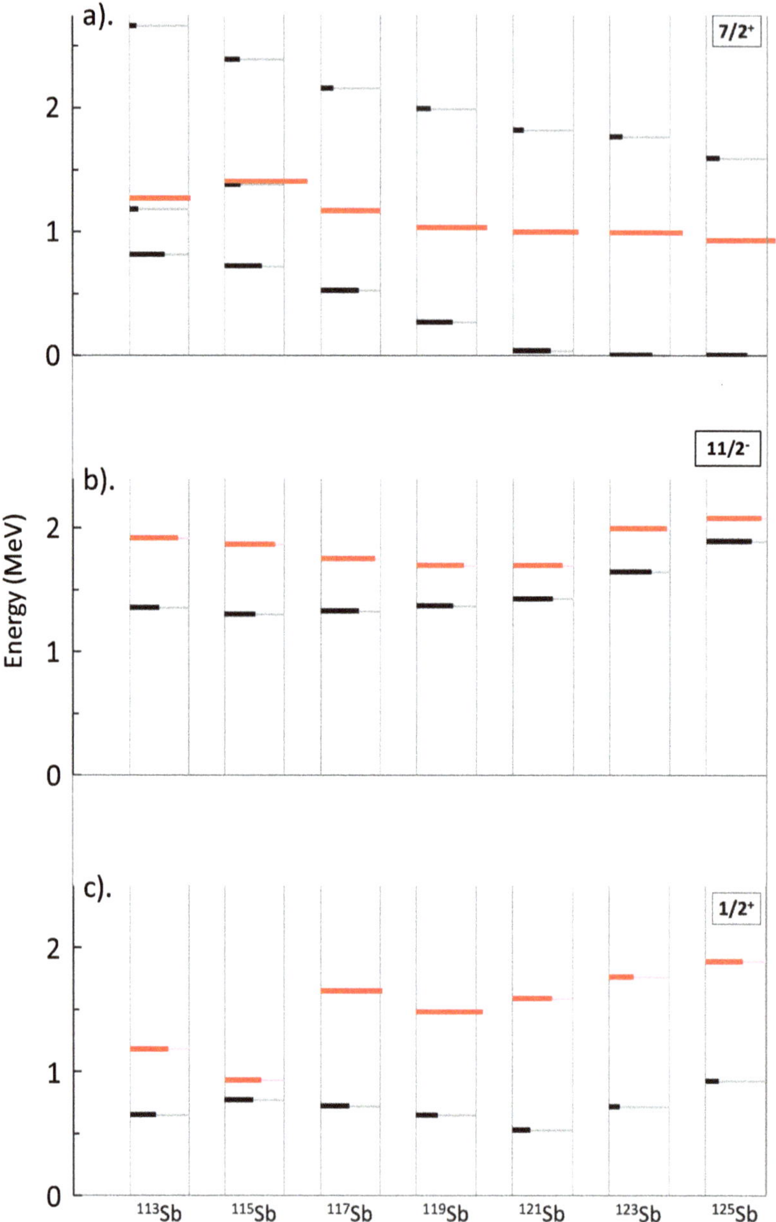

Figure 2.14. Population strengths for the one-proton addition reaction ^{A}Sn(^{3}He,d)$^{A+1}$Sb leading to selected states in $^{113-127}$Sb. (a) States populated with $L = 4$ angular distributions. (b) States populated with $L = 5$ angular distributions. (c) States populated with $L = 0$ angular distributions. Population strength is given as a fraction of the expected independent-particle strength (=1.00). The width of the heavy bars reflects this strength. Some strengths exceed 1.00. The red bars represent the energy-weighted sums of the spectroscopic strengths. Further details are discussed in the text. The data are taken from [15].

controlled by an essential independent-particle character. But there is substantial fragmentation of strength and there is missing strength in this '$E_x < 4$ MeV view'. We present some basic ideas for exploration of what is happening in the Sb isotopes in the following.

Figure 2.15 shows systematics of high-spin states in the Sb isotopes, for the upper half of the $N = 50$–82 shell. Positive-parity states with spins greater than $J^\pi = 7/2^+$ must be due to core excitations. Similarly, negative-parity states with spins greater than $J^\pi = 11/2^-$ must be due to core excitations. Figure 2.16 shows the relationship between selected states from figure 2.15 and the lowest excited states in neighbouring Sn isotopes, which act as the 'cores' for coupling an odd proton in the Sb isotopes. Evidently there is a simple relationship of an approximate 'additivity' of the Sn core excitation energy for coupling the $j = 7/2$ spin parallel to the core spins, $J = 2, 4, 6,$ and 8, i.e. for spins $11/2^+$, $15/2^+$, $19/2^+$ and $23/2^+$. For reasons beyond the present discussion, this does not occur for the $J = 10$ core states. However, many other (lower) spin couplings are possible, where, recall, spin is a vector quantity. At present the data view is limited, but we provide some details in the following.

Figure 2.17 shows the response of ^{129}Sb to Coulomb excitation. This reveals the collective excitations in this nucleus. Figure 2.18 shows systematics of the lowest excited states in $^{119–131}$Sb and some selected higher excited states. Figure 2.19 shows the response of ^{121}Sb and ^{123}Sb to Coulomb excitation, again revealing collective excitations in these nuclei. A distinctive pattern is suggested, albeit without an obvious interpretation. What is observed is that the independent proton configuration behind the lowest $7/2^+$ states and the lowest $5/2^+$ states, which can be interpreted as having shell model 'parentage', $1g_{7/2}$ and $2d_{5/2}$, retain separate groups of collective excitations, i.e. in ^{121}Sb and ^{123}Sb the collective excitation strengths are

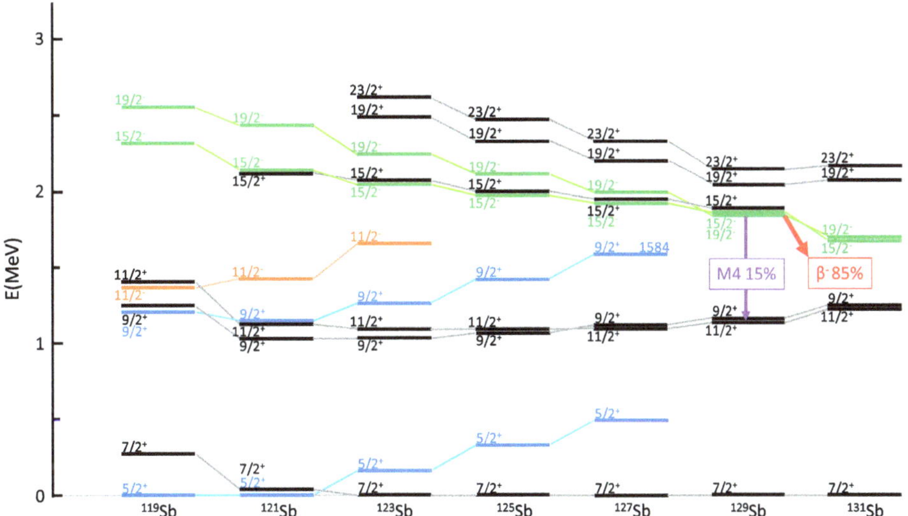

Figure 2.15. Systematics of selected high-spin states in the odd-mass Sb isotopes. Note the 'spin gap' that occurs in ^{129}Sb where the electromagnetic decay is restricted to an M4 multipolarity, with the result that beta decay is also a favoured decay mode. The data are taken from ENSDF.

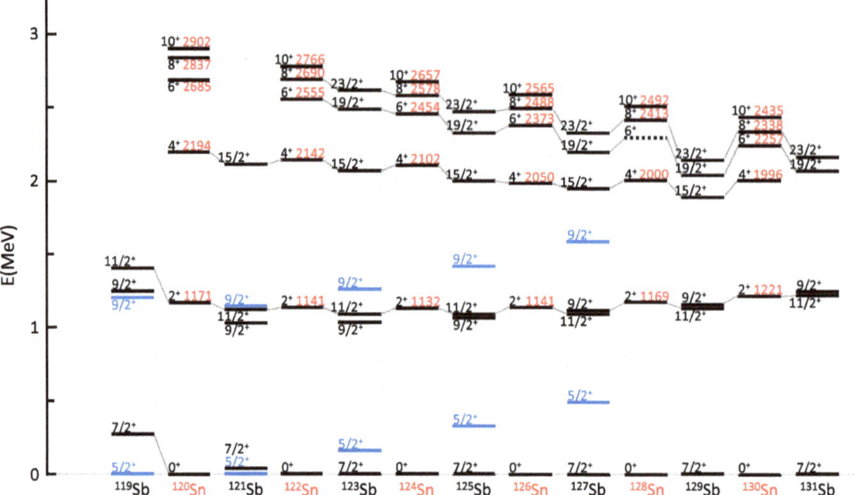

Figure 2.16. Selected states in the odd-mass Sb isotopes matched to Sn 'core' states. This illustrates a 'weak coupling', i.e. only minor energy differences between the ^{A}Sn core excitation and the $^{A+1}$Sb excitation. The couplings shown are for $J_{\mathrm{core}} + j_{\mathrm{particle}}$ in maximum possible alignment. The data are taken from ENSDF.

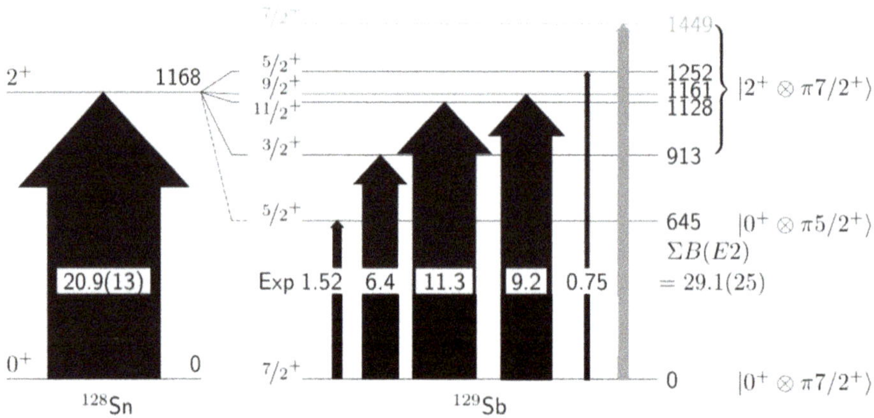

Figure 2.17. Population of the collective multiplet, $\pi 1g_{7/2}^{+1} \times ^{128}\mathrm{Sn}(2_1^+)$ in ^{129}Sb by Coulomb excitation. The deduced E2 excitation strengths are given in Weisskopf units (upwards). Reprinted with permission from [16]. Copyright (2020) by the American Physical Society.

predominantly confined to sets of states $J = j - 2, j + 1, j + 2$, with only weak transition strengths between the two groups. This would be indicative of an interaction acting on spatial degrees of freedom. Such an interaction will not couple a 'spin-flip' such as exists between $1g_{7/2}$ ('spin down') and $2d_{5/2}$ ('spin up'). The specific confinement of collective strength to the $J = j - 2, j + 1, j + 2$ states, presumably dominated by the Sn, $J = 2$ core excitation, and absence of strength to $J = j - 1, j$ states is noted, but is beyond the present level of discussion. Some further systematic views of the Sb isotopes are explored in the exercises.

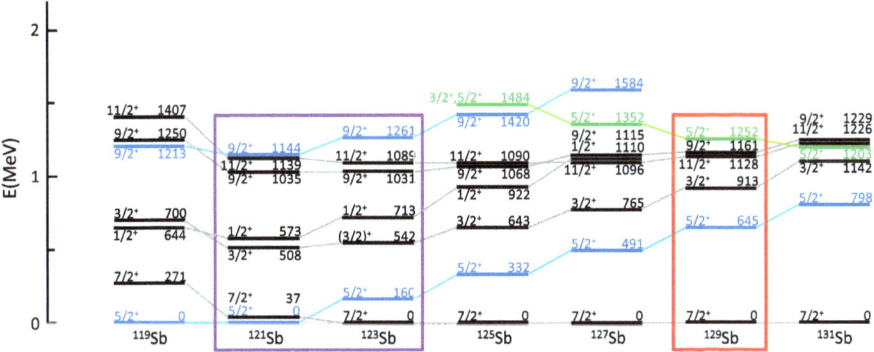

Figure 2.18. Systematics of the low-energy states in the odd-mass Sb isotopes. The boxed isotopes are the focus of E2 transition strengths in the previous and following figures. The data are taken from ENSDF.

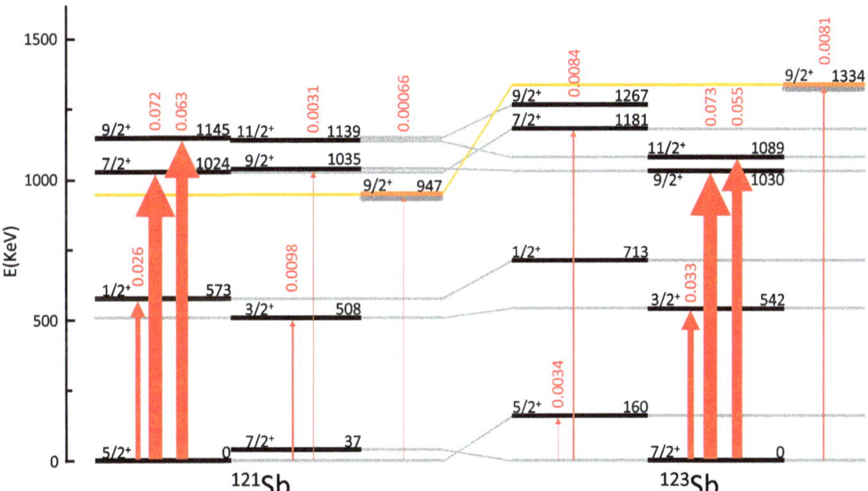

Figure 2.19. The E2 transition strengths (in e^2b^2) in ^{121,123}Sb. Note the change in the ground-state configuration. The states depicted in orange are intruder states, cf figure 2.13. Details are discussed in the text. The data are taken from ENSDF.

Magnetic and spectroscopic quadrupole moment data for the odd-mass Sb isotopes are shown in figures 2.20 and 2.21. The dramatic discontinuity between ^{121}Sb and ^{123}Sb is due to the interchange of ground-state configurations. There has been a widely held view that this interchange is between $2d_{5/2}$ and $1g_{7/2}$ spherical shell model states. However, the emerging view in this chapter is that these states are probably weakly deformed, thus placing the structures at the research frontier and beyond the scope of the present narrative.

The systematic features of the odd-mass Sb isotopes add further insight into the survival/fragmentation of shell model configuration strengths adjacent to a closed shell, but they reveal some features that demand research. Weak coupling to Sn core excitations is suggested for high-spin states but not for coupling of ground-state

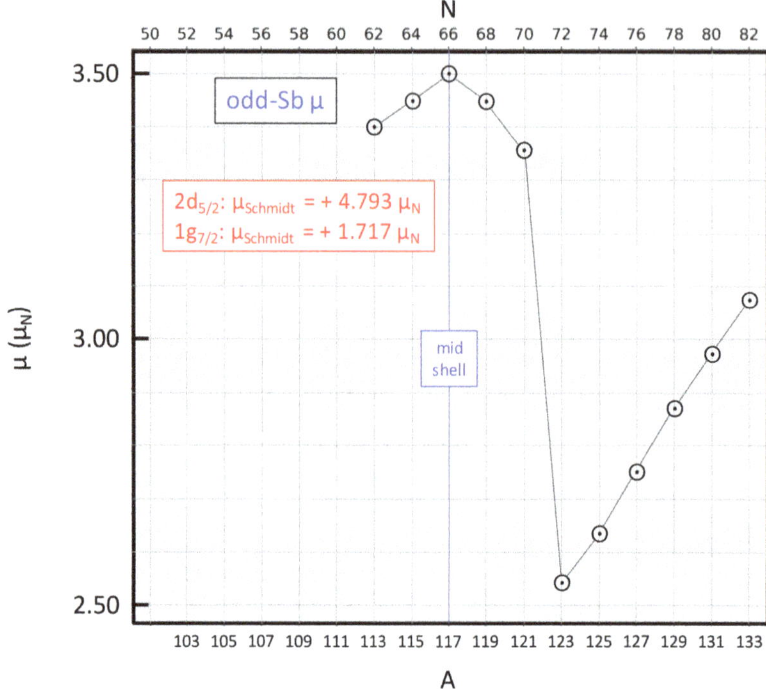

Figure 2.20. Magnetic moment data for the odd-mass Sb isotopes, given in nuclear magnetons. The extreme independent-particle (Schmidt) values for the shell model configurations 1g$_{7/2}$ and 2d$_{5/2}$ are indicated, but they both lie off scale. The data are taken from [18].

configurations to produce low-spin states. The simple view of emergent collectivity based on j mixing appears to be supported, where note the shell model configurations that possess positive parity in the $50 < Z < 82$ shell are 2d$_{5/2}$, 1g$_{7/2}$, 3s$_{1/2}$ and 2d$_{3/2}$ (cf figures 1.10, 1.11 and 1.29). The evidence from figure 2.19 is that the deformation-producing interaction does not involve a spin-flip, i.e. the configurations 2d$_{5/2}$ and 1g$_{7/2}$ do not mix; but 2d$_{5/2}$ and 3s$_{1/2}$ configurations can mix and 1g$_{7/2}$ and 2d$_{3/2}$ configurations can mix. This view of the Sb isotopes lies beyond the present level of discussion but can be explored using a shell-model basis with model interactions.

The systematic features of the In and Sb isotopes illustrate the range of structural types manifested in data for nuclei adjacent to singly closed shells. Other examples are found in isotopes and isotones with $N, Z = 7, 9, 19, 21, 27, 29, 81, 83$ and $N = 49, 51, 125, 127$. Organizing spectroscopic information into identification of particle and hole states, intruder states, particle–core and hole–core coupling states is a major data handling task. The challenge of identifying shell model states from the fragmentation manifested in one-nucleon addition and removal reactions demands expert knowledge, and often such data are inadequate to make decisions regarding locations of energy centroids of shell model configurations. Identification of intruder states is handled in [19]. States resulting from particle–core coupling are

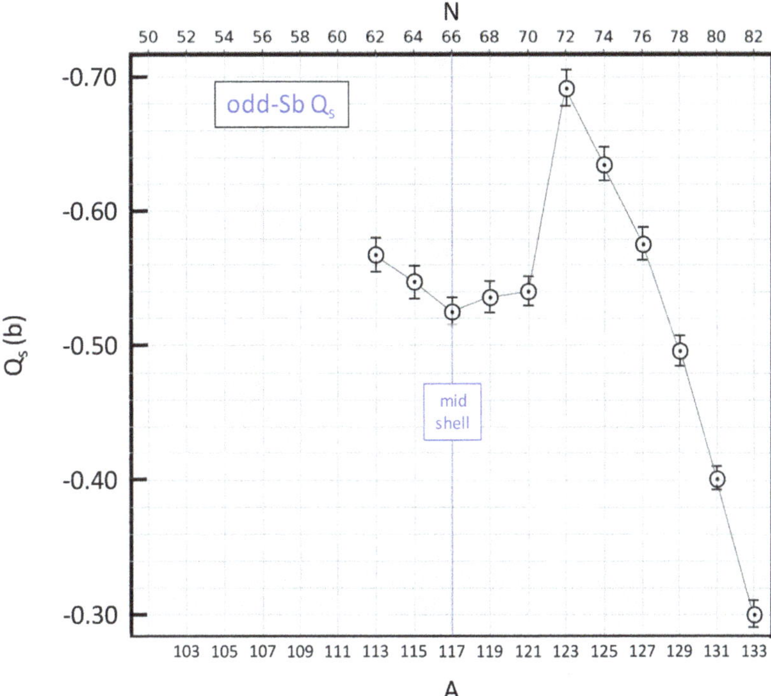

Figure 2.21. Spectroscopic quadrupole moment data for the odd-mass Sb isotopes given in barns. The data are taken from [18].

often not easy to identify and recognizing patterns for such 'multiplets' is a major research task which may require new experiments [16, 17]. Herein, we relegate further details of shell-model state identification in nuclei adjacent to singly closed shells to the following exercises.

2.3 Singly closed shell nuclei (a brief note)

Elucidating the underlying role of independent-particle motion in nuclei at singly closed shells, beyond a qualitative view, is a complex task. The reason is that the structure of singly closed shell nuclei is heavily dominated by pairing correlations. These correlations result in the emergence of so-called seniority coupling schemes. The pattern of seniority-dominated excitations in nuclei can be simple, as shown in figure 2.16. To understand the pattern—essentially that it is independent of the number of particles—requires quantum mechanics in the guise of so-called 'quasispin' which is at a level of difficulty beyond the present treatment. Indeed, the investment of effort weighed against the number of nuclei encountered that exhibit these patterns is not justified here.

For the general reader, excitation patterns in singly closed nuclei have received attention in section 4.2 of [19]. An introduction to the quantum mechanics of pairing is sketched in section 5.2 of [19]. A detailed development of pairing in many-fermion systems is given in chapter 4 of [20], wherein the algebraic structure of quasispin is thoroughly treated.

2.4 Exercises

The exercises provided in this chapter contain a theme which emphasizes topics that border on the research frontier. Specifically, the approach taken for organizing data into the nuclear structure perspectives provided in sections 2.1 (indium isotopes) and 2.2 (antimony isotopes), is extended. The exercises organize data into a structure perspective for other isotopic and isotonic series of nuclei adjacent to single-closed shells.

Note that not all nuclei adjacent to single-closed shells are considered in these exercises. Notably, light nuclei (adjacent to closed shells with N, $Z = 2$, 8, 20, 28) and the Tl ($Z = 81$) and Bi ($Z = 83$) isotopes are not discussed. The reasons are complexity due to intruder states coexisting at low energy and (for the lightest nuclei) very few shell model configurations, confined to a limited systematic view. In summary, these nuclei lie too close to the research frontier for a pedagogical treatment, such as being pursued in the present focus.

2.4.1 $N = 49$ isotones

2-1 Figure 2.22 shows selected states in the odd-mass $N = 49$ isotones. The states highlighted in red are intruder states. The excitation of the 2_1^+ states in the $N = 50$ closed shell $A + 1$ nuclei are shown as green bars with stars. By comparison with the In ($Z = 49$) isotopes, using data in ENSDF, as far as possible explore the assignment of excited states in the $N = 49$ isotones to produce patterns similar to that established for the In isotopes, cf exercises 2-2 and 2-3.

2-2 Figures 2.23(a) and (b) show states populated in the one-neutron addition reactions (a) ^{82}Se(d,p)^{83}Se, (b) ^{84}Kr(d,p)^{85}Kr. There are some multiple occurrences of positive-parity states with the same spin. Using data in ENSDF and the decoupled band view of the intruder states in the odd-mass In isotopes, as far as possible classify states in ^{83}Se and ^{85}Kr into a structural pattern similar to the In isotopes.

2-3 Figure 2.24 show states in ^{87}Sr populated in the ^{86}Sr(d,p)^{87}Sr and (^{3}He,α)^{87}Sr reactions. Figure 2.25 shows states populated in the ^{87}Sr(p,p′)^{87}Sr reaction. Using data in ENSDF, explore classification of states in ^{87}Sr in a structural pattern similar to the In isotopes. The angular distributions for the ^{87}Sr(p,p′)^{87}Sr reaction are shown in figure 2.26 to provide support for assignment of ^{87}Sr states as resulting from coupling of the $1g_{9/2}$ neutron–hole configuration to core states with spin-parity 2^+ and 3^-.

2-4 The intruder states in the $N = 49$ isotones give rise to isomerism in some nuclei. By comparing with the examples of isomerism observed in the In isotopes, cf figure 2.3, identify what isomeric decay features have been assigned for $N = 49$ nuclei, using data in ENSDF. Note: there are some

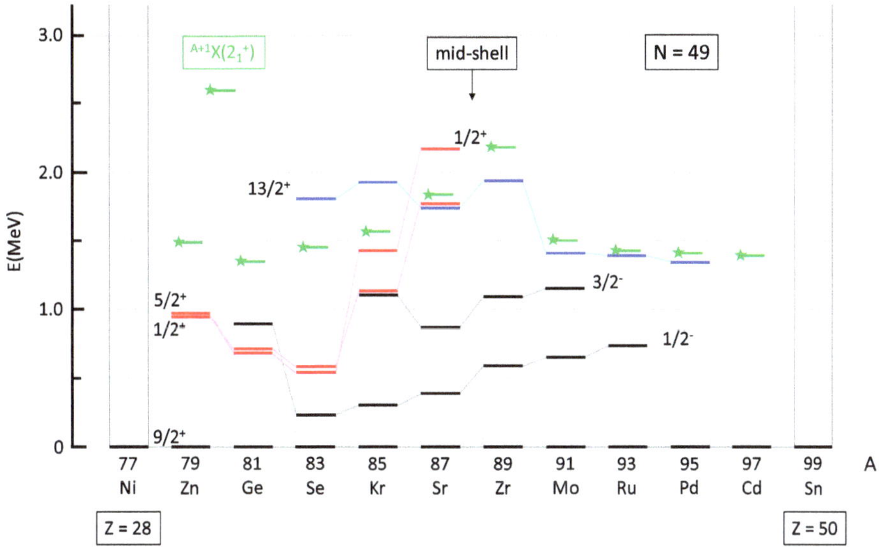

Figure 2.22. Systematic view of energies of low-lying states in the $N = 49$ isotones. The ground states, with spin-parity $9/2^+$, match the expectation of a shell model $1g_{9/2}$ neutron–hole configuration with respect to the $N = 50$ closed shell. The low-lying excited state with spin-parity $1/2^-$ matches the expectation of a shell model $2p_{1/2}$ neutron configuration. The states highlighted in red are interpreted as intruder states. The evidence for the intruder state assignments comes from single-neutron transfer reaction spectroscopy which is shown in figures 2.23 and 2.24. By comparison with the In ($Z = 49$) isotopes, these states are assigned to be a deformed band involving the Nilsson configuration $1/2^+[431]$. The proximity of the $1/2^+$ and $5/2^+$ band members and a comparison with figure 4.8 suggests the decoupling parameter lies in the range $+3 < a < +5$. The 2_1^+ excitations in the closed shell even–even core nuclei with mass $A + 1$ are shown as green lines with a green star. There are states with spin-parity $13/2^+$ that appear to be good candidates for the coupling $1g_{9/2}^{-1} \times {}^{A+1}X$. The data are taken from ENSDF, but there are ambiguities in identification and existence of some low-lying states so the present view is a subjective one on the part of the Authors. The data for ^{79}Zn are taken from [2].

remarkable developments in ^{79}Zn [2] that epitomize a view of the structure of nuclei, in an extreme neutron-rich region, through isomerism.

2.4.2 $N = 51$ isotones

2-5 Figure 2.27 shows selected states in the odd-mass $N = 51$ isotones. The excitation of the 2_1^+ states in the $N = 50$ closed shell $A - 1$ nuclei are shown as green bars with stars. By comparison with the Sb ($Z = 51$) isotopes, using data in ENSDF, as far as possible explore the assignment of excited states in the $N = 51$ isotones to produce patterns similar to that established for the Sb isotopes, cf exercises 2-6 and 2-7.

2-6 Figure 2.28 shows states in ^{91}Zr populated in the ^{92}Zr(p,d)^{91}Zr and ^{90}Zr(d,p)^{91}Zr reactions. Figure 2.29 shows states populated in the ^{91}Zr(p,p′)^{91}Zr reaction. Using data in ENSDF, explore classification of states in ^{91}Zr in a structural pattern similar to the Sb isotopes.

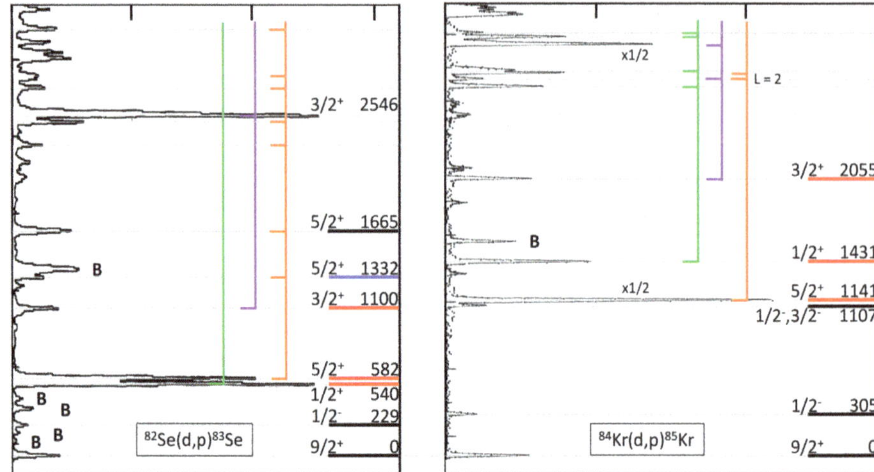

Figure 2.23. (Left) Proton spectra for the one-neutron addition reaction ^{82}Se(d,p)^{83}Se. Reprinted from [21], copyright (1978) with the permission of Elsevier. (Right) Proton spectra for the one-neutron addition reaction ^{84}Kr(d,p)^{85}Kr. Reprinted with permission from [22]. Copyright (1978) by the American Physical Society. The spin-parities and excitation energies in keV are given for selected peaks in the spectra. These reactions are expected to populate intruder states in the $N = 49$ isotones and candidates for such states are highlighted in red, cf figure 2.22. Population of more than one state with the same spin-parity is indicated by coloured vertical bars: 1/2$^+$ (green), 5/2$^+$ (red), 3/2$^+$ (violet). Peaks in the spectra due to target impurities are marked B. Other data are taken from ENSDF.

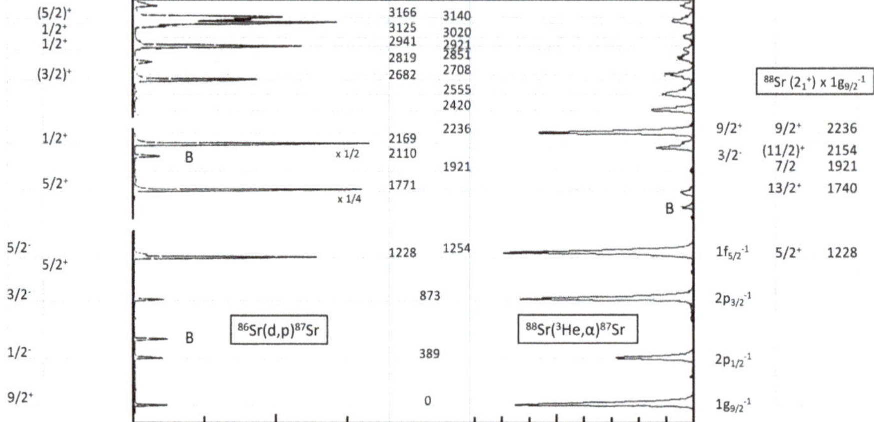

Figure 2.24. One-neutron transfer reaction spectroscopic data for ^{87}Sr: (Left) Proton spectrum from the neutron-addition reaction ^{86}Sr(d,p)^{87}Sr. Reprinted from [23], copyright (1971) with the permission of Elsevier. (Right) Alpha spectrum from the neutron-removal reaction ^{88}Sr(^{3}He,α)^{87}Sr. Reprinted with permission from [24]. Copyright (1989) by the American Physical Society. Some peaks in the spectra are labelled by the corresponding level energies (in keV), some state spin-parities are given, and on the far right, the particle–core coupling quintuplet, ^{88}Sr(2$_1^+$) $\times$ 1g$_{9/2}^{-1}$ taken from figure 2.25, is shown. Other data are taken from ENSDF.

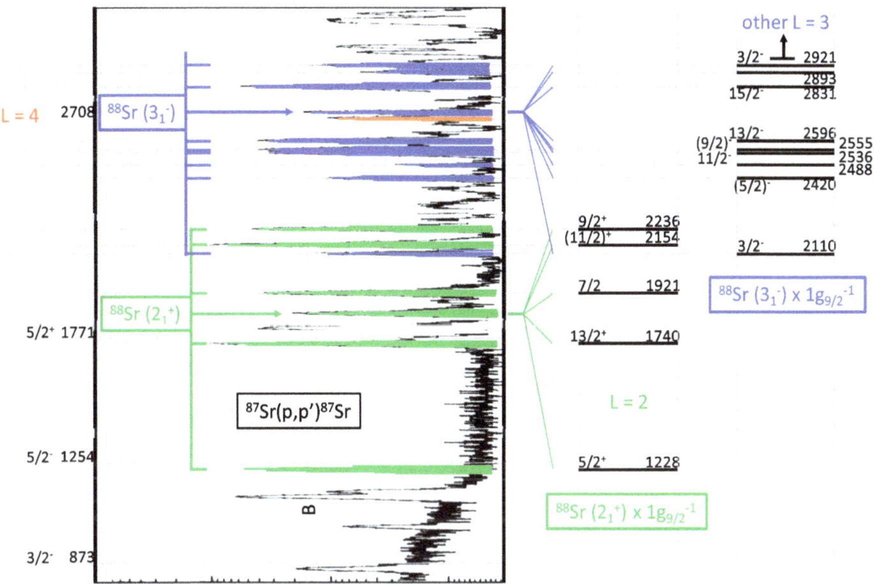

Figure 2.25. Spectrum of inelastically scattered protons from ^{87}Sr. The spin-parities and energies (in keV) are taken from ENSDF. The population of states corresponding to the particle–core coupling ^{88}Sr$(2_1^+) \times 1g_{9/2}^{-1}$ and ^{88}Sr$(3_1^-) \times 1g_{9/2}^{-1}$ are highlighted in green and blue, respectively. The target contained both ^{87}Sr and ^{88}Sr and so the ^{88}Sr2_1^+ peak (at 1836 keV) and 3_1^- peak (at 2734 keV) are present in the spectrum and their locations are indicated. The peaks corresponding to the two multiplets are distinguished by characteristic angular distributions of scattered protons with respect to the incident beam, shown in figure 2.26, for the $L = 2$ and $L = 3$ transferred angular momenta. An $L = 4$ peak is noted, highlighted in orange. Reprinted from [25], copyright (1978) with the permission of Elsevier.

2-7 For ^{89}Sr and ^{91}Zr, attempt to classify all excited states listed in ENSDF up to 2400 keV.

2-8 Based on interpretations made in exercise 2-7, attempt to classify all excited states listed in ENSDF for ^{93}Mo up to 2400 keV.

2-9 Figure 2.30 shows single-particle strengths in the $N = 51$ isotones. Using this view of the $N = 51$ isotones, attempt to classify excited states listed in ENSDF for ^{87}Kr up to 2300 keV.

2-10 As far as possible, classify states listed in ENSDF for ^{95}Ru.

2-11 Make a systematic view of long-lived isomers ($T_{1/2} >$ ns) in the $N = 51$ isotones.

2.4.3 $N = 81$ isotones

2-12 Figure 2.31 shows selected states in the odd-mass $N = 81$ isotones. Figure 2.32 shows candidate intruder states in the odd-mass $N = 81$ isotones. Make a basic shell model classification of the configurations underlying these states.

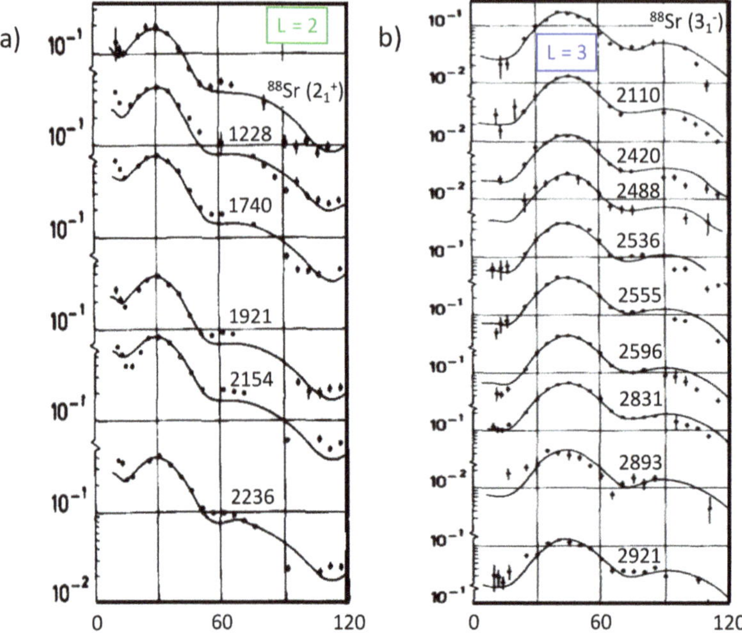

Figure 2.26. Angular distributions of inelastically scattered protons from ^{87,88}Sr. The distribution of scattered protons distinguish between angular momentum transfer of 2 ($L = 2$) in part (a) and 3 ($L = 3$) in part (b). Note the ^{88}Sr distributions are shown at the top of each figure part. The distributions for ^{87}Sr are labelled by excitation energies of the states populated. Reprinted from [25], copyright (1978) with the permission of Elsevier.

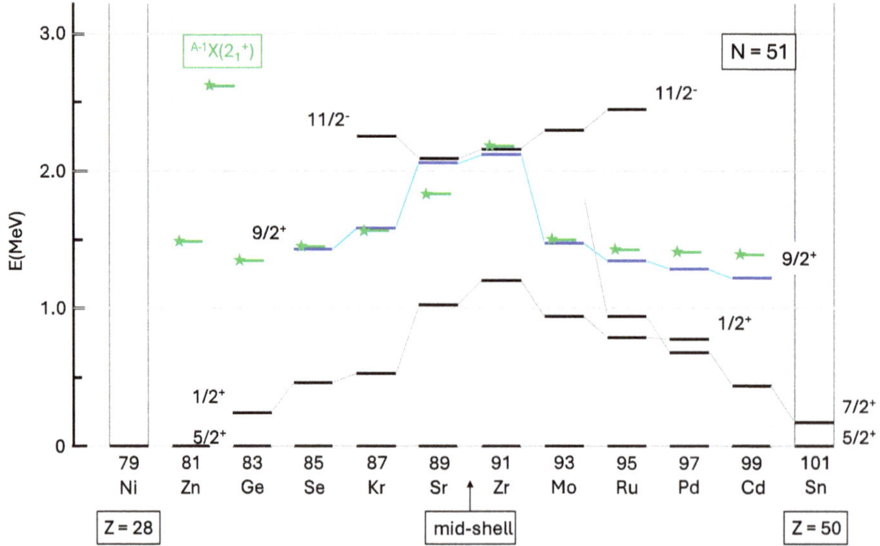

Figure 2.27. Systematic view of energies of low-lying states in the $N = 51$ isotones. Green bars with a star show the 2^+ first excited states in the $N = 50$ neighbouring $A - 1$ 'core' nuclei. The data are taken from ENSDF.

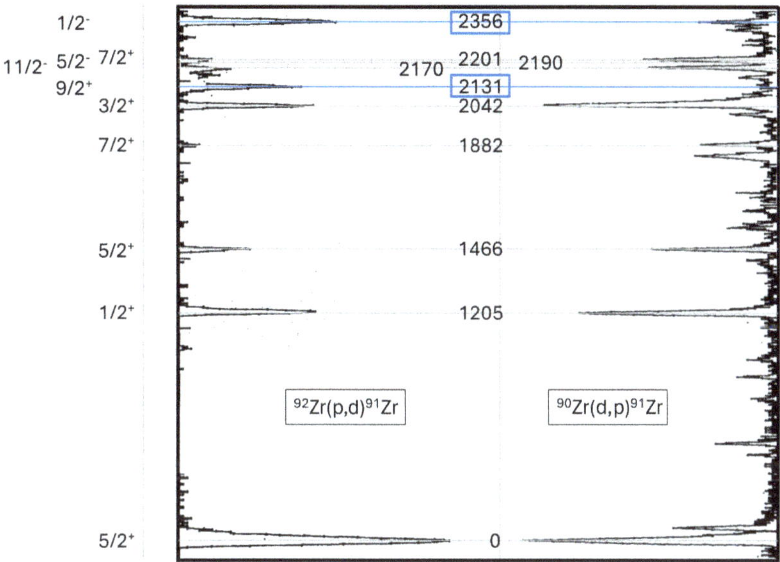

Figure 2.28. One-neutron transfer reaction spectroscopic data for ^{91}Zr. The left- and right-hand sides of the figure compare the deuteron spectrum from the neutron-removal reaction ^{92}Zr(p,d)^{91}Zr with the proton spectrum from the neutron-addition reaction ^{90}Zr(d,p)^{91}Zr, respectively. Some peaks in the spectra are labelled by the corresponding level energies (in keV), some state spin-parities are given. The states at 2131 and 2356 keV, boxed in blue, are $1g_{9/2}^{-1}$ and $2p_{1/2}^{-1}$ intruder states. Other data are taken from ENSDF. Reprinted from [26], copyright (1976) with the permission of Elsevier.

2-13 Figure 2.33 presents candidate states for the particle–core coupled structures $2d_{3/2}^{-1} \times 2_1^+(^{138}$Ba) in ^{137}Ba. Explore the excited states in ^{133}Te and ^{135}Xe with the aim of identifying particle–core coupled structures.

2-14 As a continuation of exercise 2-13, explore the occurrence of such structures in ^{139}Ce, ^{141}Nd and ^{143}Sm.

2-15 Explore the occurrence of M4 transitions in nuclei and, where known, the strengths of these transitions. (Note that they are highly converted transitions and the BrIcc website provides a means of determining internal conversion coefficients.)

2-16 What is the total internal conversion coefficient for the 65 keV M4 transition in ^{131}Sn?

2-17 Make a systematic view of long-lived isomers ($T_{1/2} >$ ns) in the $N = 81$ isotones.

2-18 Using table 2.1, make a figure (similar to figure 2.30) that presents the fragmentation of neutron–hole strength for ^{137}Ba, ^{139}Ce, ^{141}Nd and ^{143}Sm.

2-19 Using table 2.1, construct the energy-weighted centroid, $E_{EW} = \sum C^2S \times E_x / \sum C^2S$ for $l = 5$ transfer strength.

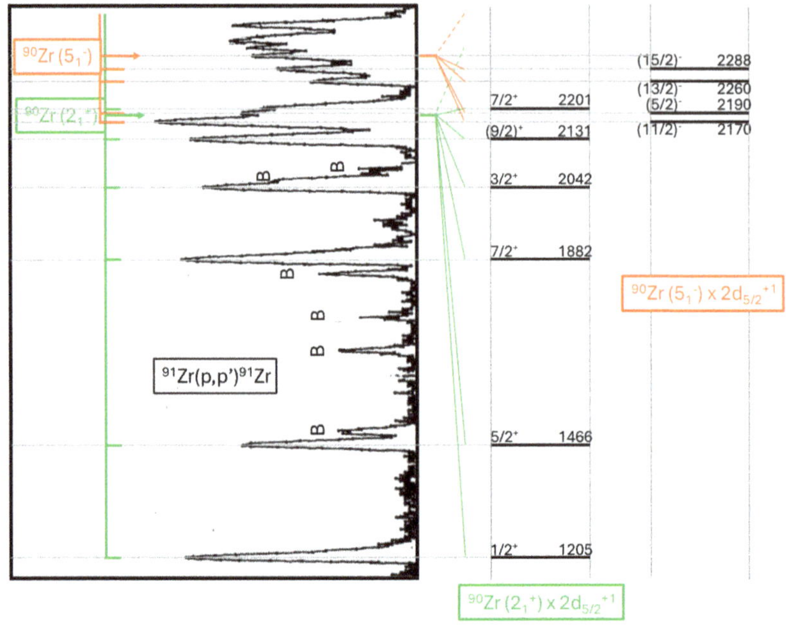

Figure 2.29. Spectrum of inelastically scattered protons from ^{91}Zr. The spin-parities and energies (in keV) are taken from ENSDF. The population of states corresponding to the particle–core coupling ^{90}Zr$(2_1^+) \times 1g_{9/2}^{-1}$ and ^{90}Zr$(5_1^-) \times 1g_{9/2}^{-1}$ are highlighted in green and orange, respectively. The target contained both ^{91}Zr and ^{90}Zr and so the ^{90}Zr 2_1^+ peak (at 2186 keV) and 5_1^- peak (at 2319 keV) are present in the spectrum and their locations are indicated. Proton spectrum peaks labelled 'B' result from target impurities. Reprinted from [26], copyright (1976) with the permission of Elsevier.

2.4.4 $N = 83$ isotones

2-20 Figures 2.34(a) and (b) show selected states in the odd-mass $N = 83$ isotones. Make a basic shell model classification of the configurations underlying these states.

2-21 Figure 2.35 presents candidate states for the particle–core coupled structures $2f_{7/2}^{+1} \times 2_1^+(^{142}Nd)$ in ^{143}Nd. Using data in ENSDF, explore the excited states in ^{141}Ce and ^{145}Sm with the aim of identifying negative-parity particle–core coupled structures.

2-22 Figure 2.36 presents particle–core coupled, $2f_{7/2}^{+1} \times {}^{A-1}X(2_1^+)$ candidate states in the $N = 83$ isotones. Using data in ENSDF, explore the excited states in ^{135}Te, ^{137}Xe and ^{139}Ba with the aim of identifying negative-parity particle–core coupled structures.

2-23 Figure 2.37 presents a systematic view of the lowest energy positive-parity low-spin states in the $N = 83$ isotones. Figure 2.38 presents a view of particle–core coupled, $2f_{7/2}^{+1} \times 3_1^-(^{142}Nd)$ candidate states in ^{143}Nd. Figure 2.39 presents states in ^{141}Ce populated in the one-neutron removal reaction ^{142}Ce(d,t)^{141}Ce and the one-neutron addition reaction ^{140}Ce(d,p)^{141}Ce. Figure 2.40 presents states in ^{143}Nd populated in the

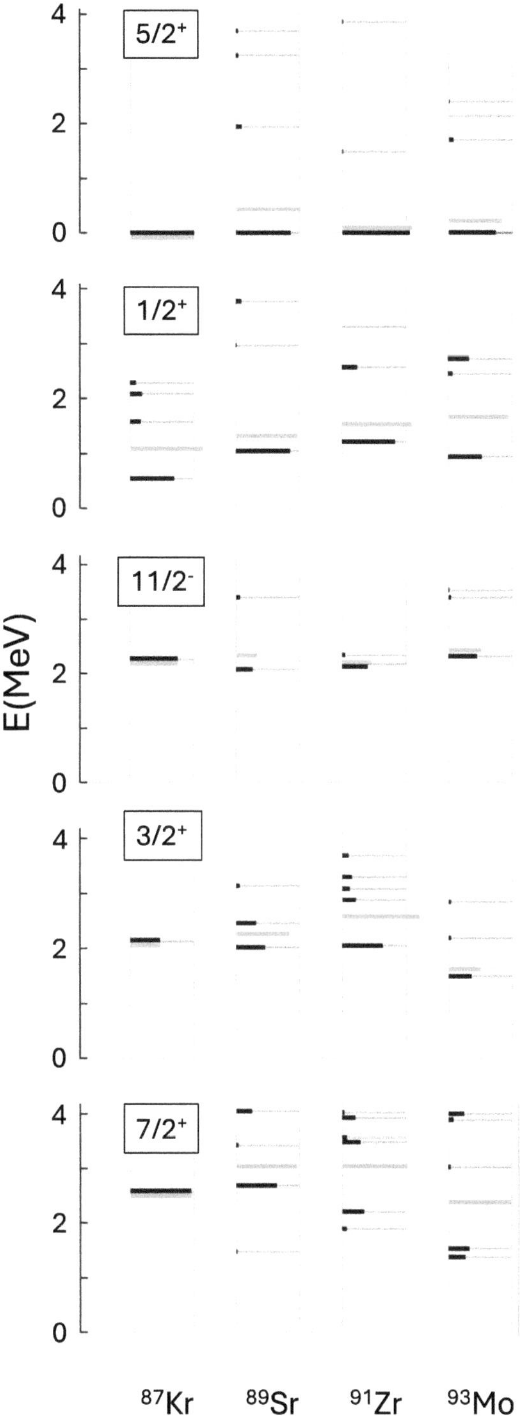

Figure 2.30. Single-particle strengths in the $N = 51$ isotones. Black bars indicate fragments of strength for $g_{7/2}$, $d_{3/2}$, $h_{11/2}$, $s_{1/2}$ and $d_{5/2}$ configurations. Green bars indicate energy-weighted centroids of these strengths. The vertical grey lines are spaced for each isotope by spectroscopic strengths of unity; thus, there is notable missing strength, particularly for $h_{11/2}$. The data are taken from [27].

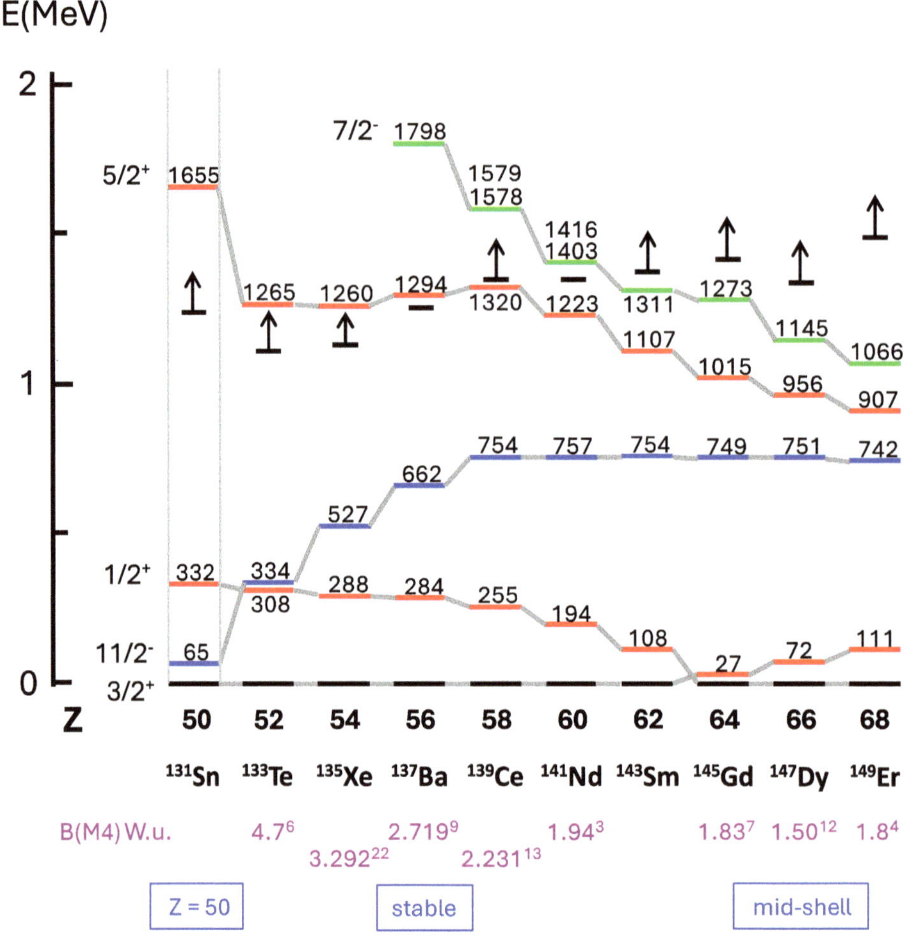

Figure 2.31. Systematic view of energies of low-lying states in the $N = 81$ isotones. Excitation energies are given in keV. The B(M4) decay strengths in Weisskopf units for the $11/2^- \rightarrow 3/2^+$ transitions are given at the bottom of the figure. Horizontal bars with vertical upward-pointing arrows indicate excitation energies above which states are omitted. The data are taken from ENSDF and XUNDL.

one-neutron removal reaction ^{144}Nd(^{3}He,α)^{143}Nd and the one-neutron addition reaction ^{142}Nd(d,p)^{143}Nd. Figure 2.41 presents the spectrum of inelastically scattered protons from ^{143}Nd and figures 2.42(a) and (b) show associated angular distributions. Using these data and data in ENSDF, attempt to make figures for particle–core coupling, similar to figures 2.35 and 2.38, for ^{141}Ce and ^{145}Sm.

2.4.5 $N = 125, 127$ **isotones**

2-24 Figure 2.44 shows selected states in the $N = 125$ isotones. Figure 2.45 shows the spectrum of tritons from the one-neutron removal reaction

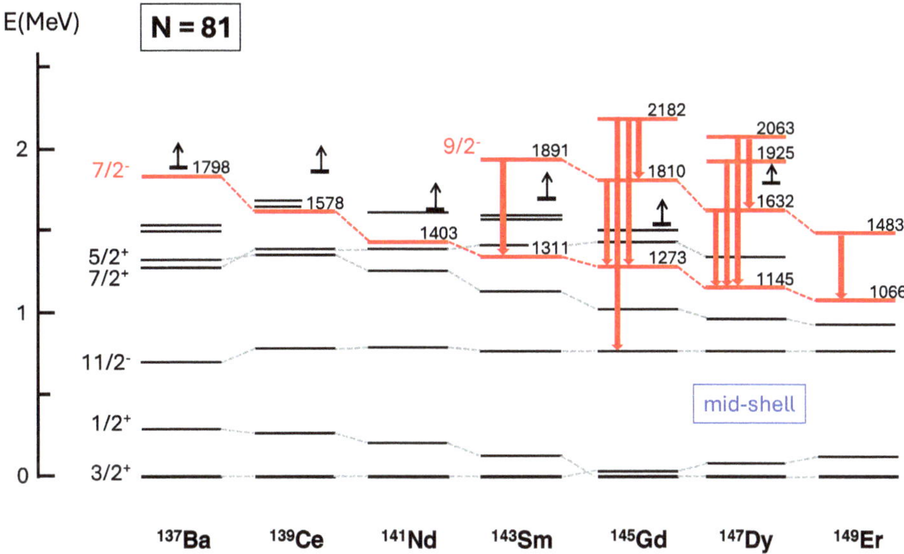

Figure 2.32. Systematic view of candidate intruder states in the heavy $N = 81$ isotones. Excitation energies are given in keV. Strong decay transitions for states that form candidate intruder bands are depicted in red. Horizontal bars with vertical upward-pointing arrows indicate excitation energies above which states are omitted. The data are taken from ENSDF.

^{210}Po(d,t)^{209}Po. Figure 2.46 shows the spectrum of protons from the one-neutron addition reaction ^{204}Po(d,p)^{205}Hg. Confirm the basic shell model classification of the states, with consultation also of the data shown in figures 1.10 and 1.12.

2-25 Figures 2.47–2.49 show various details of particle–core coupling in ^{209,211}Po, ^{211,213}Rn and ^{215}Ra. The core excitation structures are dominated at low energy by the seniority-two broken-pairs involving the shell model proton orbital 1h$_{9/2}$. The coupling is generally 'weak', i.e. the odd-mass states are localized around the energies of the core states. Suggest further excited states involving these core states coupled to excitations of the one-neutron hole (in the $N = 125$ isotones) and the one-neutron particle in the $N = 127$ isotones. Note, particularly focus on the states of highest possible spin and pursue this separately for states of negative parity and positive parity. Use data in ENSDF as necessary.

2-26 Figure 2.50 shows the systematics of the lowest energy intruder states in ^{209}Pb (q.v. Figure 1.21) and ^{211}Po and the match to the lowest states in ^{207}Pb. What would be the expected excitation of the next intruder state in ^{211}Po?

2-27 With reference to figure 2.47, using data from ENSDF, make a similar view for ^{213}Ra, ^{214}Ra, ^{215}Ra.

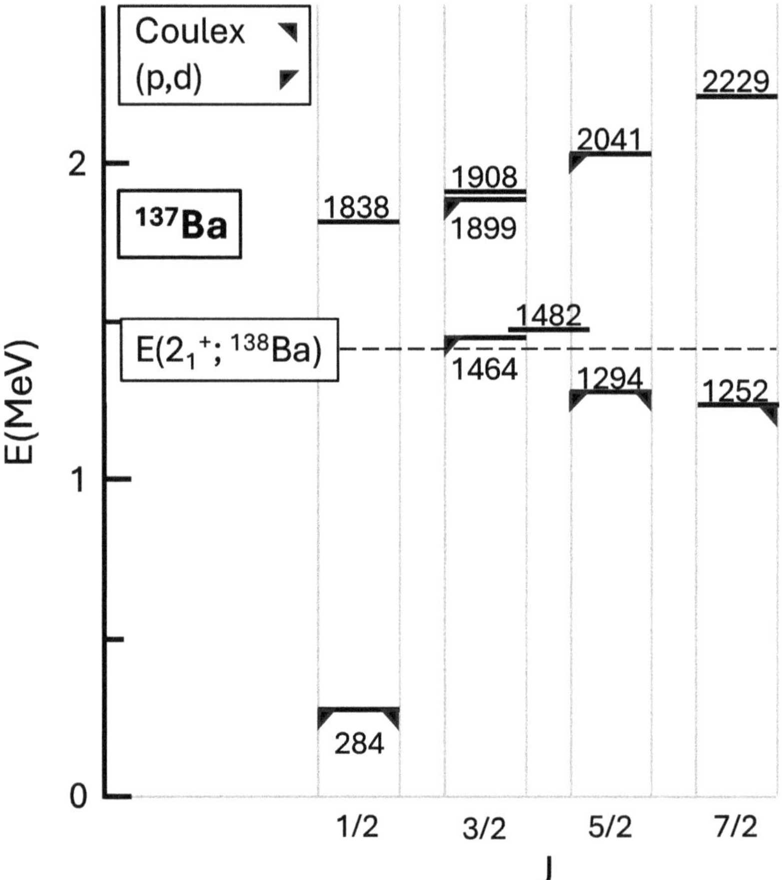

Figure 2.33. View of particle–core coupled, $2d_{3/2}^{-1} \times 2_1^+$ (^{138}Ba) candidate states in ^{137}Ba. The spectroscopic data indicated are Coulomb excitation [28] and one-neutron removal by the ^{138}Ba(p,d)^{137}Ba transfer reaction [29]. The spin assignment of the 1482 keV state is $J = 3/2$ or $5/2$. The data are taken from ENSDF.

Table 2.1. Summary of states populated in neutron-removal reactions on $N = 82$ targets, including excitation energy E, orbital angular momentum transfer l, spin parity J^π, and normalized spectroscopic factor C^2S. Excitation energies are given in MeV and are estimated to carry an uncertainty of ≈ 5 keV, rising to ≈ 10 keV in the case of ^{137}Ba at higher excitation energies. The spectroscopic factors are deduced from the (p,d) reaction for $l = 0$ and 2 transfers and from the (^{3}He,α) reaction for $l = 4$ and 5. The errors in the normalized values are typically 5% due to variation of DWBA with different input parameters, but for weaker transitions these rise where statistical errors become more significant. J^π are taken from the literature; where a J^π value is not lised, a model-dependent assumption was made that the single-particle orbitals are in the valence shell. The table is taken with permission from [30], copyright (2020) by the American Physical Society.

^{137}Ba				^{139}Ce				^{141}Nd				^{143}Sm			
E	l	J^π	C^2S	E	l	J^π	C^2S	E	l	J^π	C^2S	E	l	J^π	C^2S
0.000	2	3/2$^+$	2.56	0.000	2	3/2$^+$	2.92	0.000	2	3/2$^+$	3.04	0.000	2	3/2$^+$	3.40
0.281	0	1/2$^+$	1.86	0.252	0	1/2$^+$	1.77	0.192	0	1/2$^+$	1.69	0.110	0	1/2$^+$	1.84
0.662	5	11/2$^-$	9.42	0.755	5	11/2$^-$	8.72	0.759	5	11/2$^-$	8.99	0.758	5	11/2$^-$	10.36
1.252	4	7/2$^+$	0.26	1.321	2	5/2$^+$	2.12	1.222	2	5/2$^+$	2.44	1.100	2	5/2$^+$	3.54
1.290	2	5/2$^+$	1.01	1.347	4	7/2$^+$	0.94	1.343	4	7/2$^+$	0.62	1.362	4	7/2$^+$	1.02
1.460	2	3/2$^+$	1.17	1.598	2	(3/2)$^+$	0.40	1.565	2	(3/2)$^+$	0.23	1.533	2	(5/2)$^+$	0.17
1.840	0	1/2$^+$	0.28	1.632	2	3/2$^+$	0.12	1.597	2	(3/2)$^+$	0.06	1.708	2	(3/2)$^+$	0.40
1.900	2	3/2$^+$	0.83	1.823	2	5/2$^+$	0.09	1.822	2		0.69	1.930	2		0.07
2.040	2	(5/2)$^+$	1.08	1.889	0	1/2$^+$	0.24	1.888	0		0.14	1.990	0		0.11
2.117	2		0.07	1.911	2	(3/2)$^+$	0.69	1.968	4	7/2$^+$	0.17	2.064	2		0.47
2.230	4	7/2$^+$	0.78	2.018				2.070	2	(3/2$^+$,5/2$^+$)	0.41	2.161	4	7/2$^+$	1.01
2.271	2	(3/2$^+$,5/2)	0.04	2.090	2		0.51	2.111	2	3/2$^+$,5/2$^+$	0.14	2.274	4	7/2$^+$	0.52
2.32	5		1.35	2.143	2		0.19	2.180	0		0.05	2.450	5		1.52
2.38	2		0.07	2.251	(4)	(7/2$^+$)	0.19	2.208	5	(11/2$^-$)	1.84	2.586	5		0.87
2.44	2		0.12	2.286	5	11/2$^-$	1.63	2.31	4	7/2$^+$, (9/2$^+$)	0.83	2.662	4		0.48
2.53	2		0.14	2.362	4		0.63	2.349	4		0.50	3.05	(4)		0.64
2.61	(2)		0.02	2.426	2		0.06	2.384	4	7/2$^+$	0.20	3.13			
2.67	2		0.09	2.455	(4)		0.24	2.512				3.23			
2.75	4		0.70	2.556	4 and (0)		0.45 and 0.04	2.581	(2)		0.05				

(Continued)

Table 2.1. (*Continued*)

^{137}Ba				^{139}Ce				^{141}Nd				^{143}Sm			
E	l	J^π	C^2S	E	l	J^π	C^2S	E	l	J^π	C^2S	E	l	J^π	C^2S
2.81	(4)		0.21	2.610	(4)		0.16	2.616	(2)		0.02				
2.89	2		0.06	2.701	(4)		0.14	2.705	(2)		0.05				
2.99	5		1.24	2.800	4	$7/2^+$	0.31	2.809	(2)		0.07				
3.03	2		0.09	2.822	5	$9/2^-,11/2^-$	0.76	2.915	5		0.80				
3.12	4		0.58	2.910	2		0.11	2.939	2		0.16				
3.15	(2)		0.07	2.964	2		0.10	3.042	4		0.40				
3.21	5		0.51	3.082	(4)		0.20	3.112	4		0.56				
3.42	4		0.21	3.196	4		0.75	3.315	(2)		0.04				
3.55	4		0.43	3.282	4		0.58	3.369	2		0.19				
				3.352	4		0.30	3.407	2		0.25				

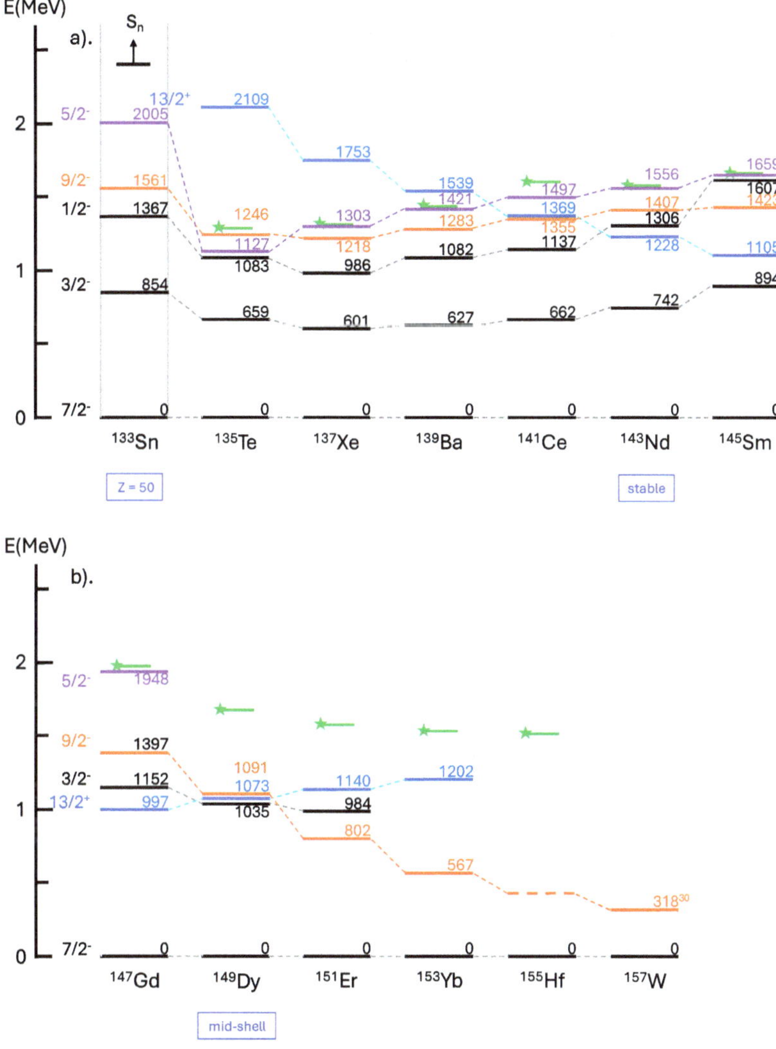

Figure 2.34. Systematic view of energies of negative-parity low-lying states in the $N = 83$ isotones; and the lowest state with spin-parity 13/2$^+$, (a) for $Z < 63$, (b) for $Z > 63$. Green bars with a star show the 2$^+$ first excited state energies in the $N = 82$ neighbouring $A - 1$ 'core' nuclei. Details are discussed in the text. The data are taken from [31] (^{135}Te), [32] (^{137}Xe), [33] (^{157}W) and ENSDF. Note that a search for the 13/2$^+$ state in ^{133}Sn supports its unbound nature [34], and the threshold for single-neutron excited states to be unbound is defined by the one-neutron separation energy, $S_n = 2402^4$ keV, as shown. The lowest lying states omitted are generally 11/2$^-$ states, for which a systematic view is presented in figure 2.36.

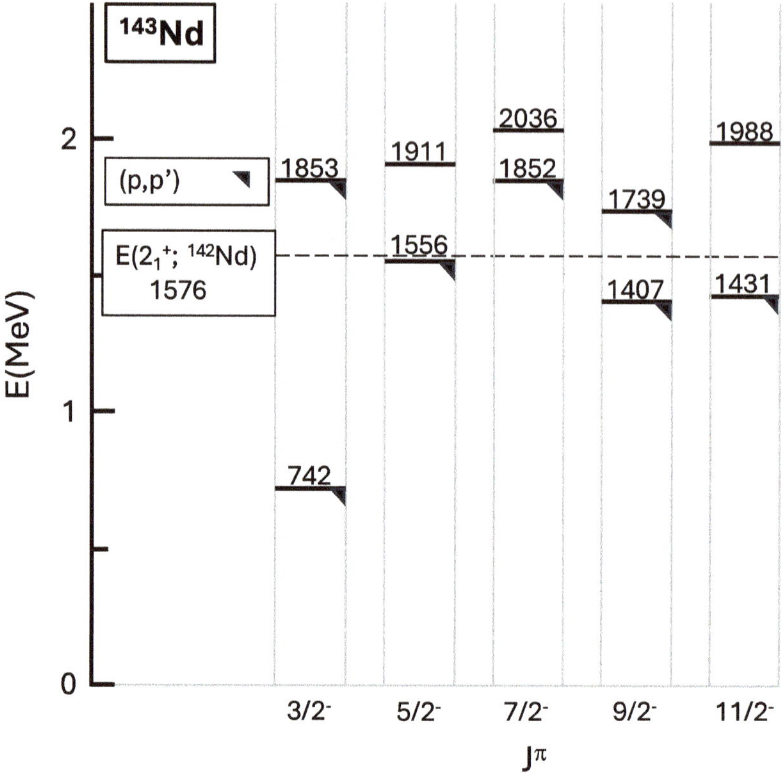

Figure 2.35. View of particle–core coupled, $2f_{7/2}^{+1} \times 2_1^+ (^{142}\text{Nd})$ candidate states in ^{143}Nd. The spectroscopic data indicate the states populated by inelastic scattering of protons [35], qv. figures 2.42 and 2.43. The data are taken from ENSDF.

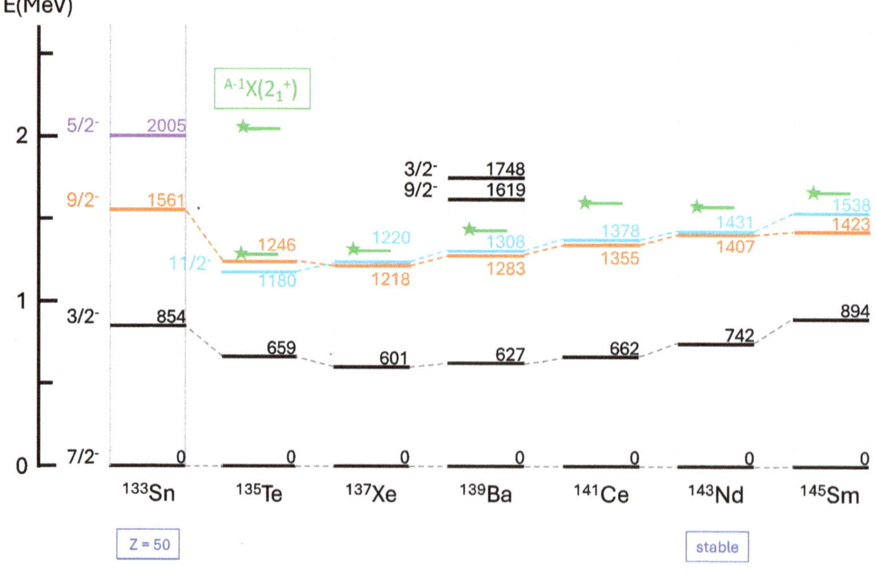

Figure 2.36. Systematic view of some particle–core coupled, $2f_{7/2}^{+1} \times 2_1^+$ ($A-1$ core) candidate states in the $N = 83$ isotones. The data are taken from ENSDF.

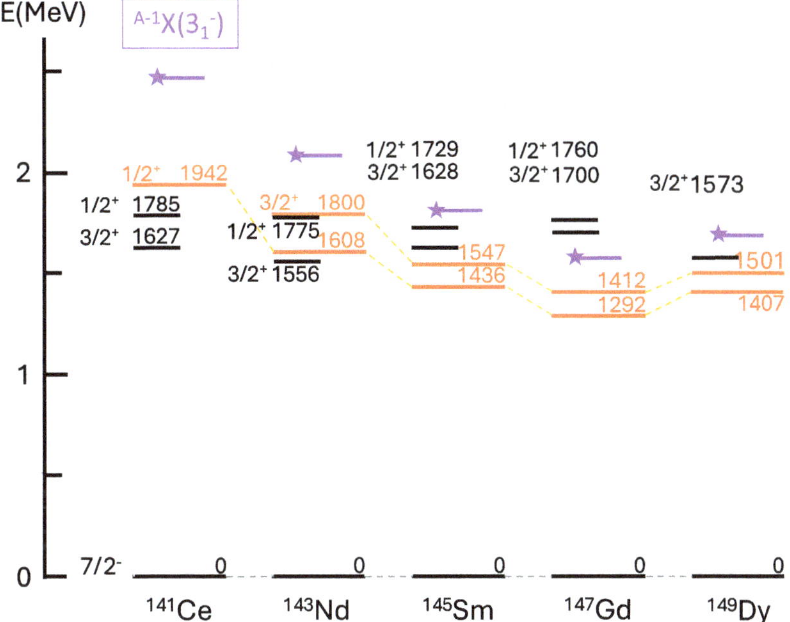

Figure 2.37. Systematic view of energies of the lowest energy positive-parity low-spin states in the $N = 83$ isotones. Violet bars with a star show the 3^- first excited states in the $N = 82$ neighbouring $A - 1$ 'core' nuclei. Details are discussed in the text. The data are taken from ENSDF.

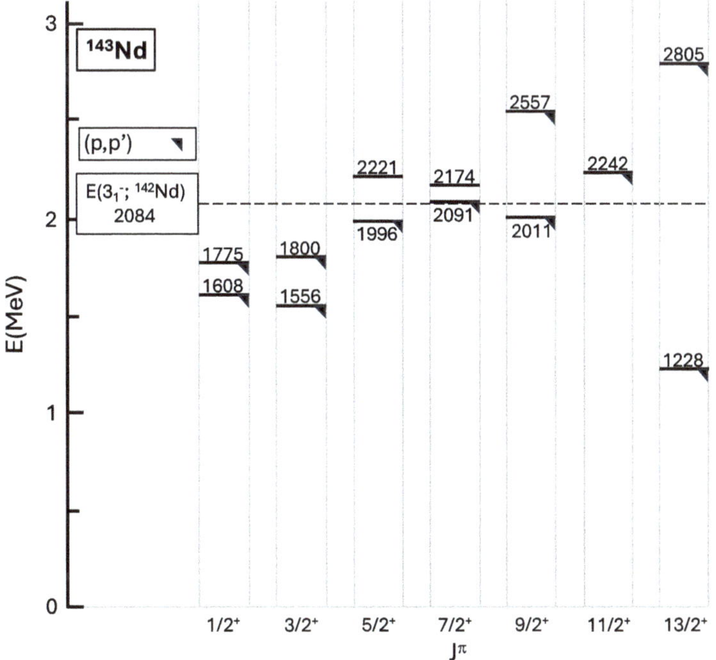

Figure 2.38. View of particle–core coupled, $2f_{7/2}^{+1} \times 3_1^-$ (^{142}Nd) candidate states in ^{143}Nd. The spectroscopic data indicate the states populated by inelastic scattering of protons [35], q.v. Figures 2.42 and 2.43. The data are taken from ENSDF.

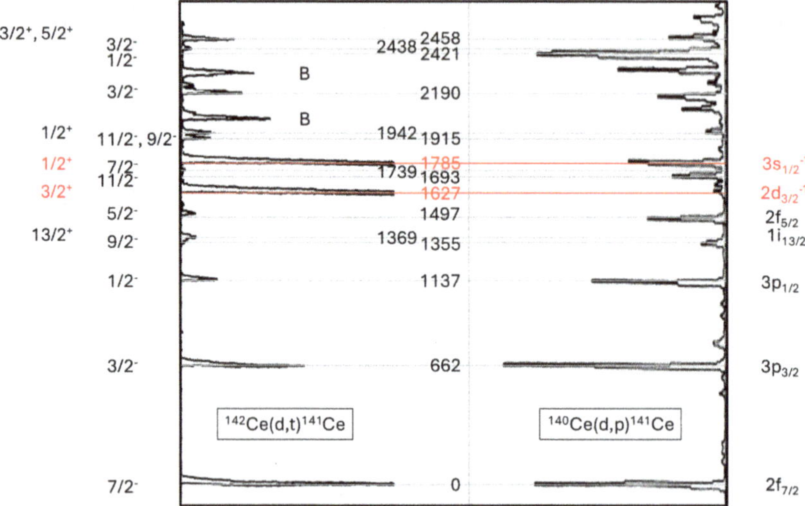

Figure 2.39. (Left) States in ^{141}Ce populated in the one-neutron removal reaction ^{142}Ce(d,t)^{141}Ce. Reprinted from [36], copyright (1979) with the permission of Elsevier. (Right) States in ^{141}Ce populated in the one-neutron addition reaction ^{140}Ce(d,p)^{141}Ce. Reprinted with permission from [37]. Copyright (1977) by the American Physical Society. Other data are taken from ENSDF.

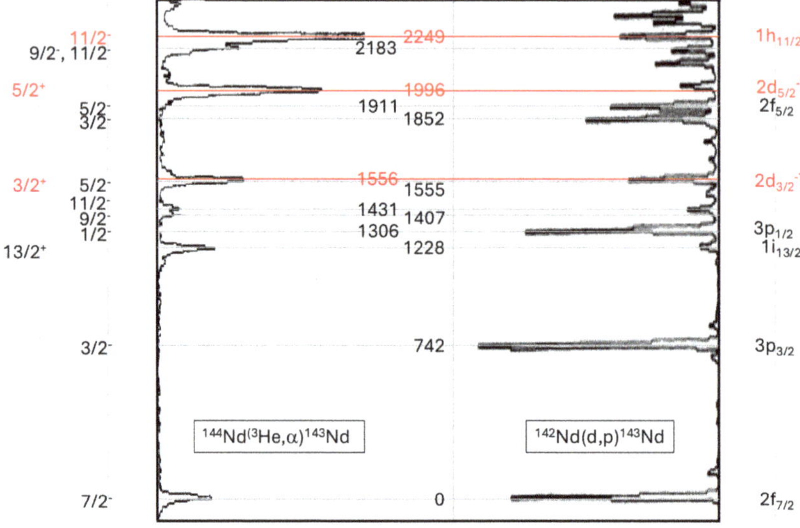

Figure 2.40. (Left) States in ^{143}Nd populated in the one-neutron removal reaction ^{144}Nd(^{3}He,α)^{143}Nd. Reprinted from [38], copyright (1988) with the permission of Elsevier. (Right) States in ^{143}Nd populated in the one-neutron addition reaction ^{142}Nd(d,p)^{143}Nd. Reprinted with permission from [37]. Copyright (1977) by the American Physical Society. Other data are taken from ENSDF.

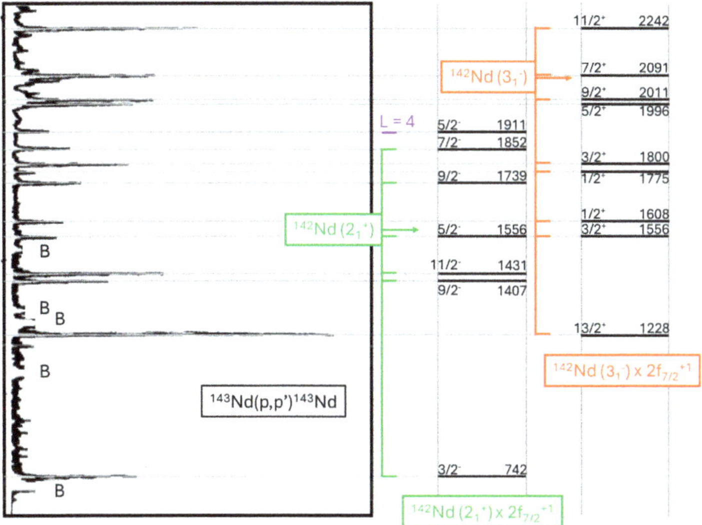

Figure 2.41. Spectrum of inelastically scattered protons from ^{143}Nd. The spin-parities and energies (in keV) are taken from ENSDF. The population of states corresponding to the particle–core coupling ^{142}Nd $(2_1^+) \times 2f_{7/2}^{+1}$ and ^{142}Nd $(3_1^-) \times 2f_{7/2}^{+1}$ are highlighted in green and orange, respectively; a state at 1911 keV populated by $L = 4$ is highlighted in violet. Proton spectrum peaks labelled 'B' result from target impurities. Compare these details with figures 2.35 and 2.38. Reprinted from [35], copyright (1989) with the permission of Elsevier.

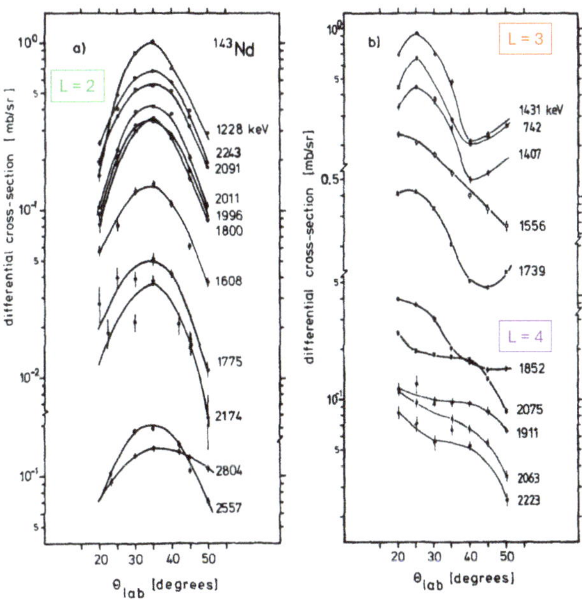

Figure 2.42. Angular distributions of inelastically scattered protons from ^{143}Nd. The distribution of scattered protons distinguish between angular momentum transfer of 2 ($L = 2$) in part (a) and 3 and 4 ($L = 3$, $L = 4$) in part (b). The distributions are labelled by excitation energies of the states populated. Note that there is an unresolved doublet of levels at 1556 keV; further note that the 1852 keV state is deduced to be populated by a mixture of $L = 2$ and $L = 4$ inelastic scattering. Reprinted from [35], copyright (1989) with the permission of Elsevier.

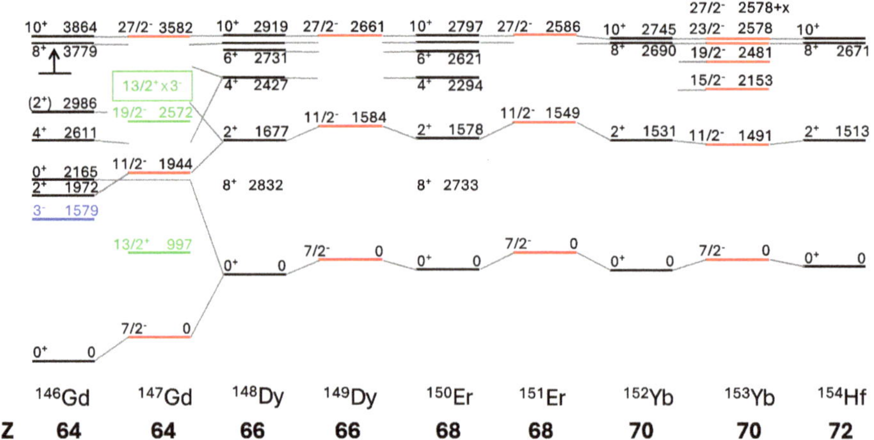

Figure 2.43. View of high-spin particle–core coupled, $2f^{+1}_{7/2}$ ×core seniority excited states for $Z > 63$. Details are discussed in the text. The data are taken from ENSDF.

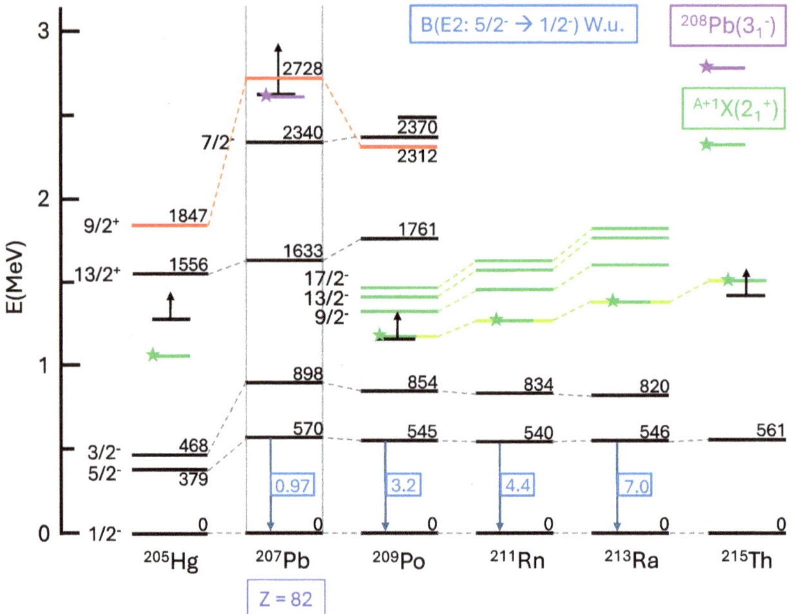

Figure 2.44. Selected systematic features of the $N = 125$ isotones. Levels coloured black are shell model candidate states; note that the energy (2370 keV) given for the $7/2^-$ state in ^{209}Po is the centroid obtained from a study of ^{210}Po(d,t)^{209}Po, see figure 2.45. Levels coloured red are intruder states. Levels coloured green are due to the ground state coupled to core excited states, see figures 2.47 and 2.48. The lowest energy core excited states with spin-parity 2^+ are indicated by green bars with a star; the 3^-_1 core excited state in ^{208}Pb is indicated by a violet bar with a star. Excitations above which levels are omitted are shown as horizontal bars with upwards-pointing arrows; this is not indicated for ^{211}Rn and ^{213}Ra because the excited state data for these nuclei are certainly very incomplete due to the difficulty of studying low-spin states in these nuclei. The E2 collective strengths between the first excited states and the ground states are shown and indicate significant onset of collectivity moving away from the $Z = 82$ shell closure [39]. Other data are taken from ENSDF.

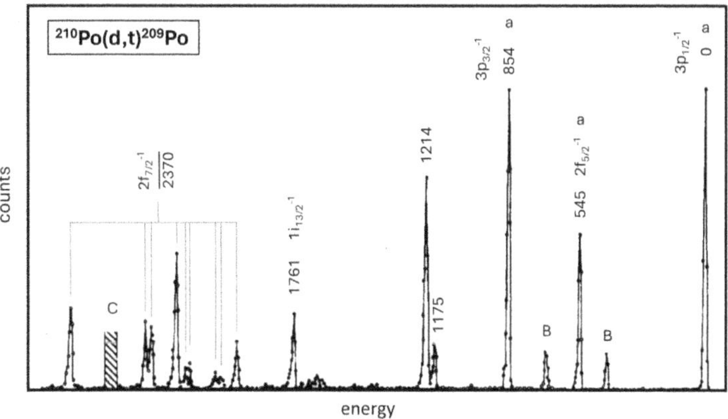

Figure 2.45. Spectrum of tritons from the one-neutron removal reaction ^{210}Po(d,t)^{209}Po. Peaks marked 'B' are due to target impurities. Peaks marked 'a' are scaled down in counts by a factor of 6. The peaks are labelled by the energies (in keV) corresponding to excitations in ^{209}Po, which are taken from ENSDF. The feature labelled 'C' is a gap in the spectrum. Shell model configurations are indicated, where note that the $L = 3$ transfer strength above 1990 keV is heavily fragmented and so a centroid value is given, which is taken from [40]. The peaks labelled 1175 and 1214 keV result from the ground-state configuration coupled to the 2_1^+ core excited state at 1181 keV. Reprinted from [40], copyright (1979) with the permission of Elsevier.

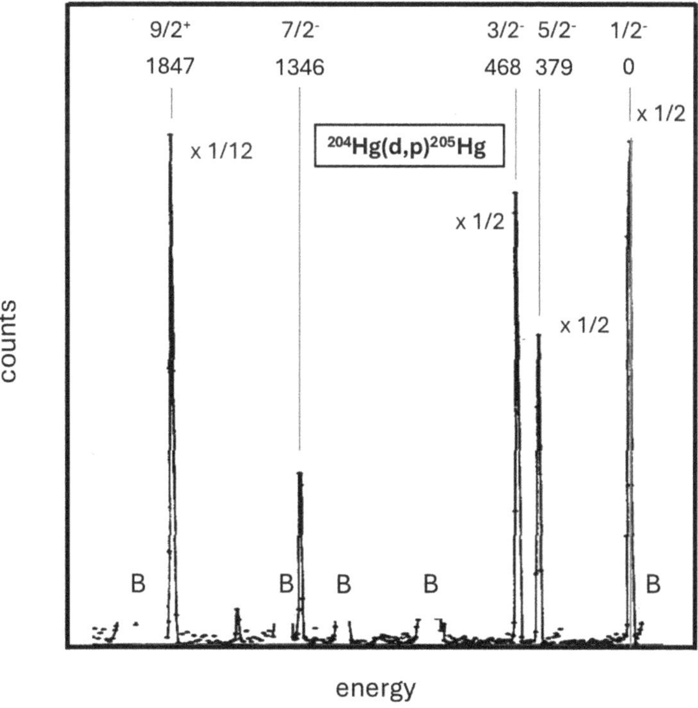

Figure 2.46. Spectrum of protons from the one-neutron addition reaction ^{204}Hg(d,p)^{205}Hg. Peaks are labelled by energies, spins and parities of corresponding states in ^{205}Hg, taken from ENSDF. Note the counts scale-down factors. There are strong background features which are blanked out and labelled 'B'. Reprinted from [41], copyright (1970) with the permission of Elsevier.

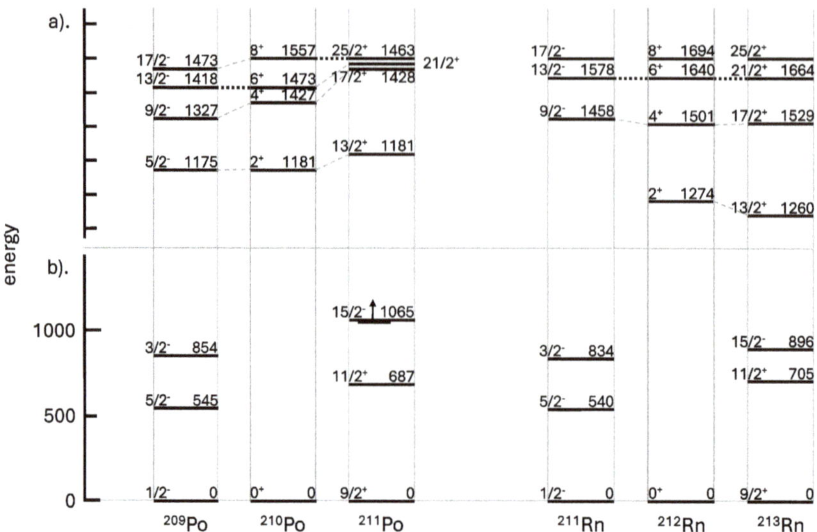

Figure 2.47. A comparison of $N = 125, 127$ excited state systematics with emphasis on particle–core coupling; the relevant even–even cores are shown. The core structures are seniority-dominated one-broken proton pair in the $1h_{9/2}$ shell model orbital. There is a clear indication that particles or holes in nuclei which are adjacent to a single-closed shell exhibit weak coupling when the even–even core structure is dominated by a high-j seniority structure and the spin coupling is 'aligned', i.e. maximal. (a) Selected high-spin states with normalization of energies indicated by heavy dotted lines; systematic patterns are emphasized by dashed lines. In ^{211}Po, states are omitted above 1051 keV and details can be found in figure 2.46. (b) The states in ^{209,211}Po and ^{211,213}Rn below 1100 keV. Note that in (a) the energy scale is expanded relative to (b) by a factor $\times 2$. The data are taken from ENSDF.

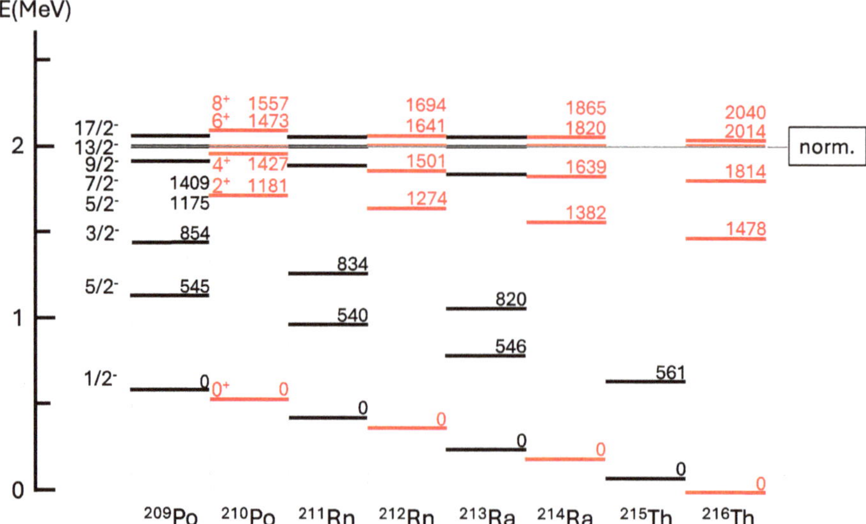

Figure 2.48. The systematics of states in the odd-mass $N = 125$ isotones with emphasis on the dominance of core seniority coupling. Energies are shown relative to the maximum spin couplings, $J = 8$ in the even–even cores and $j = 17/2$, i.e. $2 \times 8 + 1/2$ in the odd-mass nuclei. Compare this view with those shown in figures 2.47 and 2.44. The data are taken from ENSDF.

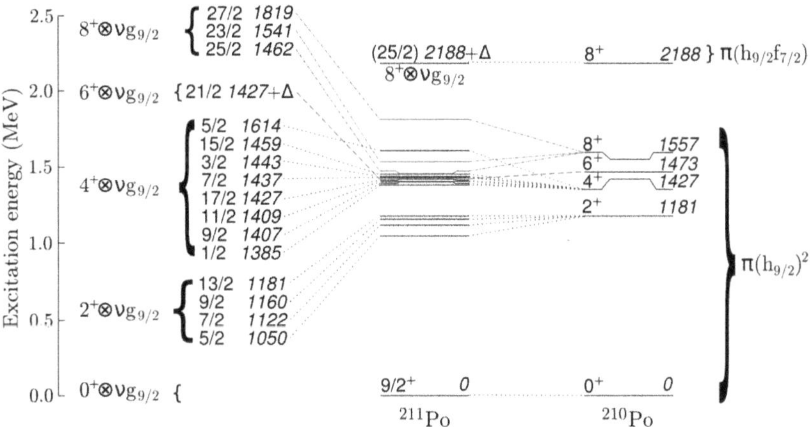

Figure 2.49. A 'complete' view of the low-energy excited states in ^{211}Po [42] and see [43]. This view shows that weak coupling between an odd nucleon and seniority-dominated core excitations applies across all spin couplings, not just 'aligned' spin coupling. Reprinted with permission from [42]. Copyright (2017) by the American Physical Society.

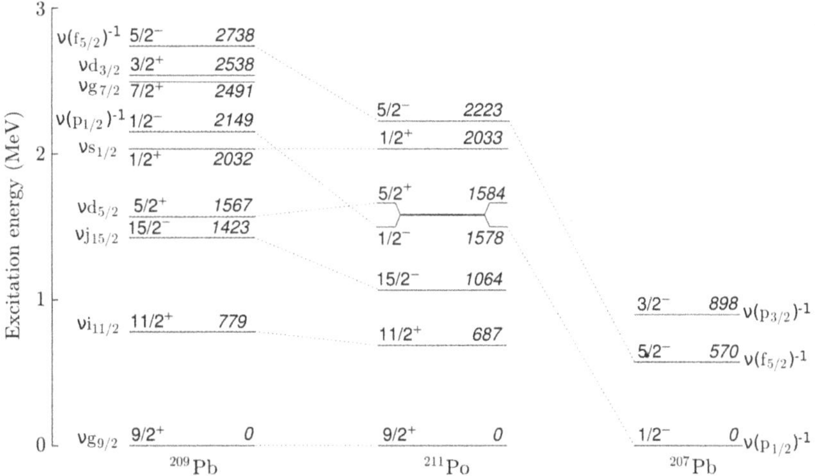

Figure 2.50. Systematics of neutron two-particle-one-hole intruder states in ^{209}Pb and ^{211}Po. Compare this view with figures 1.10 and 1.21. Reprinted with permission from [42]. Copyright (2017) by the American Physical Society.

References

[1] Mougeot M *et al* 2021 Mass measurements of $^{99-101}$In challenge *ab initio* nuclear theory of the nuclide ^{100}Sn *Nat. Phys.* **17** 1099–103

[2] Nies L *et al* 2023 Isomeric excitation energy for ^{99}In^m from mass spectrometry reveals constant trend next to doubly magic ^{100}Sn *Phys. Rev. Lett.* **131** 022502

[3] Ricketts C 2021 Fundamental nuclear properties of indium isotopes measured with laser spectroscopy *PhD Thesis* University of Manchester, Manchester

[4] Lorenz C H *et al* 2019 β decay of ^{127}Cd and excited states in ^{127}In *Phys. Rev.* C **99** 044310

[5] Saito Y *et al* 2020 Decay spectroscopy of ^{129}Cd *Phys. Rev.* C **102** 024337

[6] Vernon A R *et al* 2022 Nuclear moments of indium isotopes reveal abrupt change at magic number 82 *Nature* **607** 260–5

[7] Vernon A 2019 Evolution of the indium proton–hole states up to $N = 82$ studied with laser spectroscopy *PhD Thesis* University of Manchester, Manchester

[8] DiJulio D *et al* 2013 Coulomb excitation of ^{107}In *Phys. Rev.* C **87** 017301

[9] Vaquero V *et al* 2020 Fragmentation of single-particle strength around the doubly magic nucleus ^{132}Sn and the position of the of $0f_{5/2}$ proton–hole state in ^{131}In *Phys. Rev. Lett.* **124** 022501

[10] Bäcklin A, Fogelberg B and Malmskog S G 1967 Possible deformed states in ^{115}In and ^{117}In *Nucl. Phys.* A **96** 539–60

[11] Harar S and Horoshko R N 1972 Study of the level scheme of ^{117}In via proton transfer reactions *Nucl. Phys.* A **183** 161–72

[12] Weiffenbach C V and Tickle R 1971 Stuctre of odd-A indium isotopes determined by the (d,^{3}He) reaction *Phys. Rev.* C **8** 1668

[13] Tuttle W K, Stelson P H, Robinson R L, Milner W T, McGowan F K, Raman S and Dagenhart W K 1976 Coulomb excitation of ^{113,115}In *Phys. Rev.* C **13** 1036

[14] Karthein J *et al* 2023 Electromagnetic properties of indium isotopes elucidate the doubly magic character of 100Sn arXiv:2310.15093

[15] Mitchell A J 2012 Investigating high-*j* single particle energies in $Z = 51$ nuclei *PhD Thesis* University of Manchester, Manchester

[16] Gray T J *et al* 2020 Early signal of emerging nuclear collectivity in neutron-rich ^{129}Sb *Phys. Rev. Lett.* **124** 032502

[17] Gray T J *et al* 2024 Suppressed electric quadrupole collectivity in $_{49}$Ti *Phys. Lett.* B **855** 138856

[18] Lechner S *et al* 2023 Electromagnetic moments of the antimony isotopes $^{112-133}$Sb *Phys. Lett.* B **847** 138278

[19] Jenkins D G and Wood J L 2021 *Nuclear Data: A primer* (Bristol: IOP Publishing)

[20] Heyde K and Wood J L 2020 *Quantum Mechanics for Nuclear Structure: An Intermediate Level View* (Bristol: IOP Publishing)

[21] Montestruque L A, Cobian-Rozak M C, Szaloky G, Zumbro J D and Darden S E 1978 Study of the ^{77,79,81,83}Se level structure with the ^{76,78,80,82}Se(d,p) reaction *Nucl. Phys.* A **305** 29–62

[22] Detorie N A, Jolivette P L, Browne C P and Rollefson A A 1978 Nuclear structure information for states in ^{85}Kr from the ^{84}Kr(d,p) and ^{84}Kr($\vec{d}$,p) reactions *Phys. Rev.* C **18** 991

[23] Morton J M, Davies W G, McLatchie W, Darcey W and Kitching J E 1971 The level structures of ^{85,87}Sr from (d,p) reactions *Nucl. Phys.* A **161** 228–40

[24] Fortier S, Gales S, Hourani E, Maison J M, Massolo C P and Schapira J P 1989 Neutron-hole strengths and core coupling in ^{87}Sr via the ^{88}Sr(^{3}He,α)^{87}Sr reaction *Phys. Rev.* C **39** 82

[25] Kaptein E J, Blok H P and Blok J 1978 High resolution investigation of the ^{87}Sr(p,p') reaction *Nucl. Phys.* A **312** 97–114

[26] Blok H P, Hulstman L, Kaptein E J and Blok J 1976 Investigation of ^{91}Zr by high resolution (d,p), (p,p') and (p,d) reactions *Nucl. Phys.* A **273** 142–71

[27] Sharp D K *et al* 2013 Neutron single-particle strength outside the $N = 50$ core *Phys. Rev.* C **87** 014312

[28] Drăgulescu E, Ivaşcu M, Mihu R, Popescu D, Semenescu G, Velenik A and Paar V 1984 Coulomb excitation of levels in ^{135}Ba and ^{137}Ba *J. Phys. G Nucl. Part. Phys.* **10** 1099–114

[29] Jolly R K and Kashy E 1971 Neutron-hole-state structure in $N = 81$ nuclei II. ^{140}Ce and ^{138}Ba(p,d) *Phys. Rev.* C **4** 1398

[30] Howard A M *et al* 2020 Neutron-hole strength in $N = 81$ nuclei *Phys. Rev.* C **101** 034309

[31] Allmond J M *et al* 2012 One-neutron transfer study of ^{135}Te and ^{137}Xe by particle-γ coincidence spectroscopy: the $\nu 1 i_{13/2}$ state at $N = 83$ *Phys. Rev.* C **86** 031307

[32] Reviol W *et al* 2016 One-neutron transfer study of ^{137}Xe and systematics of $13/2_1^+$ and $13/2_2^+$ levels in $N = 83$ nuclei *Phys. Rev.* C **94** 034309

[33] Bianco L *et al* 2010 Discovery of ^{157}W and ^{161}Os *Phys. Lett.* B **690** 15–8

[34] Allmond J M *et al* 2014 Double-magic nature of ^{132}Sn and ^{208}Pb through lifetime and cross-section measurements *Phys. Rev. Lett.* **112** 172701

[35] Trache L *et al* 1989 Spectroscopy of ^{142}Nd and ^{143}Nd via inelastic proton scattering at $E_\mathrm{p} = 25$ MeV *Nucl. Phys.* A **492** 23–44

[36] Lien J R *et al* 1979 Studies of single-neutron holes in the transitional nuclei $N = 83$-89: the ^{142}Ce(d,t)^{141}Ce and ^{142}Ce(^{3}He, α)^{141}Ce pick-up reactions *Nucl. Phys.* A **324** 141–59

[37] Strömich A, Steinmetz B, Bangert R, Gonsior B, Roth M and von Brentano P 1977 (d,p) reactions on ^{124}Sn, ^{130}Te, ^{138}Ba, ^{140}Ce, ^{142}Nd, and ^{208}Pb below and near the Coulomb barrier *Phys. Rev.* C **16** 2193

[38] Løvhøiden G, El-Kazzaz S, Lien J R, Heyde K, Rekstad J, Ellegaard C, Bjerregaard J H, Knudsen P and Kleinheinz P 1988 Study of high-spin hole states in ^{143}Nd *Nucl. Phys.* A **481** 71–80

[39] Gerathy M S M *et al* 2021 Emerging collectivity in neutron-hole transitions near doubly magic ^{208}Pb *Phys. Lett.* B **823** 136738

[40] Bhatia T S, Canada T R, Barnes P D, Eisenstein R A and Ellegaard C 1979 Levels of ^{209}Po and ^{211}Po populated in one-neutron stripping and pickup from ^{210}Po *Nucl. Phys.* A **314** 101–14

[41] Andersen M L, Andersen S A, Nathan O, Bisgard K M, Gregersen K, Hansen O, Hinds S and Chapman R 1970 Neutron transfer reactions on ^{204}Hg *Nucl. Phys.* A **153** 17–31

[42] Slotte J M K *et al* 2017 In-beam γ-ray spectroscopy of low- and medium-spin levels in ^{211}Po *Phys. Rev.* C **96** 044302

[43] Cottle J R *et al* 2017 Complete spectroscopy of ^{211}Po below 2.0 MeV via the (α,n) reaction *Phys. Rev.* C **95** 064323

IOP Publishing

Nuclear Data
An independent-particle motion view
David Jenkins and John L Wood

Chapter 3

Where are Nilsson model states found in nuclei?

The manifestation of Nilsson model configurations in nuclei is introduced.

Concepts: independent-particle mean field, spheroidal potential, equipotential surface, Nilsson model Hamiltonian, Nilsson model quantum numbers, spherical basis—N, l, j, $\Omega = m_j$, π, asymptotic basis—N, n_z, Λ, Σ, Ω, π, Nilsson bands

Learning outcomes: the Nilsson model manifests extraordinary power to organize data for deformed nuclei into rotational bands built on Nilsson configurations.

The Nilsson model is an independent-particle view of nucleon motion in a deformed mean field. It is an extraordinarily successful description of a very large range of data pertaining to deformed nuclei. Indeed, a question that is not easy to answer is: 'why does it work so well?'

The most important open question regarding the applicability of the Nilsson model to nuclei is: what is the weakest deformation for which it can be regarded as useful? In this chapter, we consider mainly nuclei with strong deformation. The applicability of the Nilsson model to weakly deformed nuclei lies in the research domain, but herein we take important steps in this direction.

3.1 Basic features of the model

The Nilsson model[1] in its simplest version adopts the most elementary formulation of the shell model, cf equation (1.1), and deforms the potential to produce a spheroidal harmonic oscillator potential, i.e. a cylindrically symmetric oscillator potential with a spin–orbit coupling term and a well-flattening term. The kinetic energy, spin–orbit and well-flattening terms are not deformed. Thus, we consider a Nilsson model Hamiltonian of the form

$$H = \frac{p^2}{2m} + \frac{m}{2}\left[\omega_{xy}^2(x^2 + y^2) + \omega_z^2 z^2\right] + C\mathbf{l}\cdot\mathbf{s} + Dl^2, \qquad (3.1)$$

---

[1] A video-based tutorial on the Nilsson model is found in figure 3.60.

doi:10.1088/978-0-7503-5648-0ch3   3-1

where the parameters follow directly from definitions for the shell model, with the added degree of freedom that the oscillator frequency ω now has two components: ω_{xy} with respect to directions perpendicular to the symmetry axis, and ω_z in the direction of the symmetry axis.

The appearance of two oscillator frequencies, ω_{xy} and ω_z needs special handling. In the shell model, the single parameter, ω is fixed by equation (1.6), viz. $\hbar\omega = 41A^{-1/3}$ MeV, via the expectation value given in equation (1.5). This expectation value is for the observable quantity $\sum_{i=1}^{A} r_i^2$ which is a measure of the size of the nucleus and, in quantum mechanical terms, is a measure of the confinement of a particle of mass m, the mass of the proton and neutron (which, recall, are nearly identical). This is a fundamental step in the quantization of the nucleus and sets the energy scale for all nuclear systems. Thus, the shell model energy-shell ordering and energy scale are determined within the model. The presence of deformation and two oscillator frequency parameters requires some thinking which is subtle and is presented below.

The primary measure of nuclear deformation, in terms of an observable, is the quadrupole moment. Other moments may be non-zero, e.g. a hexadecapole moment. Recall that an octupole moment is not an observable (odd-rank spherical tensor operators have zero diagonal matrix elements), but a nucleus can have an octupole shape (recall the asymmetry of fission). We proceed in a manner that is particular to quadrupole moments and implied spheroidal deformation. The issue of triaxial shapes (ellipsoidal deformation) does not appear to be important for the strongly deformed nuclei considered herein.

The observable termed the 'quadrupole moment' is an electric quadrupole moment. But the mean-field for an independent-particle description of a nucleus of mass A is for the combined influence of $A - 1$ nucleons, protons plus neutrons, on an individual nucleon. It is assumed that the spheroidal shape of the nuclear mass distribution is the same as that implied for proton distribution in the nucleus, based on the electric quadrupole moment. There is the possibility of making a model distinction between spheroidal mean fields for protons and for neutrons: this is done within the parameters defined above, notably C and D in equation (3.1) (but with no allowance made for the mass difference between protons and neutrons—this is too small vis à vis the comparison of the model with data). Specifics of parameter values are discussed shortly.

The key step in arriving at the parameterization of the deformed Nilsson model potential is to define an equipotential surface for the nucleus. This step is ad hoc: it enables a connection to be made between ω_{xy}, ω_z and observed quadrupole moments. Thus, we introduce the relationships, cf equation (3.1),

$$\omega_{xy}^2 R_x^2 = \omega_{xy}^2 R_y^2 = \omega_z^2 R_z^2, \tag{3.2}$$

where R_x, R_y, R_z are radial distances to a sharply defined nuclear surface. Then, via

$$\omega_{xy} := \omega_0 \exp\{\varepsilon/3\}, \quad \omega_z := \omega_0 \exp\{-2\varepsilon/3\}, \tag{3.3}$$

a constant volume, proportional to $R_x R_y R_z$, follows. The imposition of a constant nuclear volume under changing deformation is assumed, based on the observed (near) incompressible nature of nuclear matter, recall $R = 1.2A^{1/3}$ fm. The parameter, ε is connected to a quadrupole deformation via the relationship

$$V = \frac{m\omega_0^2 r^2}{2} - \frac{4}{3}\frac{\sqrt{\pi}}{5}\varepsilon m\omega_0^2 r^2 Y_{20}(\theta, \phi), \tag{3.4}$$

where $Y_{20}(\theta, \phi)$ is the quadrupole spherical harmonic with axial symmetry, and recall,

$$r^2 Y_{20}(\theta, \phi) = \sqrt{\frac{5}{16\pi}}(2z^2 - x^2 - y^2). \tag{3.5}$$

equation (3.4) defines

$$V_{\text{def}} := -\frac{4}{3}\frac{\sqrt{\pi}}{5}\varepsilon m\omega_0^2 r^2 Y_{20}(\theta, \phi), \tag{3.6}$$

and the diagonal eigenvalues of V_{def} are given by

$$E_{\text{def}} = \frac{2}{3}\varepsilon(N + 3/2)\hbar\omega_0 \left[\frac{3\Omega^2}{4j(j+1)} - \frac{1}{4}\right], \tag{3.7}$$

where $\Omega := j_3$ (body-frame directional component of j). For a single j subshell, the energies, E_{def} are depicted in figure 3.1. A useful semi-classical view of E_{def} in terms of orbital alignments within the deformed mean field of the nucleus is given in figure 3.2. Recall the more narrowly that a quantum particle is confined, the higher is its zero-point energy.

Equation (3.7) expresses the diagonal matrix elements of V_{def}. The term V_{def} also has off-diagonal matrix elements: these cause mixing of j values but preserve Ω for the model. Because the deformed potential is reflection symmetric, parity is also conserved. In consequence the mixing matrix that must be diagonalized involves a subspace of Hilbert space that is characterized simply by configurations with the same Ω, π values. This mixing is depicted schematically in figure 3.3. Energy eigenstates are expressed in a spherical basis as

$$|\Omega\pi\rangle = \sum_{jl} C_{jl} |Nj\, l\Omega\pi\rangle. \tag{3.8}$$

The influence of V_{def} can extend to mixing $|\Omega\pi\rangle$ configurations differing in N by 2, so-called $\Delta N = 2$ mixing. For present purposes this only occurs in very isolated cases and so discussion of this is limited to a few brief notes later in the narrative.

A conceptual view which forms a basic part of the Nilsson model formulation is the so-called 'asymptotic' solution. This corresponds to the extreme cylindrically deformed limit where the harmonic oscillator part of the Hamiltonian leads to energy eigenvalues, viz.

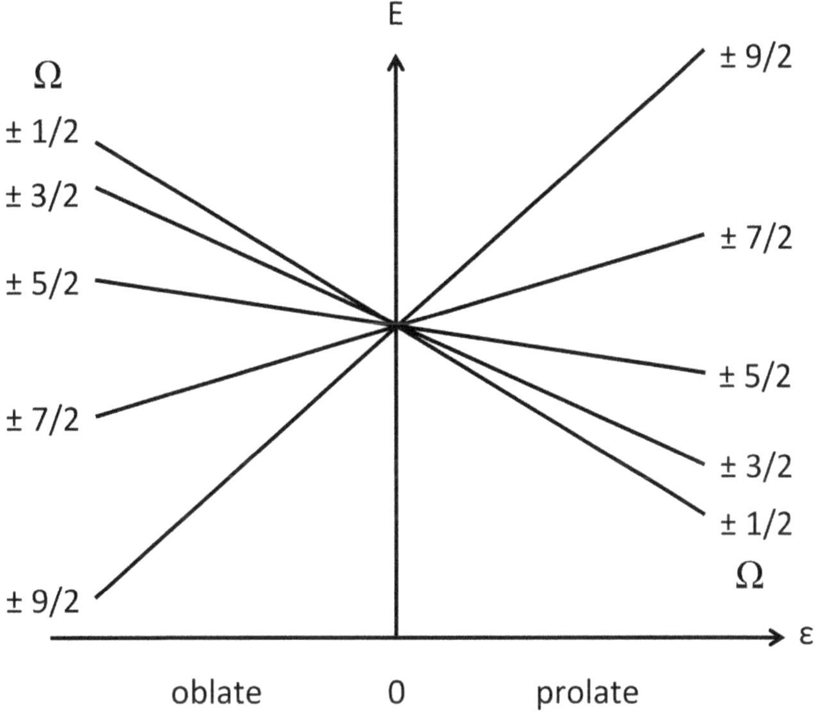

Figure 3.1. The energy spectrum for a single-j orbital in a spheroidal harmonic oscillator potential, cf equation (3.7). The energies are linear in the deformation parameter, ε and quadratic in the Nilsson quantum number, Ω. The difference between states in a prolate deformed potential and an oblate deformed potential is a sign change in ε and a reversal in the energy order of the Ω values: nuclei with large ground-state oblate deformation are not yet identified; there are a few weakly deformed nuclei that are oblate, but they appear to be axially asymmetric. Each orbital, labelled by Ω, is two-fold degenerate, i.e. the states $+\Omega$ and $-\Omega$ are indistinguishable: this is the result of the plane of reflection symmetry through the centre of mass of the nucleus at right angles to the axis of rotational symmetry of the spheroidal shape. Reproduced from [1], copyright IOP Publishing Ltd. All rights reserved.

$$E(N, n_z) = (N - n_z + 1)\hbar\omega_{xy} + (n_z + 1/2)\hbar\omega_z. \qquad (3.9)$$

If this expression were the end of the model story, one sees that by making the elongation axis become infinitely 'stretched' one allows many particles in such a potential to take on a total energy that could approach an arbitrarily small value. Nuclei do not do this: a constraint must be imposed such that the 'shape' of the potential matches the shape of the nucleus. 'Shape' here equates to the quadrupole moment of the nucleus. Hence, one might view equations (3.2) and (3.3) as motivated by this constraint.

The asymptotic quantum numbers are the popular and universally used labels of configurations. The spherical basis is the universally preferred basis for diagonalization. The spherical basis plays an essential role in the coupling of Nilsson model configurations to rotational degrees of freedom and is presented in chapter 4. The spherical basis also plays an essential role in the interpretation of one-nucleon transfer reactions. The reason for this is that the reaction takes place in the

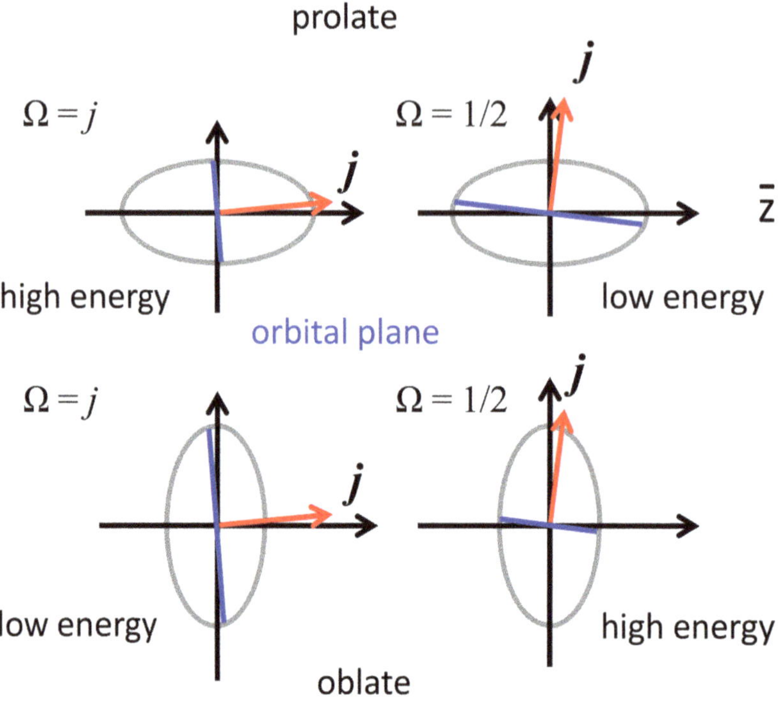

Figure 3.2. A schematic view of orbital alignments in spheroidal deformed potentials. A simple interpretation of the energies associated with this view is to recall that for a particle of mass m confined in a one-dimensional infinite square-well potential of width L, subject to the laws of quantum mechanics, the energies are given by $n^2\hbar^2/2mL^2$, where $n = 1, 2, 3, \ldots$, i.e. 'narrowing' the confinement raises the energy of independent-particle states. Reproduced from [1], copyright IOP Publishing Ltd. All rights reserved.

laboratory frame in which the spherical basis provides good quantum numbers. In practical terms, the C_{lj} coefficients in equation (3.8) take on physical significance in the definition of one-nucleon transfer strengths leading to final states in a deformed odd-mass nucleus, so-called 'spectroscopic factors'. Some details are given later.

The asymptotic basis leads to only diagonal energies for V_{def}. It is the 'spherical' terms, $C\mathbf{l}\cdot\mathbf{s}$ and Dl^2 in the Hamiltonian that mix this basis. The main value of the asymptotic basis is as a set of approximate labelling quantum numbers. Because the spin–orbit term is a mixing term, the asymptotic basis adopts $\Lambda := l_z$ and $\Sigma := s_z$ as approximately good quantum numbers, viz. $|Nn_z\Lambda\Sigma\rangle\Omega\pi$. Discussion of ambiguities, with respect to parameterization of the Nilsson model Hamiltonian, between the choice of basis in which diagonalization is carried out is beyond the scope of the present discussion[2].

---

[2] There is an extensive literature that explores Nilsson model parameterizations, even coordinate system representations such as deformation-stretched coordinates and presents Nilsson model energy level diagrams and tables of expansion coefficients, cf equation (3.8). We note that effects such as rotation-particle coupling render fine-tuning at the independent-particle deformed mean-field stage ambiguous. Thus, we only present the most basic details of the Nilsson model. Rotation-particle coupling is the focus of chapter 4.

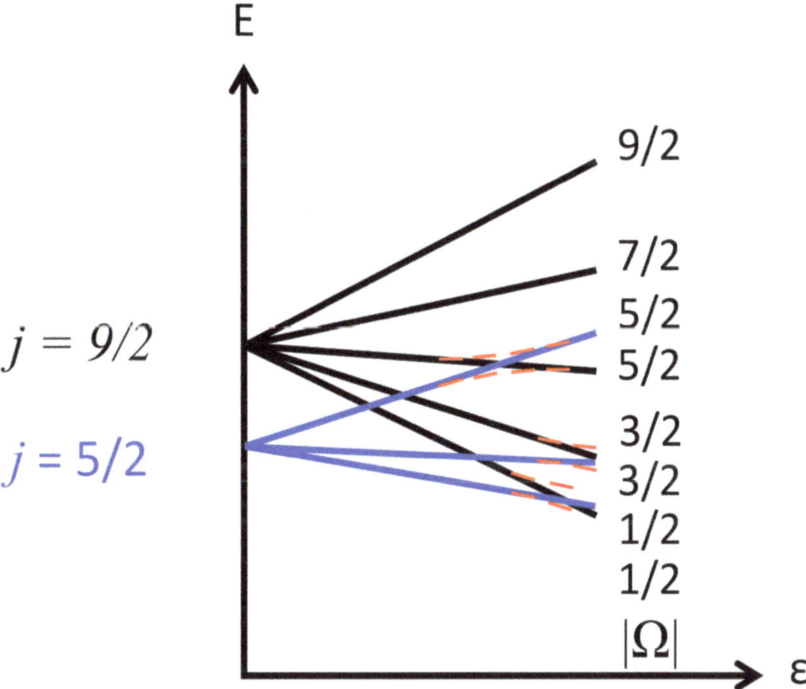

Figure 3.3. A schematic view of the mixing of $|j\Omega\rangle$ configurations (shell model configurations) in a spheroidal deformed potential. This results from the influence of V_{def}, e.g. Equation (3.6), which produces the energy shifts illustrated in figure 3.1, and mixes configurations differing in j by two units; parity is conserved. This leads to the expansion of Nilsson states $|\Omega\pi\rangle$ over $|Nj\,l\Omega\pi\rangle$ spherical basis states, cf equation (3.8). The mixing implied in the figure is artificially weak to help visualize the effect. Reproduced from [1], copyright IOP Publishing Ltd. All rights reserved.

An example of a so-called Nilsson diagram is depicted in figure 3.4. This is shown for prolate deformation which totally dominates the lanthanide and actinide mass regions. In using a Nilsson diagram, one needs to know the number of nucleons, e.g. for Lu, $Z = 71$, and an estimate of the deformation parameter, e.g. from a quadrupole moment of the nucleus of interest or of a neighbouring nucleus. Thus, figure 3.5 shows the identification of Nilsson states to be expected in ^{175}Lu. Figure 3.6 shows observed low-energy Nilsson bands in ^{175}Lu. The diagrams, as depicted, are only approximate with respect to level order for a given deformation and should not be viewed as quantitative[3]. The primary reason for lack of quantitative 'predictions' is because of rotation-particle coupling effects, other deformation effects, and pairing correlations. Some details of other deformation

---

[3] Note that there is a scaling across j values, cf equation (3.7), which is not easy to see in the plot, namely that the energy scale depends inversely on $j(j + 1)$. For example, the energy scale difference for the dependence $\Omega^2/j(j + 1)$, between $(\Omega, j) = (5/2, 5/2)$ and $(7/2, 7/2)$, is 5/7 cf 7/9, i.e. 0.71 cf 0.78 as manifested in the Nilsson states 5/2[402] and 7/2[404]. Thus, they exhibit an energy separation that is nearly independent of ε, which is evident in the plot.

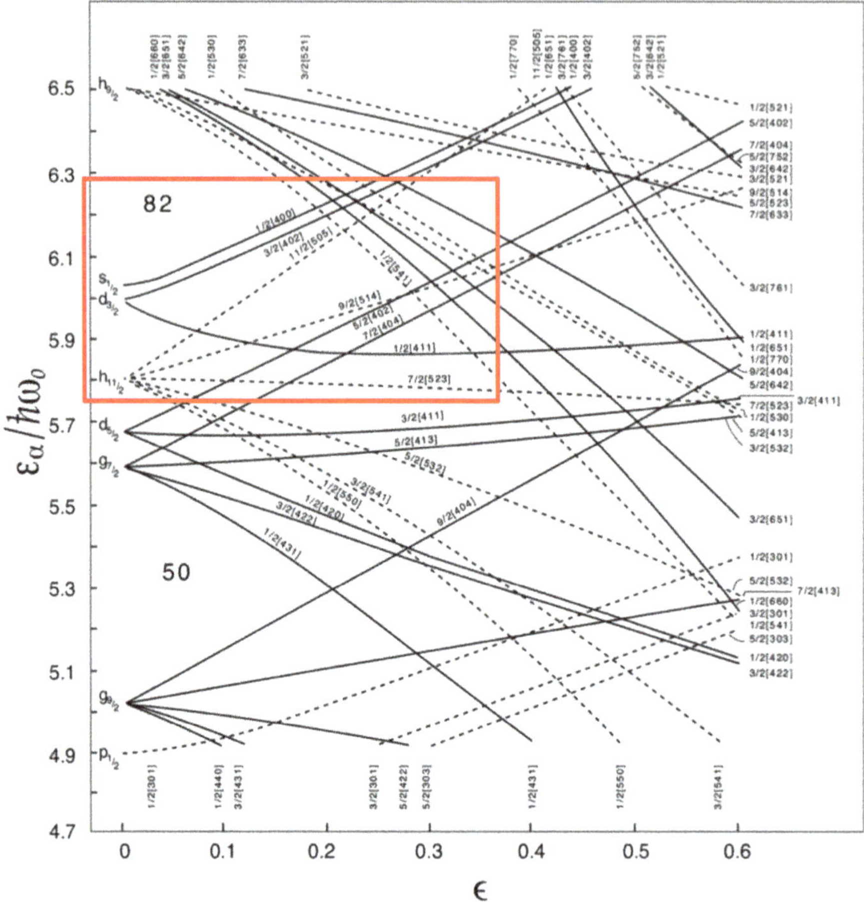

Figure 3.4. A Nilsson diagram for odd-proton nuclei in the rare-earth (lanthanide) region. The states are labelled by the asymptotic quantum numbers $\Omega[Nn_z\Lambda]$, the relationships $\Omega = \Lambda + \Sigma$ and $\pi = (-1)^N$ are implicit, positive (negative) parity states in the diagram are indicated by solid (dashed) lines. The units of the energy scale are given by $\hbar\omega_0 = 41A^{-1/3}$ MeV $(A \sim 170)$ $\sim$7.5 MeV for the middle of the rare-earth region. The energy ordering of the asymptotic states can be understood from equation (3.9) for the large deformation limit where $\hbar\omega_{xy} \gg \hbar\omega_z$, cf figure 3.2, i.e. highest values of n_z are the most energetically favoured. A useful empirical relationship is $n_z + \Lambda = l$. Note, e.g. the near-parallel energies of the 5/2[402] and 7/2[404] states; this is explained in the text. The region in the red box is expanded in figure 3.5. Reproduced from [1], copyright IOP Publishing Ltd. All rights reserved.

effects are presented below; rotation-particle coupling effects are presented in chapter 4. Basic details of pairing correlation effects are introduced progressively in the narrative as needed.

A more specialized topic is the incorporation of deformations beyond Y_{20}, e.g. Y_{22}, Y_{30}, Y_{40}. A semi-classical view of the effect of Y_{40} deformation on Nilsson states is shown in figure 3.7. Octupole and ellipsoidal, Y_{22} deformations are considered briefly later in the narrative. Note that nuclear quadrupole deformation based on

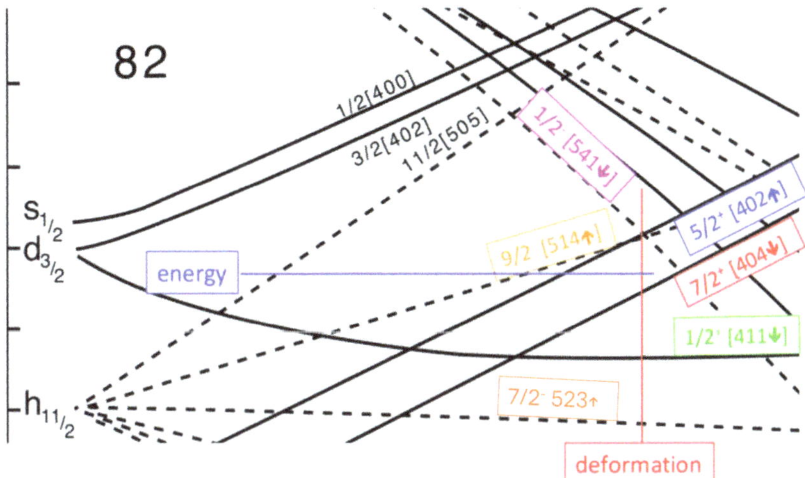

Figure 3.5. Nilsson diagram selected (red-boxed region in figure 3.4) for interpretation of states in the lutetium ($Z = 71$) isotopes. For present purposes, details of deformation and possible fine-tuning of parameters is unnecessary: it is sufficient to count from $Z = 82$ to 70, by twos as each Nilsson state is 'crossed' to arrive at the likely deformation that matches observed Nilsson band heads in the Lu isotopes, e.g. see figure 3.6. The energy scale, indicated by notches, is about 0.75 MeV per inter-notch interval. Reproduced from [1], copyright IOP Publishing Ltd. All rights reserved.

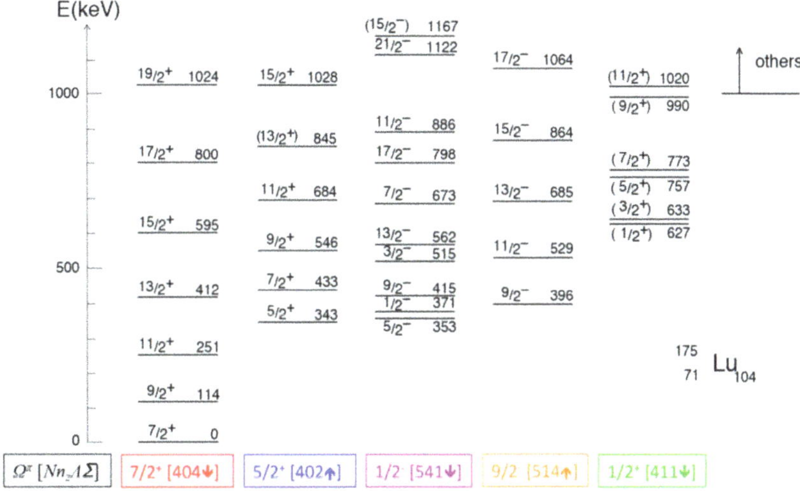

Figure 3.6. Nilsson bands in the low-energy region of excitation in ^{175}Lu. The labelling quantum numbers are given at the bottom of the figure. Usually, band-head spin $= \Omega = K$. The irregular energy spacings of the bands labelled $1/2^-[541]$ and $1/2^+[411]$ are characteristic of $K = 1/2$ bands and are discussed in chapter 4. The data are taken from ENSDF. Reproduced from [1], copyright IOP Publishing Ltd. All rights reserved.

electromagnetic properties is expressed in terms of a parameter β_2 which is slightly different from ε, the relationship between them is $\varepsilon := \varepsilon_2 = 3\beta_2\sqrt{5}/16\pi \sim 0.94\beta_2$. (The lowest-order term in the conversion between ε_4 and β_4 is $\varepsilon_4 \sim -0.85\beta_4$.)

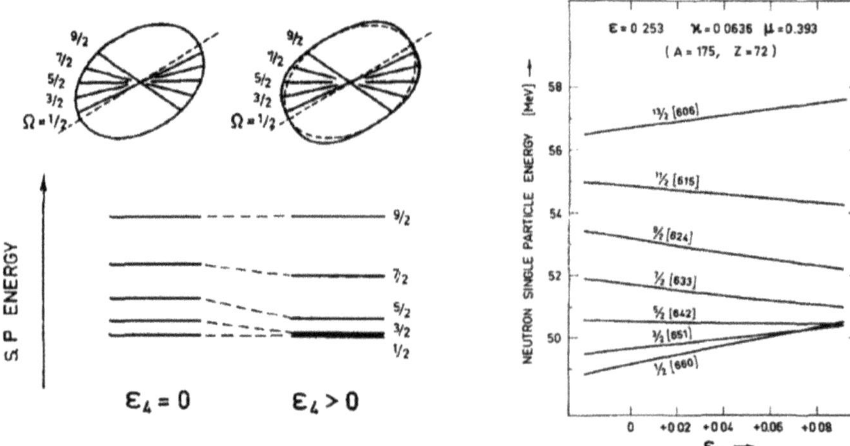

Figure 3.7. This figure presents a schematic view of the effect of a hexadecapole component to the independent-particle deformed mean-field. There is strong evidence for hexadecapole deformation in nuclei and this will influence the energy spacing of Nilsson states. The view is of orbital alignments, like the idea expressed in figure 3.2. Specifically, the effect on a single j orbital in a prolate spheroidal deformed potential is shown for positive values of the hexadecapole deformation parameter, ε_4. Note that the left-hand side of the figure is for relative energies of a $j = 9/2$ orbital and the right-hand side is for absolute energies of a $j = 13/2$ orbital. Reproduced from [1], copyright IOP Publishing Ltd. All rights reserved.

3.2 What constitutes a Nilsson state?

The identification of Nilsson model states is often tied to the identification of the rotational bands built on these states. This in turn often depends on assigning states to rotational bands and then deducing the Nilsson configuration for each band by identifying the lowest spin-parity state in the band and equating this value to an $|\Omega\pi\rangle$ configuration. This needs some explanation which follows below. We then address other spectroscopic signatures for identifying Nilsson configurations.

Nilsson diagrams usually provide a strong guide for systematic assignment of Nilsson configurations to band heads. This has led to a network of assignments of Nilsson configurations across broad mass ranges of deformed nuclei with respect to changing neutron number, i.e. isotopic sequences, and changing proton number, i.e. isotonic sequences. An example of fixed proton number (Ho, $Z = 67$) with changing neutron number is shown in figure 3.8. With such a network, knowledge of the deformation of a nucleus is not critical; indeed, the ordering of the individual Nilsson states may provide a fair estimate of the deformation of the nucleus under study.

Two key features are responsible for the simplicity manifested in figure 3.8. First, Nilsson states are two-fold degenerate, i.e. $\pm\Omega$. When a nucleus possesses a plane of reflection symmetry passing through the centre of mass at right angles to the symmetry axis, two-fold degeneracy is mandated. Second, an attractive pairing force acts to cause Nilsson states to fill pairwise. Above 1 MeV of excitation so-called broken-pair states are encountered, and this leads to an escalating complexity (this is

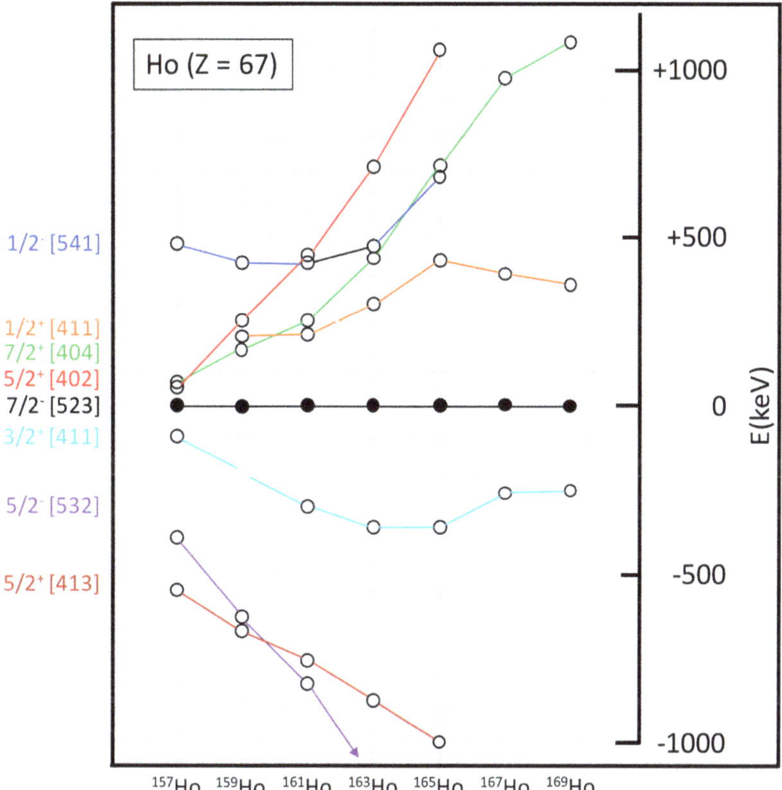

Figure 3.8. The experimental energy systematics of Nilsson states in the holmium ($Z = 67$) isotopes. Energies are shown relative to the Nilsson state 7/2⁻[523], which is the ground-state configuration of nearly all these isotopes (the exception, in ^{155}Ho which has $N = 88$, is in a region of rapidly decreasing deformation and this is beyond the scope of the present discussion). The data are taken from ENSDF.

the focus of chapter 5). The energy of 1–2 MeV that separates these excitation modes from the lowest non-broken-pair states is termed the 'pairing energy gap'.

Nilsson configurations can be assigned to states in nuclei by observed properties other than sorting excited states into rotational bands and inspecting band-head spin-parities. These observed properties are detailed below.

A leading type of spectroscopic information for assigning Nilsson configurations is one-nucleon transfer reaction spectroscopy, e.g. (d,p), (d,t), (d,^{3}He), (^{3}He,d), (t,α), there are many possibilities. Figures 3.9(a)–(c) show one-proton transfer reaction spectroscopy data for ^{165,167,169}Ho obtained using the reaction (t,α) on targets of ^{166,168,170}Er. There are recurring patterns of band-member population for each Nilsson configuration which are detailed below. Figure 3.10 shows the rotational band information for ^{165}Ho. By comparing the spectroscopic peaks in figure 3.9(a) with rotational band structures in figure 3.10, it is evident that there are distinct patterns for population of members of a rotational band by one-nucleon transfer: some details follow.

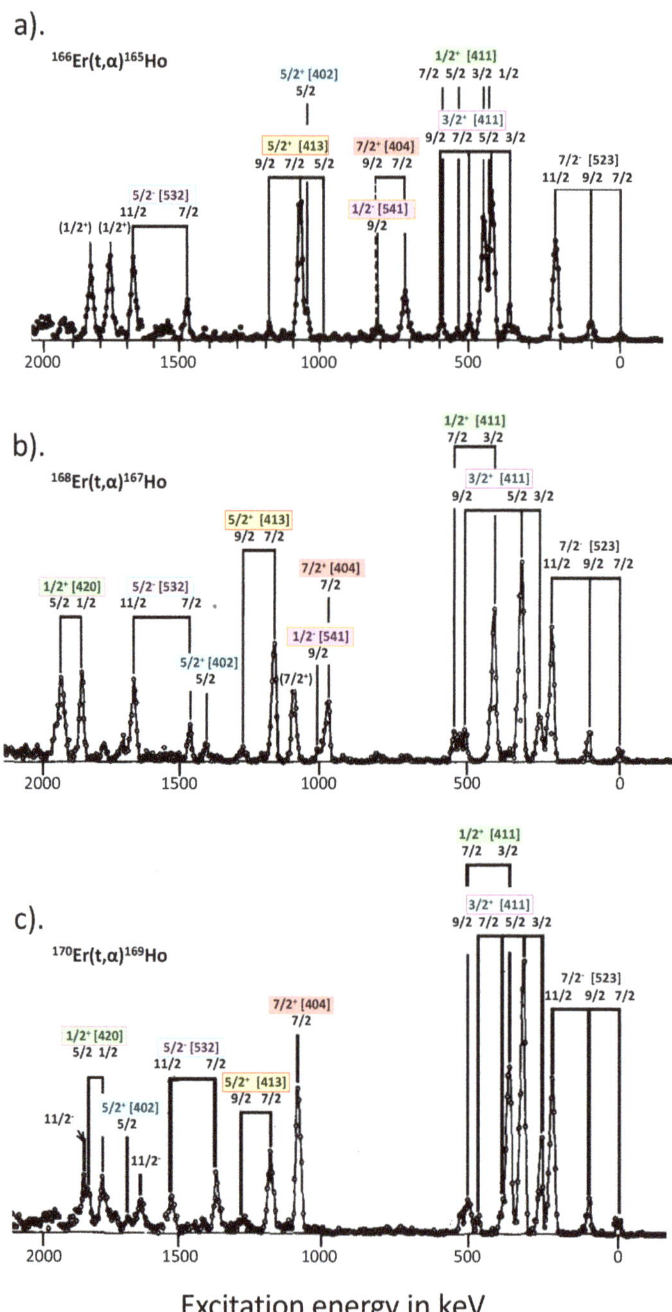

Figure 3.9. Spectra of helions (^{4}He nuclei $= \alpha$) for one-proton removal from (a) ^{166}Er, (b) ^{168}Er, (c) ^{170}Er using the (t, α) reaction, shown as counts versus energy in keV. The peaks in the spectrum are identified by rotational band assignments. The patterns of population for states in each band are a distinctive characteristic of the transfer reaction process and are discussed in the text. Reproduced from [1], copyright IOP Publishing Ltd. All rights reserved, based on figures that appear in [2, 3].

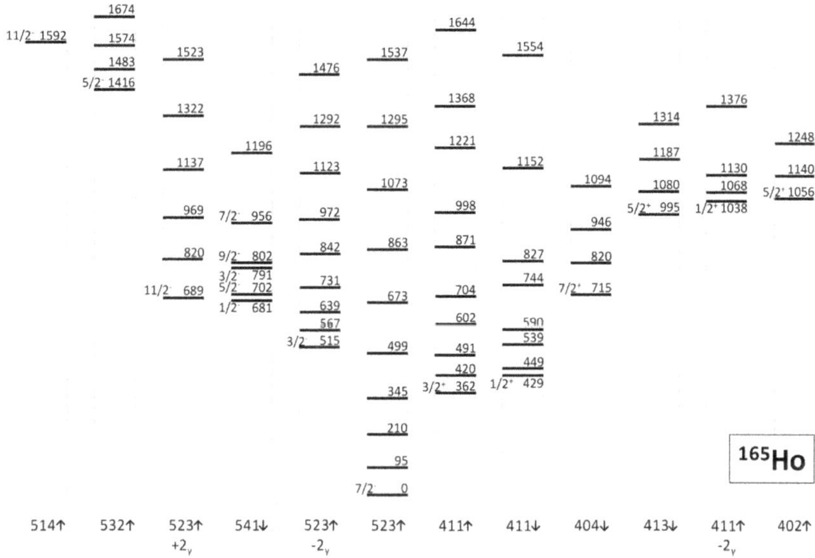

Figure 3.10. Low-lying states in ^{165}Ho arranged into Nilsson bands and labelled by the quantum numbers $Nn_z\Lambda\Sigma$ (note, $\Sigma = +1/2$ is designated as ↑ and $\Sigma = -1/2$ as ↓). The data are taken from ENSDF.

Table 3.1. Population of Nilsson band members in ^{165}Ho by the ^{166}Er(t,α)^{165}Ho one-proton-transfer reaction, cf figure 3.9(a). The j values of dominant spherical configurations are taken from figure 3.4. Excitation energies] of states are taken from figure 3.10.

Band	Spin of dominant band member populated	j value of dominant spherical configuration	Excitation energy of state (keV)
$7/2^-[523\!\uparrow]$	11/2	11/2	209.8
$3/2^+[411\!\uparrow]$	5/2	5/2	419.5
$1/2^+[411\!\downarrow]$	3/2	3/2	449.3
$7/2^+[404\!\downarrow]$	7/2	7/2	715.3
$1/2^-[541\!\downarrow]$	9/2	9/2	802.3
$5/2^+[402\!\uparrow]$	5/2	5/2	1055.8
$5/2^+[413\!\downarrow]$	7/2	7/2	1079.6
$5/2^-[532\!\uparrow]$	11/2	11/2	1674

An inspection of the right-hand most set of alpha particle groups in figure 3.9(a) shows that it is the spin-11/2 member of the $7/2^-[523\!\uparrow]$, ground-state band that is most strongly populated. Moving from right to left, i.e. increasing in excitation energy, such associations in the other bands are made in table 3.1. Other band members in ^{165}Ho are populated only weakly or not observably: notably, cf figure 3.9 (a), the bands labelled $523\!\uparrow - 2_\gamma$ and $523\!\uparrow + 2_\gamma$ with band-head energies of 515.5

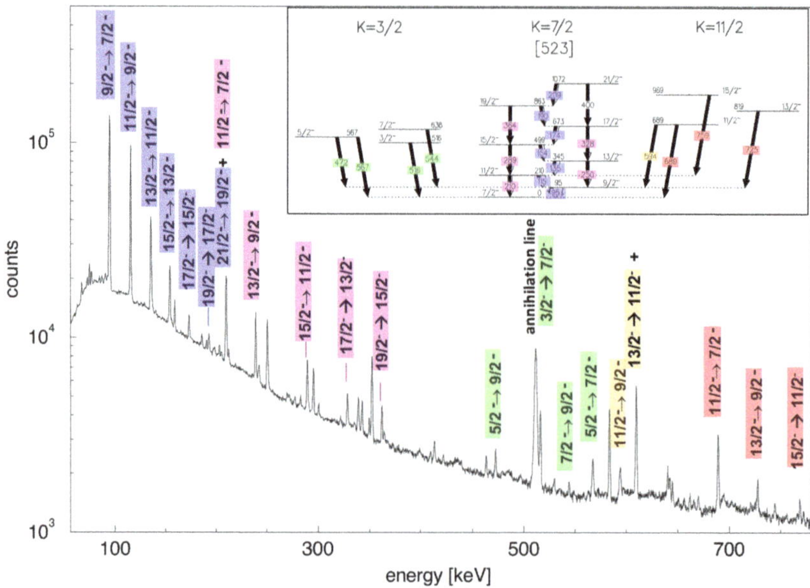

Figure 3.11. Gamma-ray spectrum from the Coulomb excitation of a ^{165}Ho target by a beam of ^{40}Ar ions. The process selects predominantly states connected to the ^{165}Ho ground state, via multiple steps of E2 excitation, viz. the ground-state $7/2^-[523]$ band and two bands resulting from the couplings $K = 3/2, 7/2^-[523] - 2_\gamma$ and $K = 11/2, 7/2^-[523] + 2_\gamma$ with band heads at 515 and 689 keV, respectively, as shown in figure 3.10. The 2_γ core excitation is usually called the gamma band and occurs in ^{166}Er at an excitation of 786 keV and in ^{164}Dy at 762 keV. A discussion of these excitation energies is beyond the scope of the present focus. Reprinted from [4], copyright IOP Publishing, all rights reserved.

and 688.8 keV, respectively. These bands are populated in Coulomb excitation, as shown in figure 3.11 and are interpreted as the ground-state Nilsson configuration coupled to the core excitation termed the 'gamma' band. The coupling of the gamma band $K = 2$ degree of freedom to the $7/2^-[523\uparrow]$ ground-state configuration results in bands with $K = 3/2$ and $K = 11/2$ (recall, K is defined with respect to a fixed axis, the nuclear symmetry axis, and so the coupling is for the collinear vectorial sum or difference.) A more subtle feature is that the (t,α) transfer reaction probes proton hole states. Thus, Nilsson states such as $7/2^+[404\downarrow]$, $1/2^-[541\downarrow]$ and $5/2^+[402\uparrow]$ are populated progressively less strongly with distance from the Fermi energy which is located at $Z = 68$ (Er). A deeper look at such details is made in chapter 4: these details require the role of pairing correlations to be introduced.

A more limited type of spectroscopic information is provided by beta decay strengths, and even alpha decay strengths in selected mass regions (notably in actinide nuclei). Thus, the strongest observed decay branches may connect specific Nilsson configurations. In the rare earth region, the cases of enhanced beta decay strength involve the pairs of configurations $\nu[523\downarrow]-\pi[523\uparrow]$, $\nu[514\downarrow]-\pi[514\uparrow]$. Examples of so-called selection rules for Nilsson quantum numbers, as encountered in alpha decay and beta decay, are addressed in the exercises.

3.3 Survey of Nilsson model states in odd-mass nuclei

Nilsson model states occur widely across the nuclear mass surface. By far most occurrences are centred on the lanthanide (rare-earth) region with $N > 89$, $Z < 78$ and the actinide region with $N > 133$, $Z > 85$. Other mass regions which exhibit Nilsson states are narrowly defined. In the following, a selective survey is presented. Figures 3.12(a)

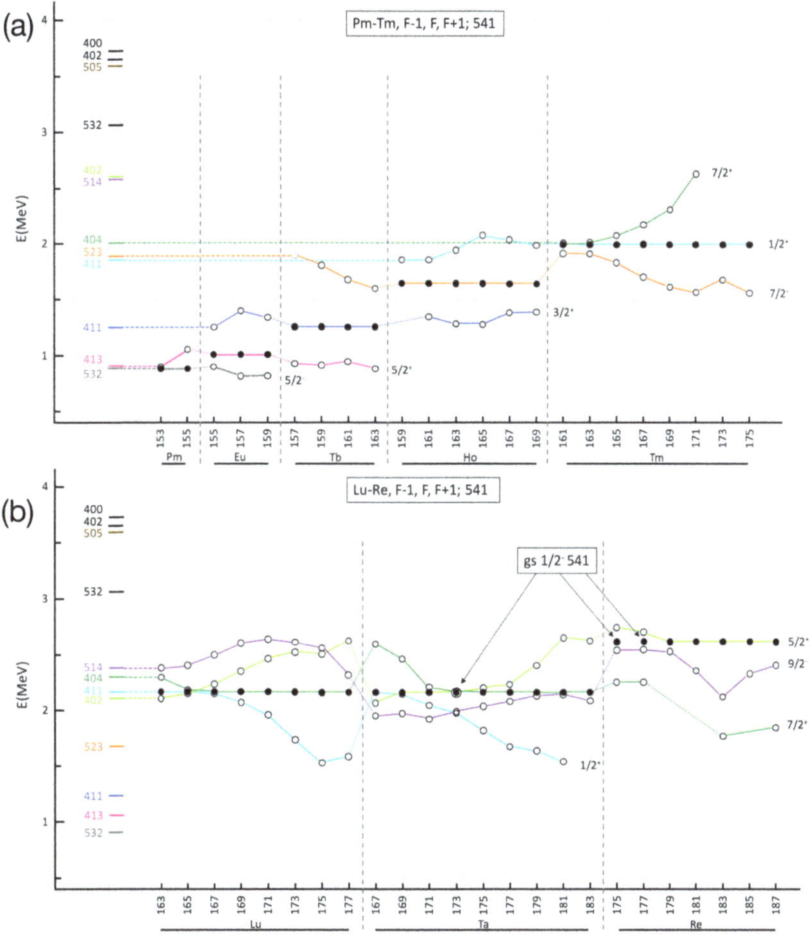

Figure 3.12. Systematic energy trends in selected Nilsson states for odd-proton configurations in the rare-earth region, (a) Pm ($Z = 61$) to Tm ($Z = 69$), (b) Lu ($Z = 71$) to Re ($Z = 75$). Ground states are designated by solid dots and excited states by open circles. The energies shown for ground states are all the same within an isotopic series and are arbitrary between isotopes of the different elemental species. Data are limited to the first one or two Nilsson configurations identified above and below the energy of the ground-state configuration in the Nilsson diagram shown in figure 3.4. The configuration 1/2⁻[541] is omitted, even though it is the ground state in ^{173}Ta and ^{175,177}Re. The reason is that this state is an intruder structure and follows the characteristic parabolic energy trend exhibited by such structures. Thus, it does not follow an energy pattern characteristic of non-intruder Nilsson states. Details are given explicitly in figure 4.31 in [8]. Note that the ground states of the Lu and Ta isotopes are both dominated by the configuration 7/2⁺[404], i.e. there is an inversion in the energy order of the configurations 7/2⁺[404] and 9/2⁻[514] with the result that the configuration 9/2⁻[514] is not observed as the (expected) ground state in the Ta isotopes. The data are taken from ENSDF.

and (b) present a survey of proton Nilsson states in the lanthanide region. The states selected are the ground states, shown as black dots, and one or two Nilsson states from above and below these states in the Nilsson diagram, shown as open circles. The systematic variations in relative energies can be interpreted as resulting from changing deformation of the mean field; but other factors are addressed in the exercises. Similar figures for neutron Nilsson states in the lanthanide region and for proton and neutron Nilsson states in the actinide region are available in review articles [5–7]; they can also be assembled from data in ENSDF in a straightforward manner. Nilsson states for neutrons in the lanthanide region and for states in the actinide region are handled in the exercises. Nilsson states in other, more narrowly defined regions are also handled in the exercises.

3.4 Exercises

The exercises provided in this chapter expand the view of Nilsson states, developed for odd-proton nuclei in the rare earth/lanthanide region, into odd-neutron nuclei in the rare earth/lanthanide region and generally into the actinide region. Also, some attention is given to a view of Nilsson states in other mass regions and to more specialized aspects of Nilsson states.

3.4.1 Rare earth/lanthanide odd-neutron nuclei

Figures 3.13–3.18 provide a basic view of Nilsson states for odd-neutron nuclei in the rare earth/lanthanide region. While the systematic features involve a larger number of Nilsson states and more nuclei than odd-proton nuclei (the open shell, $82 < N < 126$ is 'larger' than $50 < Z < 82$), the pattern establishes that the Nilsson model is an excellent qualitative guide to organizing a view of data for all strongly deformed odd-neutron nuclei in this mass region.

The key summary view is provided by figure 3.18, but this is for selected states, e.g. Figure 3.17 illustrates the $N = 105$ isotones and the more extended view, into ^{181}Os, ^{183}Pt, ^{185}Hg and ^{187}Pt. We immediately note that the Pt, Hg and Pb isotopes are in a region of shape coexistence, and this is dealt with in later exercises. Here, we focus on ^{181}Os and the change in energy ordering of the Nilsson states. Figure 3.13 reveals that the $1/2^-[521\downarrow]$ state is 'downsloping' with increasing deformation, whereas the $5/2^-[512\uparrow]$ and $7/2^-[514\downarrow]$ states are 'upsloping', so a decrease in deformation could explain the energy reordering. Figure 3.13 reveals that a decreasing deformation should result in the appearance at low energy of the $7/2^-[503\uparrow]$ state, and further the $9/2^-[505\downarrow]$ and $11/2^+[615\uparrow]$ states.

    3-1 Using data in ENSDF, organize the low-energy excited states in ^{181}Os into bands, cf the view in figure 3.15, to explore how complete the view of ^{181}Os is with respect to a change in deformation. (Note: the view is very incomplete.)

    3-2 A further view of the Nilsson states in figure 3.13, in the direction of issues raised in exercise 3-1, can be obtained by inspection of data for the $N = 107, 109, 111, 113$ isotones: using data in ENSDF, organize the low-energy excited states in ^{181,183,185,187}W, ^{183,185,187,189}Os into bands, cf

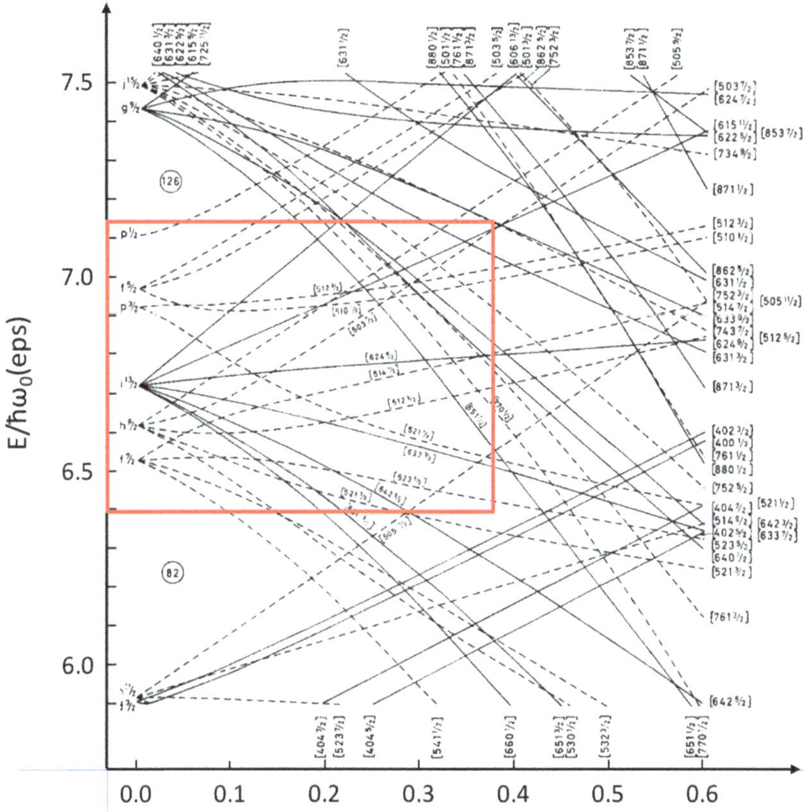

Figure 3.13. A Nilsson diagram for odd-neutron nuclei in the rare-earth (lanthanide) region. Reprinted from [9], copyright (1969) with the permission of Elsevier.

the view in figure 3.15; and produce views as in figure 3.17 for these isotone numbers. Note: this view reveals systematic details for the $7/2^-[503\uparrow]$, $9/2^-[505\downarrow]$ and $11/2^+[615\uparrow]$ states. Especially note the feeding of the $11/2^+[615\uparrow]$ states by 'inverted' spin sequences, viz. $3/2^+ \rightarrow 7/2^+ \rightarrow 11/2^+$, as revealed in the tables of transitions in ENSDF and XUNDL for ^{183,185,187}W and ^{189}Os, and similar patterns of behaviour in other isotopes in this region. This is a characteristic of a particle coupled to an axially asymmetric rotor and is looked at in later exercises.

3-3 Exercise 3-2 encounters the 'high-mass border' of the strongly deformed region for the rare earth/lanthanide nuclei. The 'low-mass border' for isotones is $N = 89$. As far as possible, using data in ENSDF, organize the low-energy excited states in ^{151}Sm, ^{153}Gd and ^{155}Dy into bands, similar to figure 3.15.

The other limitation to the view provided by figure 3.18 is that only a few Nilsson states are shown for each isotope. A leading question is: 'above what excitation

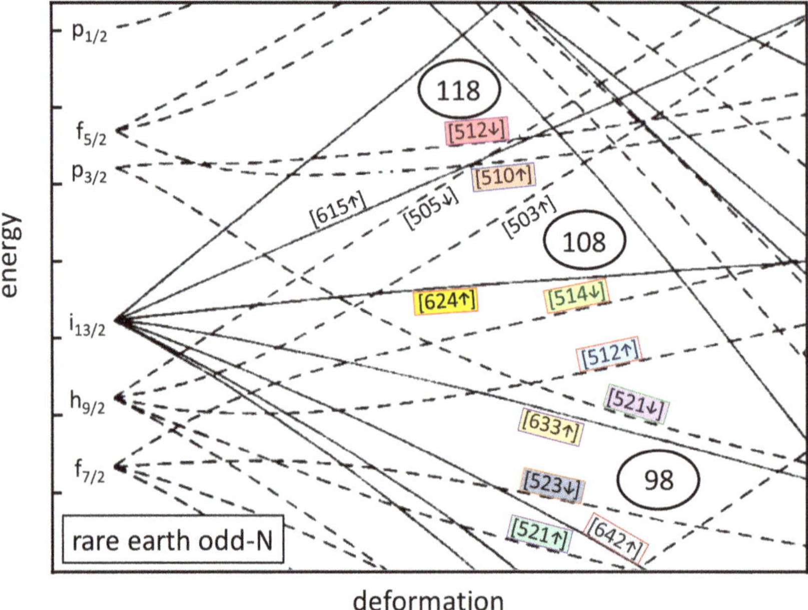

Figure 3.14. Nilsson diagram selected (red-boxed region in figure 3.13) for interpretation of states in the $N \sim 105$ isotones. The colour coding is to assist identification of Nilsson states in figures 3.15, 3.16 and 3.19. The energy scale and deformation scale are shown in figure 3.13. Reprinted from [9], copyright (1969) with the permission of Elsevier.

energy does one encounter degrees of freedom that lie outside of the Nilsson band classification?' For example, figure 3.11 illustrates a view, from Coulomb excitation, of excitations in ^{165}Ho that lie beyond this simple organizational framework, namely that involve so-called 'gamma-band' degrees of freedom. Such excitations are identified by inelastic scattering and the example of ^{167}Er is shown in figure 3.19. The population of $5/2^{+}[642\uparrow]$ and $9/2^{+}[624\uparrow]$ bands occurs via strong Coriolis mixing of the members of the Nilsson multiplet which stems from the high-j, $i_{3/2}$ shell model parent configuration.

    3-4 Using data in ENSDF, organize the low-energy excited states in ^{167}Er into bands, cf the view in figure 3.15.

    3-5 A comparison of the energies of the 'gamma' bands in ^{165}Ho (figures 3.10 and 3.11) and ^{167}Er (figure 3.18) with gamma band energy systematics for even–even rare earth nuclei, shown in figure 3.20, suggests that such degrees of freedom should be expected at relatively low energy ($E_x < 800$ keV) in mass regions centred on $A \sim 165$, also $A \sim 185$. Using data in ENSDF, explore other occurrences of such excitations below 800 keV in: ^{163}Ho, ^{163}Dy, ^{165}Er, ^{185}Re.

Beta decay provides connections between odd-N and odd-Z nuclei, manifested in radioactive decay scheme spectroscopy. There are wide variations in beta decay rates which are controlled primarily by spin and parity changes. There are two

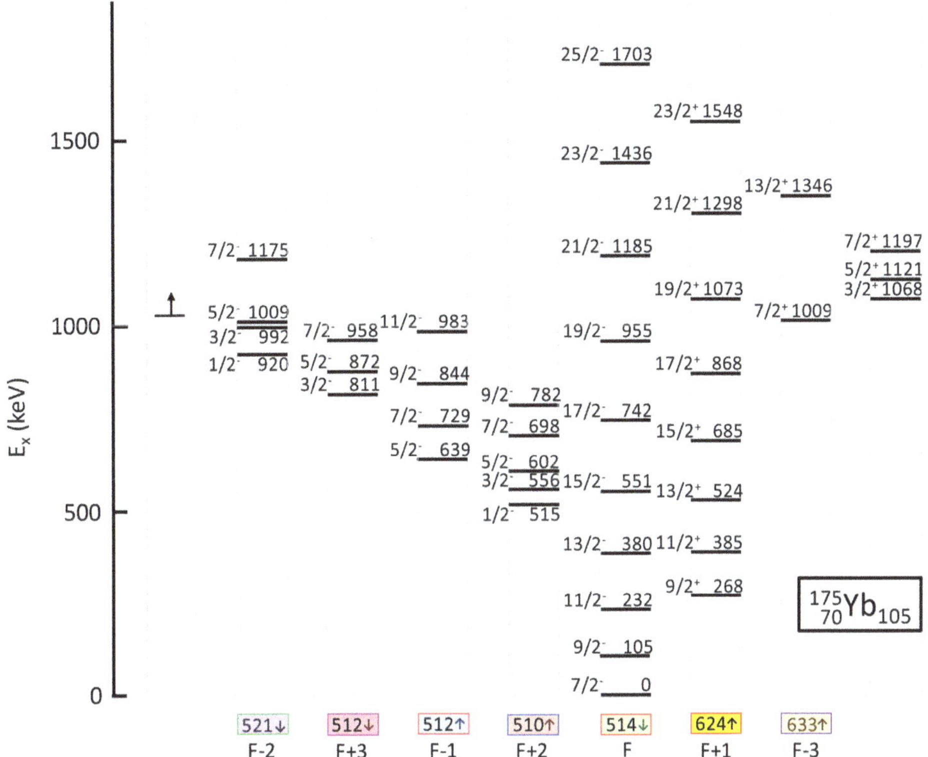

Figure 3.15. Low-lying states in ^{175}Yb arranged into Nilsson bands and labelled by the quantum numbers $Nn_z\Lambda\Sigma$ (note, $\Sigma = +1/2$ is designated as ↑ and $\Sigma = -1/2$ as ↓). The labels involving 'F' are a guide to Nilsson states, ordered by energy, above ($F + 1$, $F + 2$, ...) and below ($F - 1$, $F - 2$, ...) the Fermi energy (F). The horizontal bar with the vertical arrow indicates the excitation energy above which levels have been omitted. The data are taken from ENSDF. (Note: the use of parentheses for spin assignments in ENSDF has been removed.)

fundamental categories termed 'Fermi' and 'Gamow–Teller' decays. In the rare earth region, specific Nilsson state combinations stemming from the neutron configuration $h_{9/2}$ and the proton configuration $h_{11/2}$ lead to fast Gamow–Teller, GT decays.

> 3-6 By identifying beta decay pairs where fast GT transitions can occur, e.g. neutron $5/2^-$ [523↓] and proton $7/2^-$[523↑], explore Nilsson state assignments that can be made in daughter nuclei by this spectroscopic 'fingerprint'. [Starting hint: look at the ^{167}Ho $\rightarrow$ ^{167}Er decay scheme. Further suggestion: look at ^{181}Os $\rightarrow$ ^{181}Re decay.]

3.4.2 Actinide odd-mass nuclei

Figures 3.21 and 3.22 show Nilsson diagrams for the actinide region. Figures 3.23 and 3.24 provide a basic view of Nilsson states observed for odd-mass nuclei in the actinide region. To a considerable degree, the Nilsson states manifested in the

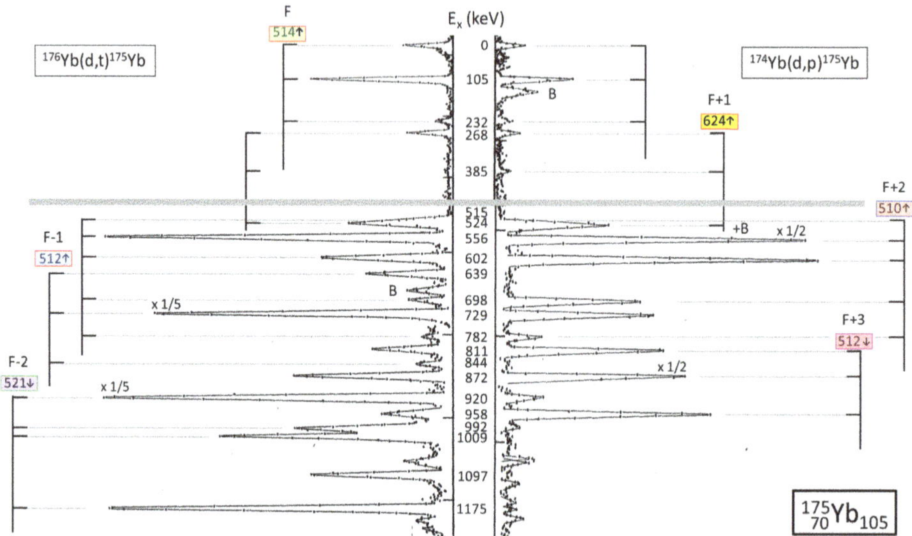

Figure 3.16. Triton and proton spectra for the ^{176}Yb(d,t)^{175}Yb and ^{174}Yb(d,p)^{175}Yb one-neutron removal and one-neutron addition reactions shown matched to illustrate the population of states in ^{175}Yb. The ground state is at the top of the figure. Note that population of states at low energy occurs in both reactions, but with increasing excitation energy there is a 'separation' into hole states (neutron removal) and particle states (neutron addition). The heavy grey horizontal line demarcates a change in the energy scale, which is expanded above 500 keV to facilitate labelling as the line density gets higher. Peaks marked 'B' indicate events due to target impurities. The labels $F-2, F-1, F, \ldots, F+3$ are explained in the caption to figure 3.15. The energies are taken from ENSDF. Reprinted with permission from [10]. Copyright (1979) by the American Physical Society.

actinide region closely follow the patterns observed in the rare earth/lanthanide region. The major difference is the higher density of Nilsson configurations due to the larger open shells arising in the actinide region. Indeed, the 'high-mass boundary' may not yet have been reached. These exercises introduce a few basic perspectives of odd-mass actinide isotope structure, e.g. via one-nucleon transfer reaction spectroscopy; and especially point to alpha decay schemes where favoured alpha decay between identical Nilsson states is an important tool for extending structural interpretations, especially towards superheavy elements.

One-nucleon transfer reactions provide a foundational view through which Nilsson states have been mapped in the actinide region. Figure 3.25 illustrates such data for the ^{242}Pu(d,p)^{243}Pu and ^{244}Pu(d,t)^{243}Pu reactions; figure 3.26 illustrates such data for the ^{250}Cf(α,t)^{251}Es. The einsteinium isotopes represent a useful 'gateway' to building α-decay chain systematics for odd-proton nuclei descending from the so-called 'superheavy element', SHE region of the Chart of the Nuclides. Figures 3.27 and 3.28 illustrate Nilsson state identifications in some mendelevium, Md isotopes and a lawrencium isotope, ^{255}Lr. Figure 3.29 provides a gateway view to alpha decay chains for heavy odd-neutron actinide isotopes. Alpha decay hindrance factors are the key indicator of whether the decay involves a change in

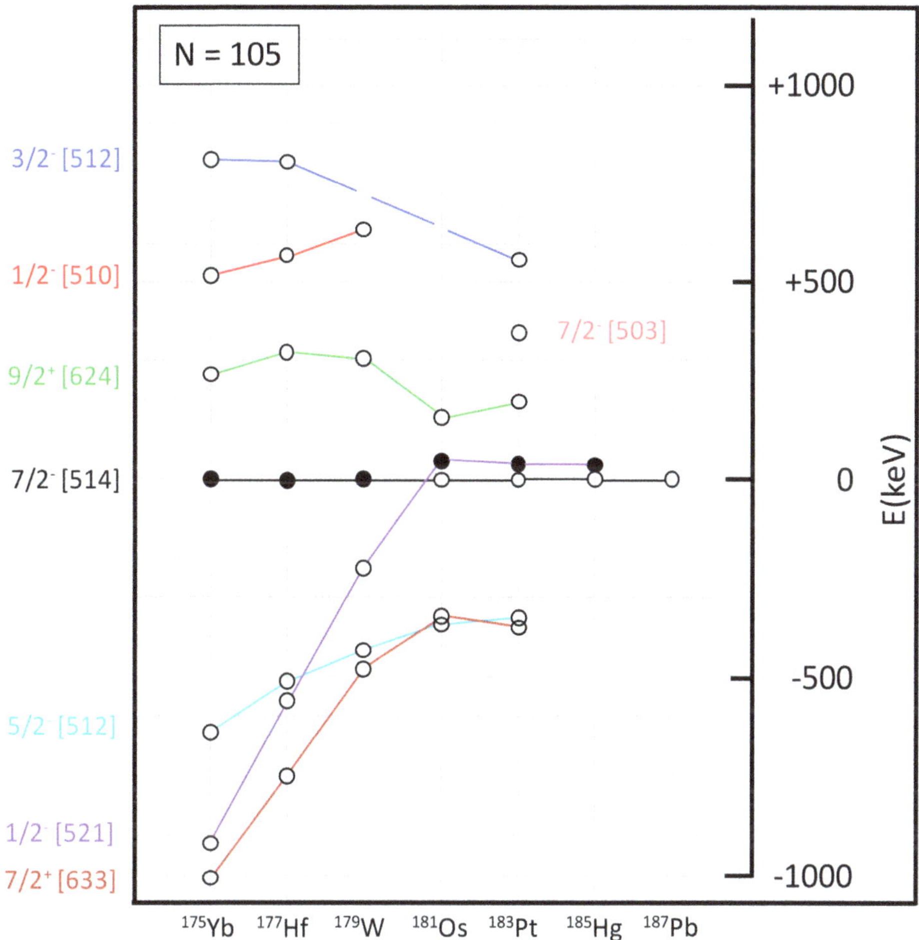

Figure 3.17. The experimental energy systematics of Nilsson states in the $N = 105$ isotones.

Nilsson state between parent nucleus and daughter or no change. No change is indicated by a small hindrance factor, HF (ideally HF = 1). We refer persons with specialized interests to the original work of John Rasmussen [12].

   3-7 Set up, cf figure 3.29, and explore such a 'gateway' for odd-neutron SHE exploration. Note that the states in ^{243}Pu populated in the (d,p) reaction, cf figure 3.25, provide a useful guide to Nilsson states that can be expected at low energy up to $N = 164$.

   3-8 Sometimes, the α decays of interest have been established but long ago and so the details lack key information such as coincident γ rays, needed to determine to which excited state in the daughter nucleus the decay occurs. Identify some cases of established α decays where γ-ray spectroscopy is needed to make Nilsson state identification of parent decays.

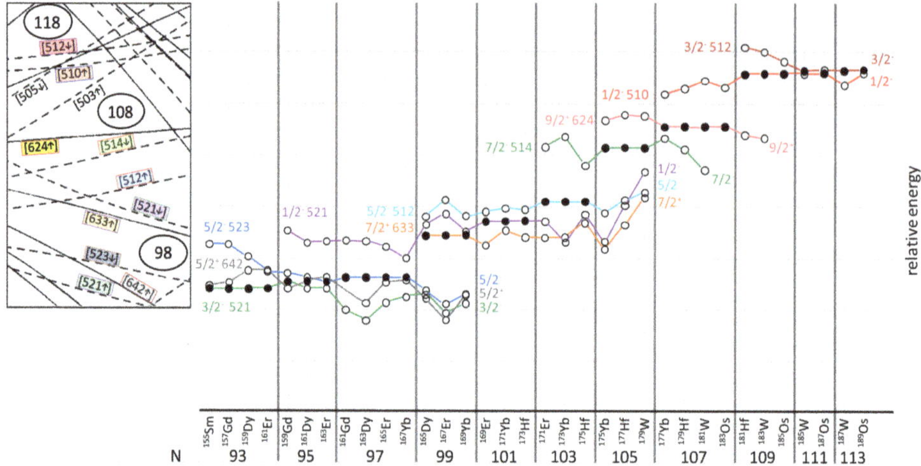

Figure 3.18. Global systematics of Nilsson states for odd-neutron nuclei in the rare earth (lanthanide) region. The heavy black dots are ground states: from isotone series to isotone series their energies are set to minimize major shifts in energy. The states selected are limited to setting a useful framework for assessing the smoothness with which the Nilsson states are filled across the rare earth region: a selected portion of figure 3.14 is embedded as an inset to facilitate this view. The data are taken from ENSDF.

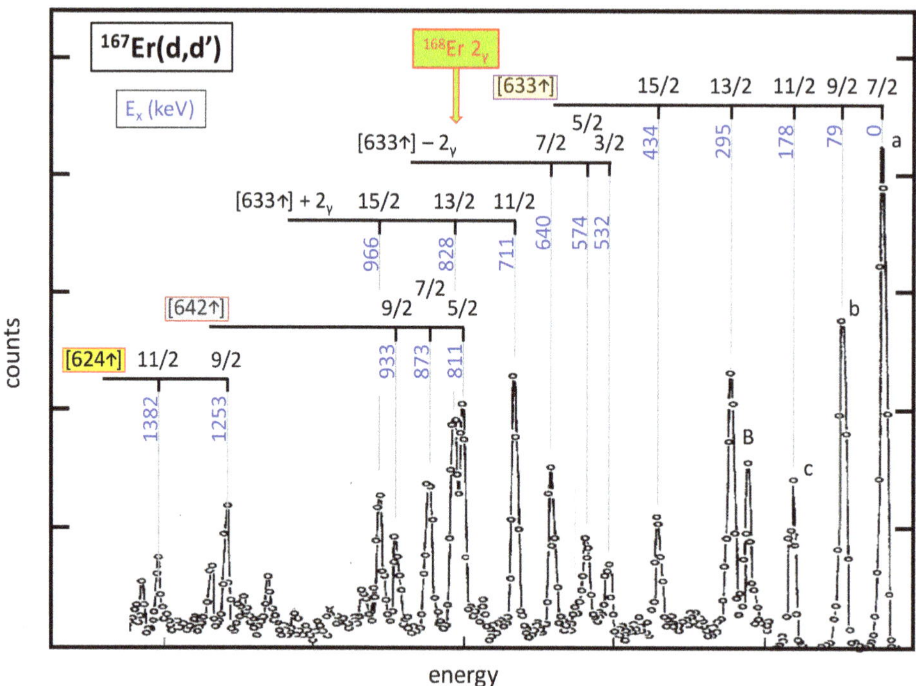

Figure 3.19. Deuteron spectrum from inelastic scattering of 12 MeV deuterons on a target of ^{167}Er. Peaks are assigned to bands and energies are given in keV. The peak labelled 'B' is due to target isotopic impurity. Reduction of counts scale is indicated for the lowest states, a— × 1/200, b— × 1/40, c— × 1/20. Note that the coupling $3/2^+$ { $7/2^+$ [633↑] × 2_γ }, $E_x = 532$ keV and $11/2^+$ { $7/2^+$ [633↑] × 2_γ }, $E_x = 711$ keV both lie well below the 2_γ core excitation (indicated for ^{168}Er at 821 keV, and similarly in ^{166}Er at 786 keV). The population of the $5/2^+$[642↑] and $9/2^+$[624↑] bands is discussed in the text. Other data are taken from ENSDF. Reprinted from [11], copyright (1973) with the permission of Springer Nature.

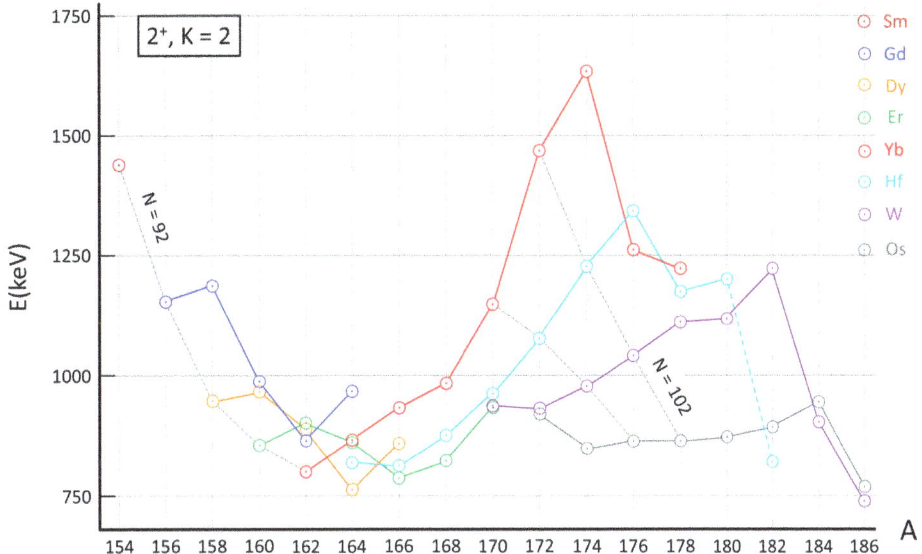

Figure 3.20. Systematics of gamma band energies in even–even rare earth nuclei. The figure is reproduced from [8], copyright IOP Publishing Ltd, all rights reserved.

3-9 Following on from exercise 3-8, even when γ-ray decays following γ decay have been established, parity changes may be involved: consider in your answer to exercise 3-8 how conversion-electron spectroscopy can resolve this issue. The BrIcc website will be useful to address this exercise.

3-10 The Nilsson configurations observed at low energy in the vicinity of ^{251}Es ($Z = 99$) show some commonalities with those observed at low energy in the vicinity of ^{167}Er ($N = 99$). Explore the extent of this similarity, e.g.

    (a) which Nilsson states appear in a different energy order?

    (b) what differences in energy separations between similar Nilsson configurations exist?

3.4.3 $Z > 50$, $N < 82$ region odd-mass nuclei

The identification of Nilsson states in the $Z > 50$, $N < 82$ region is limited. The primary reason is that the occurrence of strong deformation and the observation of strongly coupled rotational bands is limited; and there are no stable isotopes, to provide targets for one-nucleon transfer reaction spectroscopy, at the centre of this region. Historically, support for the Nilsson model in a given mass region usually starts by 'imposing' a Nilsson model interpretation on observed low-energy states. Eventually, a broad and consistent view emerges for the region. This process is 'underway' in the neutron-deficient odd-mass nuclei centred on $Z = 60$, $N = 70$. An early and detailed study, for ^{133}Nd, is illustrated in figure 3.30. Notably, conversion-electron spectroscopy and lifetime measurements were made, and this established E1 isomerism with associated non-rotational structures, i.e. candidate Nilsson states. A

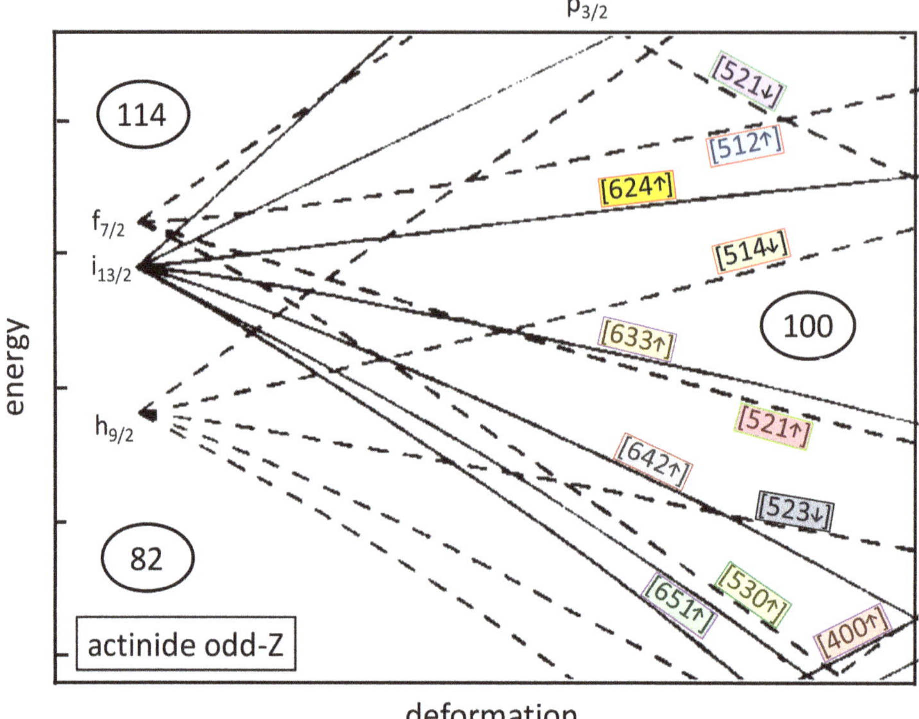

Figure 3.21. Nilsson diagram for odd-proton actinide nuclei. The colour coding is to assist in the navigation of Nilsson-state identifications in some of the following figures. The quantum numbers labelling the states are $[Nn_z\Lambda\Sigma]$, where $\Sigma = +1/2(\uparrow)$ or $-1/2(\downarrow)$ and Ω^π quantum number labels can be inferred via $\Omega = \Lambda + \Sigma$ and $\pi = (-1)^N$. Solid lines are even-parity states, dashed lines are odd-parity states. The shell model 'parent' states are labelled l_j; where recall $l = 0, 1, 2, 3, 4, 5, 6, 7, 8$ are labelled s, p, d, f, g, h, i, j, k, respectively, are the standard encoding of angular momentum values and $j = l \pm 1/2$. Note the useful circumstantial relationship $l = n_z + \Lambda$. Reprinted from [9], copyright (1969) with the permission of Elsevier.

useful Nilsson diagram for these isotopes is provided in figure 3.31. A limited systematic view has emerged for the $N = 73$ isotones; this is shown in figure 3.32.

We strongly emphasize, establishing a structural type in a mass region necessitates detailed systematic study across the mass surface. Eventually, a network of interconnected structures—by systematics of nuclear spins and moments deduced from atomic beam magnetic resonance and optical hyperfine atomic spectroscopy, from alpha and beta decay schemes, by one-nucleon transfer reaction spectroscopy, by isomeric structures and their decays—emerges. This may require long-term, dedicated programs of nuclear structure investigation. At present, such a detailed picture for the $Z > 50$, $N < 82$ region remains incomplete.

    3-11 Using data in ENSDF and XUNDL, attempt to make a systematic view of Nilsson bands in:

        (a) the $N = 71$ isotones ^{127}Ba, ^{129}Ce, ^{131}Nd, ^{133}Sm;

        (b) the $N = 69$ isotones ^{125}Ba, ^{127}Ce, ^{129}Nd.

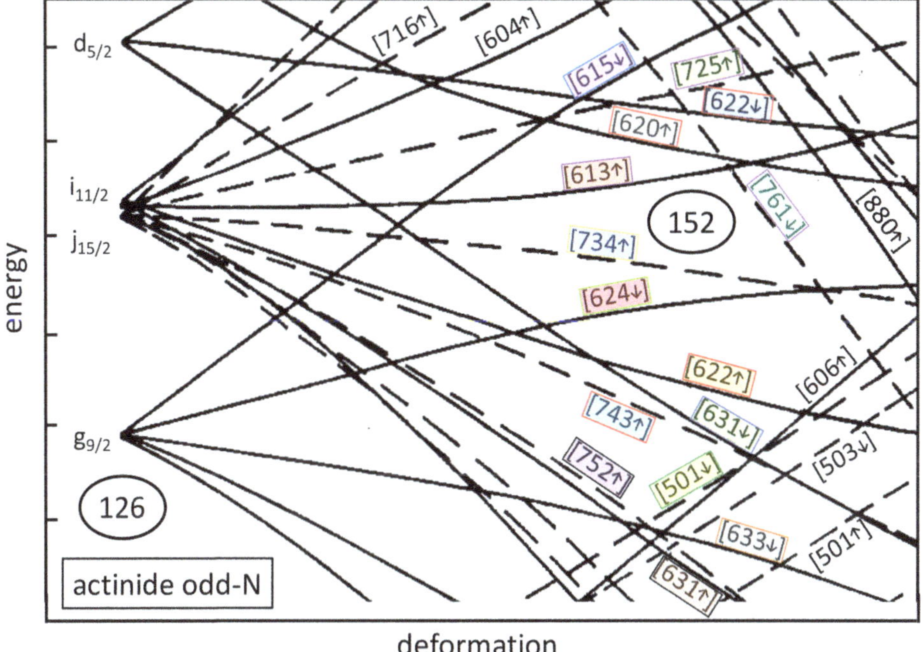

Figure 3.22. Nilsson diagram for odd-neutron actinide nuclei. The colour coding is to assist in the navigation of Nilsson-state identifications in some of the following figures. See the caption to figure 3.21 for details of the quantum number labels. Caution, the asymptotic quantum numbers are in use and the positions of the labelling are sometimes misleading with respect to the spherical shell model parent quantum numbers. For example, the [613↑] and [624↓] states are dominated by mixing of the shell model configurations $g_{9/2}$ and $i_{11/2}$: the line emerging from $g_{9/2}$ is labelled '[624↓]', which is correct asymptotically but at the location of the label (note the curvature in the lines) the states so labelled are near 50:50 admixtures of these configurations, q.v. Figure 3.3. Reprinted from [9], copyright (1969) with the permission of Elsevier.

3-12 The neutron numbers involved in exercise 3-11 match the proton numbers for the Lu and Tm isotopes. What similarities exist between Nilsson states in these two regions for these 'mirrored' pairs of nucleon numbers?

3-13 Using data in ENSDF and XUNDL, explore the identification of Nilsson bands in odd-proton nuclei in the $Z \sim 60$, $N \sim 70$ region. The Nilsson diagram in figure 3.4 can be used for this exercise.

The problem with nuclei in this region is that, where detailed data are available, there are too many states at low excitation energy relative to a Nilsson model view. This is indicated in figures 3.30 and 3.32. A more detailed view for ^{129}Ba is presented in figure 3.33 which shows states populated in the ^{130}Ba(d,t)^{129}Ba one-neutron removal reaction. A more detailed view for ^{131}Ce is presented in figure 3.34 which shows states populated in the beta decay of the high-spin isomer in ^{131}Pr. A view of how to accommodate the extra states is presented in figure 3.35 for states in ^{129}Ba. Essentially, these nuclei can be described as a particle coupled to an axially

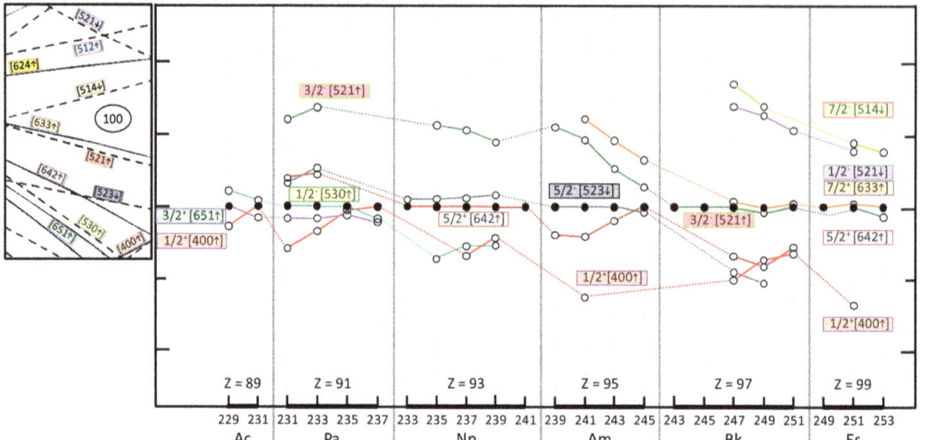

Figure 3.23. Systematics of selected Nilsson states in the actinide region for odd-Z nuclei. Ground states are shown as solid black dots, states which are predominantly of particle character are located above the ground states and hole states are located below. The energy notches mark 500 keV intervals. The Nilsson diagram relevant to these nuclei, adapted from figure 3.21, is inset as a guide. The quantum numbers labelling states in the main body of the figure are $\Omega^{\pi}[Nn_{z}\Lambda\Sigma]$, for other details and the inset see the caption to figure 3.21. The design of the figure follows the style of similar figures found in reference [6], which, in turn, uses a style adopted in an earlier review of this topic for the lanthanide region [5].

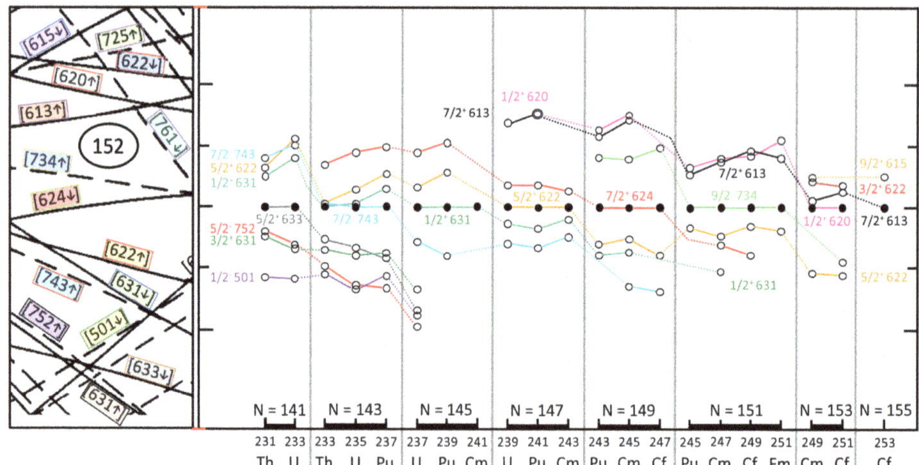

Figure 3.24. Systematics of selected Nilsson states in the actinide region for odd-N nuclei. The energy notches mark 500 keV intervals. The Nilsson diagram relevant to these nuclei, adapted from figure 3.22, is inset as a guide. Note that the more complete quantum number labelling used in figure 3.23 is not employed here because of the spatial limitations. Also, the colour coding used in figure 3.22 (cf the inset) is not used. The design of the figure follows the style of similar figures found in reference [6]. For other details, see the caption to figure 3.23.

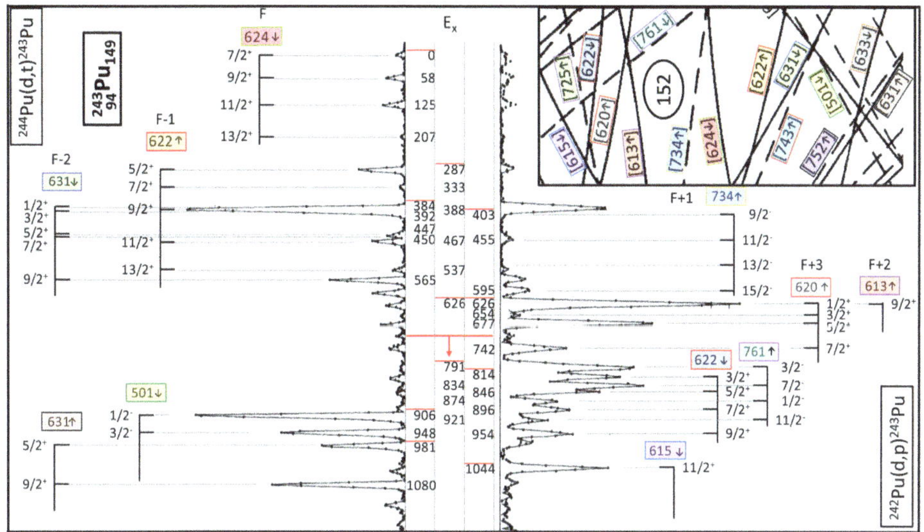

Figure 3.25. Spectra of protons and tritons from the one-neutron addition and removal reactions ^{242}Pu(d,p)^{243}Pu and ^{244}Pu(d,t)^{243}Pu. The spectrum peaks are matched with the energies in keV of the levels in ^{243}Pu to which they correspond, given in the central columns. Nilsson band assignments are shown and the band-head energies are indicted with red horizontal lines above these energies. The energy above which states (seen either in these reactions or other spectroscopic probes) are omitted is indicated by a heavy red bar and a downwards pointing arrow. Higher-spin band members are not observably populated and are not indicated. The Nilsson diagram relevant to ^{243}Pu, adapted from figure 3.22, is inset as a guide, with matched colour coding. Energies, spin-parities and band assignments are taken from ENSDF. Reprinted with permission from [13]. Copyright (1976) by the American Physical Society.

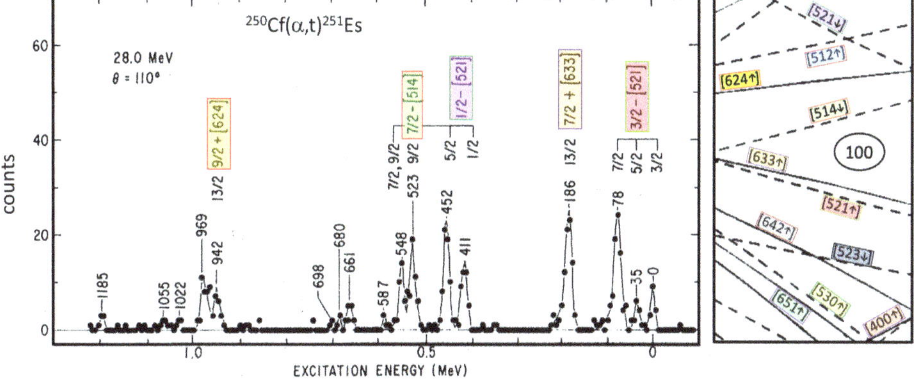

Figure 3.26. Spectrum of tritons from the one-proton addition reaction ^{250}Cf(α,t)^{251}Es. The spectrum peaks are labelled with the energies (in keV) of the levels in ^{251}Es (Z = 99) to which they correspond; Nilsson band assignments are also indicated, with the band member spin assignments. The Nilsson diagram relevant to ^{251}Es, adapted from figure 3.21, is inset as a guide, with matched colour coding. Reprinted with permission from [14]. Copyright (1977) by the American Physical Society.

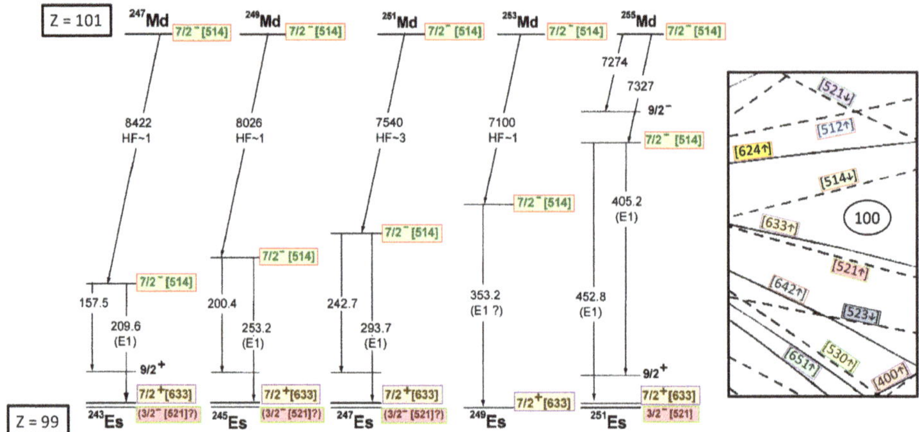

Figure 3.27. Alpha decay schemes for the parent nuclei ^{247,249,251,253,255}Md, showing Nilsson state assignments. Alpha decay energies in keV and α-decay hindrance factors (see text) are indicated. The Nilsson diagram relevant to these nuclei, adapted from figure 3.21, is inset as a guide. The quantum numbers labelling states in the decay schemes are $\Omega^{\pi}[Nn_z\Lambda]$ and in the inset are $[Nn_z\Lambda\Sigma]$, where $\Sigma = +1/2(\uparrow)$ or $-1/2(\downarrow)$, $\Omega = \Lambda + \Sigma$ and $\pi = (-1)^N$. Reprinted from [15], copyright (2005) with the permission of Springer Nature.

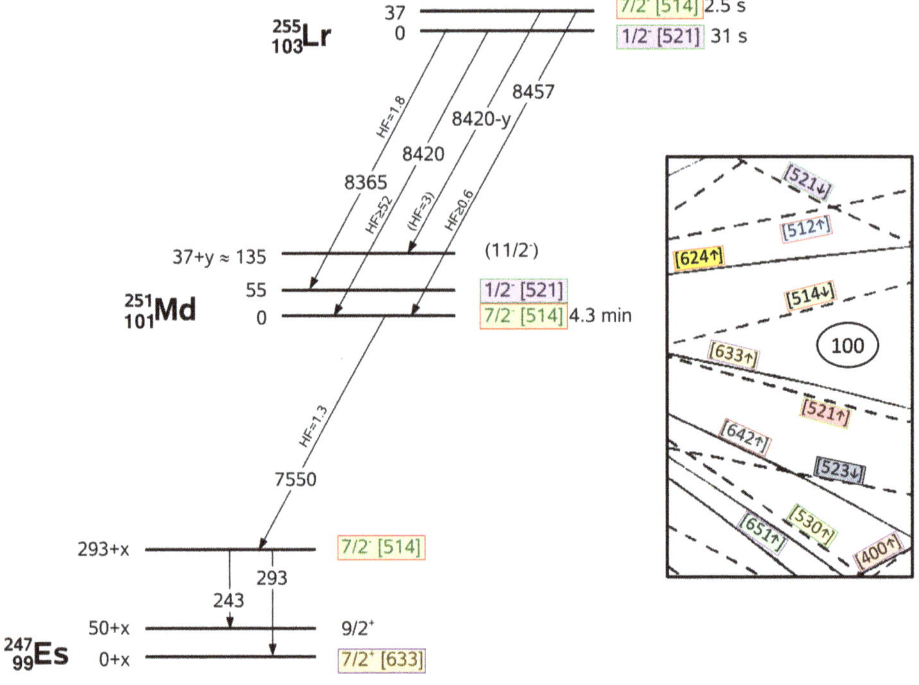

Figure 3.28. Alpha decay schemes for the chain ^{255}Lr $\rightarrow$ ^{251}Md $\rightarrow$ ^{247}Es, showing Nilsson state assignments. The Nilsson diagram relevant to these nuclei, adapted from figure 3.21, is inset as a guide. Alpha decay energies in keV and α-decay hindrance factors (see text) are indicated. Reprinted from [16], copyright (2006) with the permission of Springer Nature.

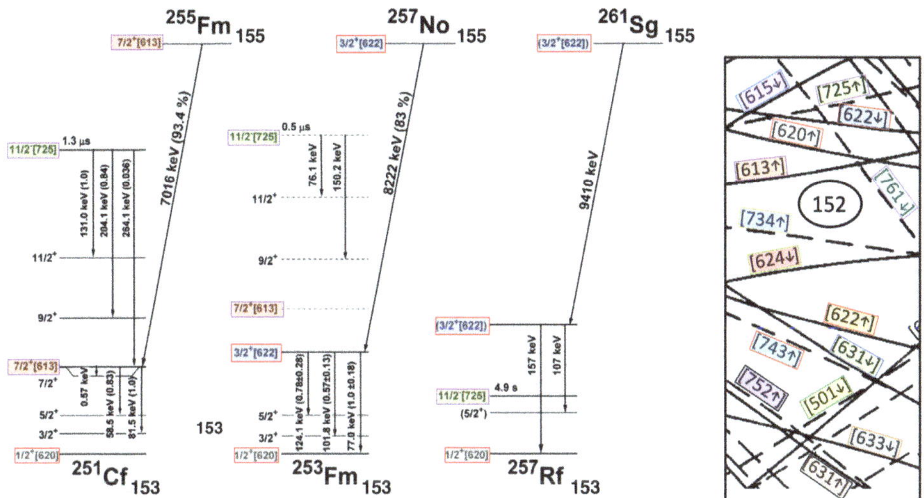

Figure 3.29. Alpha decay schemes for the parent nuclei ^{255}Fm, ^{257}No and ^{261}Sg, and isomerism in the daughter nuclei ^{251}Cf, ^{253}Fm and ^{257}Rf, showing Nilsson state assignments. The Nilsson diagram relevant to these nuclei, adapted from figure 3.22, is inset as a guide. Reprinted from [17], copyright (2010) with the permission of Springer Nature.

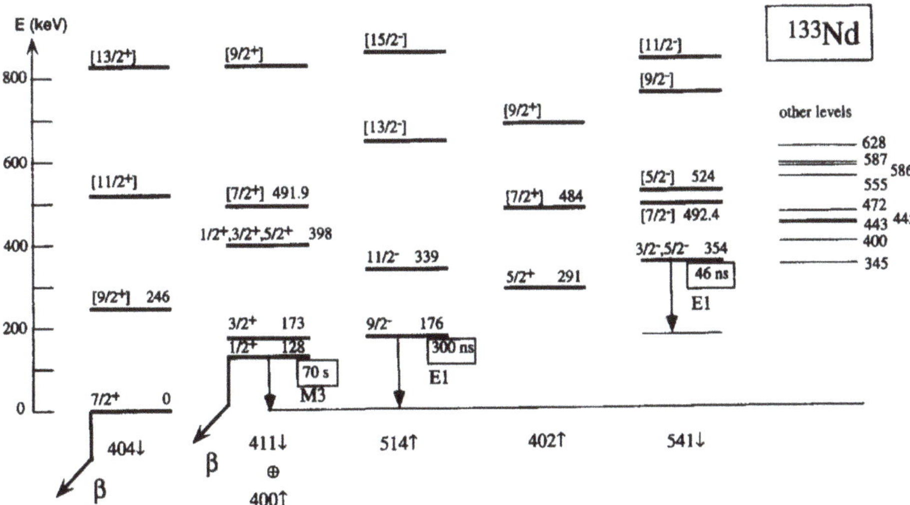

Figure 3.30. Nilsson bands assigned in ^{133}Nd. Note that there are two isomeric states that decay by E1 transitions and an isomeric state that undergoes beta decay and also decays by an M3 transition. These isomeric decays are very restrictive with respect to the possible Nilsson states on which the rotational bands are built. Particularly note that E1 transitions involve a parity change and, as revealed in Nilsson diagrams for this mass region, Nilsson states with negative-parity are not abundant in this region. Further, note that below 700 keV there are many states that lack band assignments and associated Nilsson configurations. This is discussed in the text. Reprinted from [18], copyright (1995) with the permission of Elsevier.

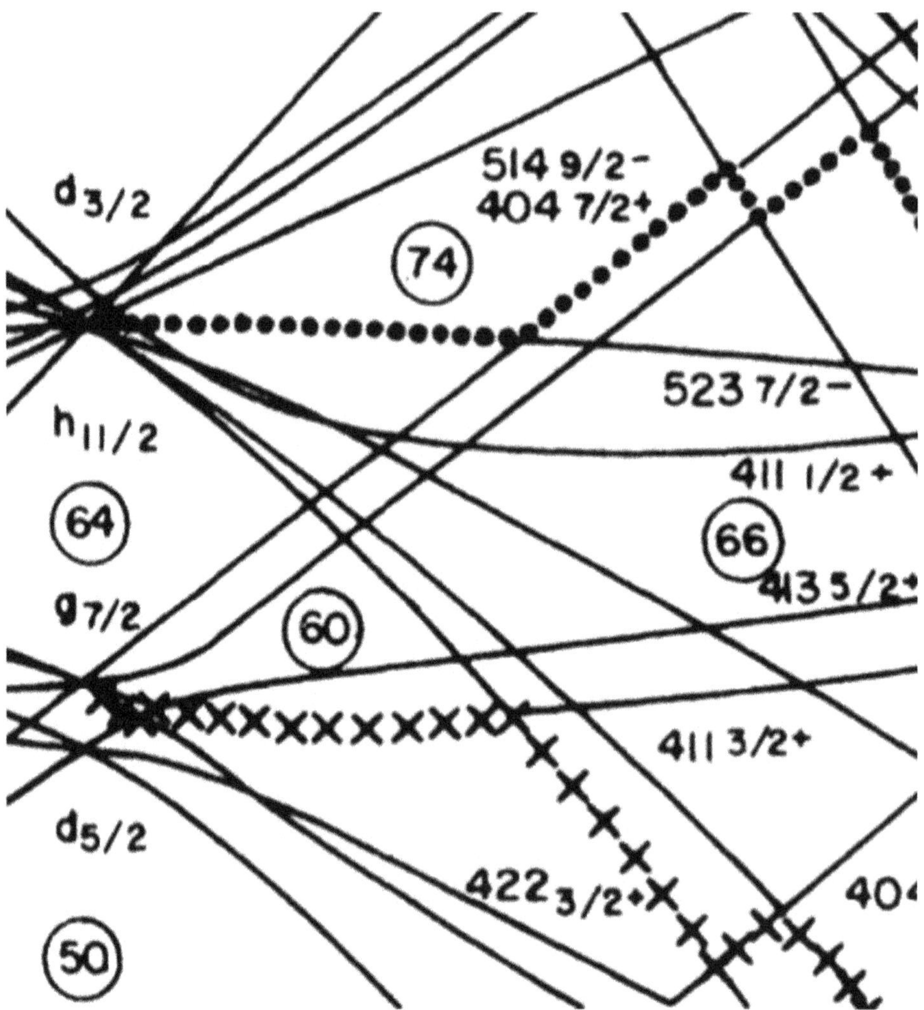

Figure 3.31. Nilsson diagram for odd-neutron nuclei in the $Z \sim 60$, $N \sim 70$ region. Note the paucity of Nilsson states with negative parity which severely constrains the interpretation of Nilsson bands in ^{133}Nd. Reprinted with permission from [19]. Copyright (1974) by the American Physical Society.

asymmetric rotor, where the extra states result from rotations about the so-called unfavoured axis of deformation of a triaxial core.

The view presented in figure 3.35 was pioneered by Juergen Meyer-ter-Vehn [22, 23]. The most extensive example of this view from a data perspective is presented in figure 3.36 for ^{125}Xe. The quantum mechanical details are beyond the present level of discussion, but some guidance is provided in this figure. The model is for a single high-j particle coupled to an axially asymmetric (triaxial) rotor. Rotations are possible about a favoured axis of rotation (the axis with respect to which the moment of inertia is largest, and so the rotational constant $\hbar^2/2I_{\max}$ is smallest and this results in lowest rotational energies), and rotations are possible about an

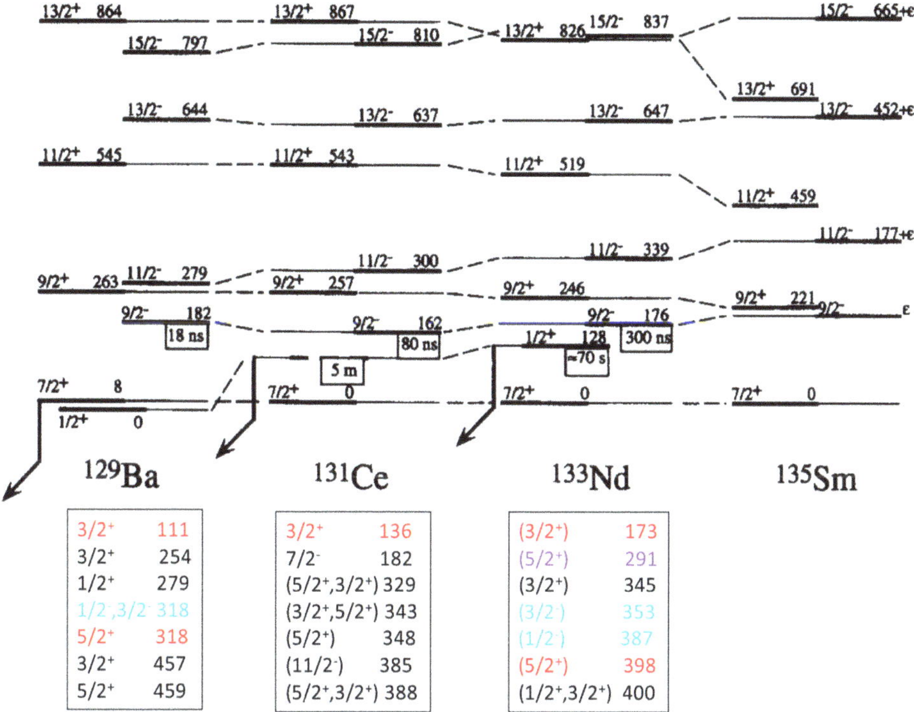

129Ba	
$3/2^+$	111
$3/2^+$	254
$1/2^+$	279
$1/2^-,3/2^-$	318
$5/2^+$	318
$3/2^+$	457
$5/2^+$	459

131Ce	
$3/2^+$	136
$7/2^-$	182
$(5/2^+,3/2^+)$	329
$(3/2^+,5/2^+)$	343
$(5/2^+)$	348
$(11/2^-)$	385
$(5/2^+,3/2^+)$	388

133Nd	
$(3/2^+)$	173
$(5/2^+)$	291
$(3/2^+)$	345
$(3/2^-)$	353
$(1/2^-)$	387
$(5/2^+)$	398
$(1/2^+,3/2^+)$	400

Figure 3.32. Nilsson bands assigned to the $N = 73$ isotones with $Z = 56 - 62$. The heavy arrows identify beta decaying isomers in these isotopes. States that lack band assignments or Nilsson configurations in ^{129}Ba, ^{131}Ce and ^{133}Nd are shown in boxes below these isotopes; colour coding is to match possible corresponding states. This is discussed further in the text. Reprinted from [18], copyright (1995) with the permission of Elsevier.

unfavoured axis of rotation (with respect to which the moment of inertia component is a minimum). The unfavoured axis of rotation is the broken axis of symmetry, i.e. the quantum numbers Ω and K are no longer good quantum numbers, and no longer (necessarily) equal. But excitations are organized about approximate quantum numbers with the appearance of Ω and K, here labelled $\tilde{\Omega}$ and $\tilde{K}$.

In figure 3.36, favoured rotations are manifested as sequences of states with spins $J, J + 2, J + 4, \ldots$ extending diagonally upwards to the right. The transitions between these states exhibit the maximal E2 transition strengths. The different $\tilde{\Omega}$ values are the 'shadows' of the Nilsson quantum number Ω: in ^{125}Xe these states are nearly degenerate in energy and $\tilde{\Omega} = 9/2$ is lowest in energy.

    3-14 Using data in ENSDF and XUNDL attempt an organization of data for the negative-parity states in ^{127}Xe, similar to that shown for ^{125}Xe in figure 3.36.

Caution, there are many open questions here, and the research frontier 'has been reached' with these exercises.

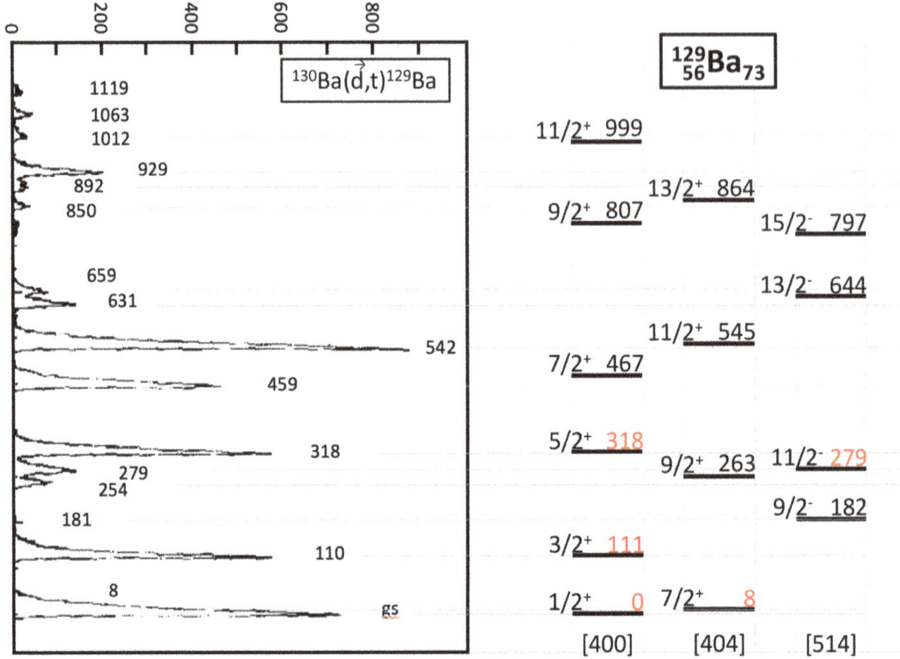

Figure 3.33. Triton spectrum observed in the one-neutron removal reaction ^{130}Ba($\vec{\text{d}}$,t)^{129}Ba using a beam of polarized deuterons, matched to assigned band structures in ^{129}Ba. Entries in red are matches to peaks in the spectrum. Reprinted from [20], copyright (1998) with the permission of Elsevier.

3.4.4 $Z < 50$, $N > 50$ region odd-mass nuclei

The identification of Nilsson states in the $Z \sim 41$, $N \sim 63$ region is beginning to be explored in some detail. Figures 3.37–3.39 provide a systematic view of Nilsson bands in odd-mass Y, Nb and Tc isotopes. Figure 3.40 provides a similar view in odd-mass $N = 61$ and 63 isotones. Nilsson diagrams for these nuclei are shown in figures 3.41 and 3.42(a) and (b). We note that, similar to the structural effect depicted in figure 1.16, there is a parallel effect for the neutron $1g_{7/2}$ orbital in the $50 < N < 82$ shell as protons fill the $1g_{9/2}$ orbital in the $28 < Z < 50$ shell. This effect lies entirely outside of the mean-field description of nuclei in this region and therefore shell model parameters must be 'forced' to give a correct description. Figures 3.42(a) and 3.42(b) illustrate the ambiguity involved in this model-based view.

The unique-parity states for odd-neutron nuclei stem from the $h_{11/2}$ shell model orbital. There are candidate states with this parentage identified across two open shells, as shown in figure 3.43 for $N = 63$. This view provides an important guide to how Nilsson bands lose their distinctive Ω, $\Omega + 1$, $\Omega + 2$, ... character. It indicates that, in turn, such distinctive character may be lost between Ru and Pd, i.e. in the Rh isotopes.

In the foregoing and following level schemes, states that lack confident band assignments are shown in boxes below the level schemes. Note that two sources of

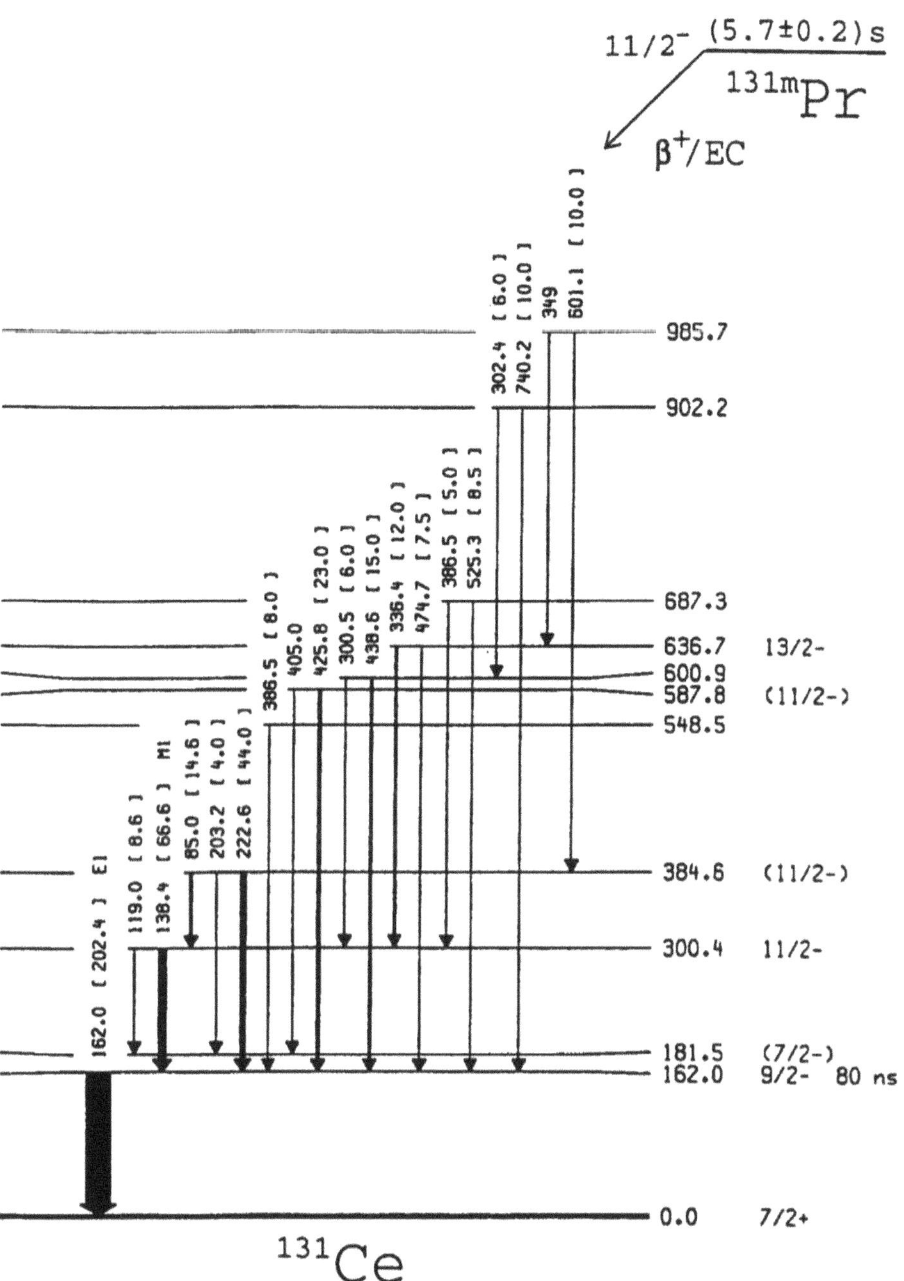

Figure 3.34. The beta decay scheme for the high-spin isomer in [131]Pr. The number of medium-spin states that decay to the $9/2^-$ isomer in [131]Ce cannot be explained by rotational bands built on Nilsson states. This is discussed in the text. Reprinted from [21], copyright (1996) with the permission of Elsevier.

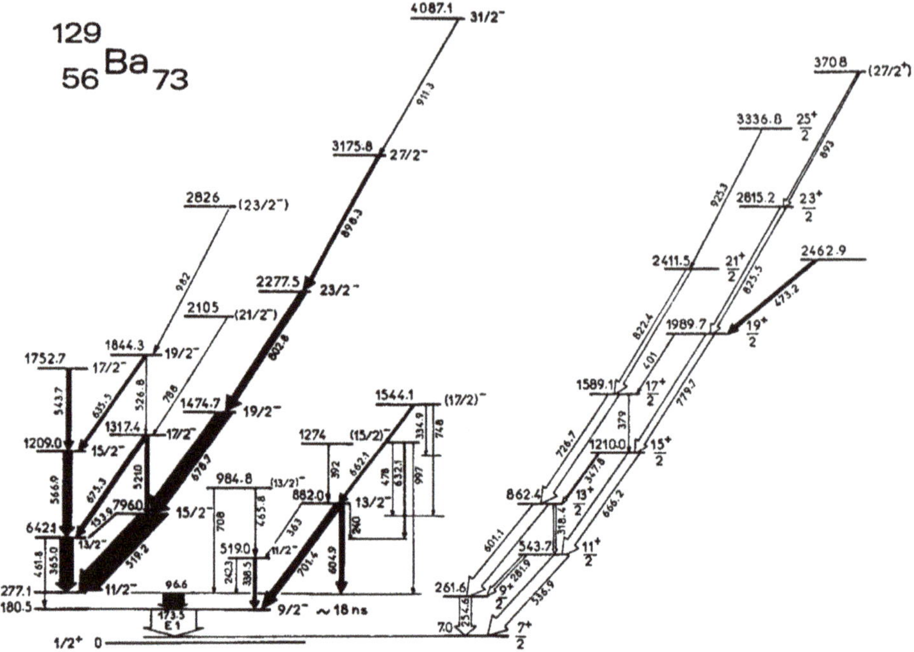

Figure 3.35. States in ^{129}Ba organized into a view developed by J Gizon, A Gizon and J Meyer-ter-Vehn. See text for details. Reprinted from [24], copyright (1977) with the permission of Elsevier.

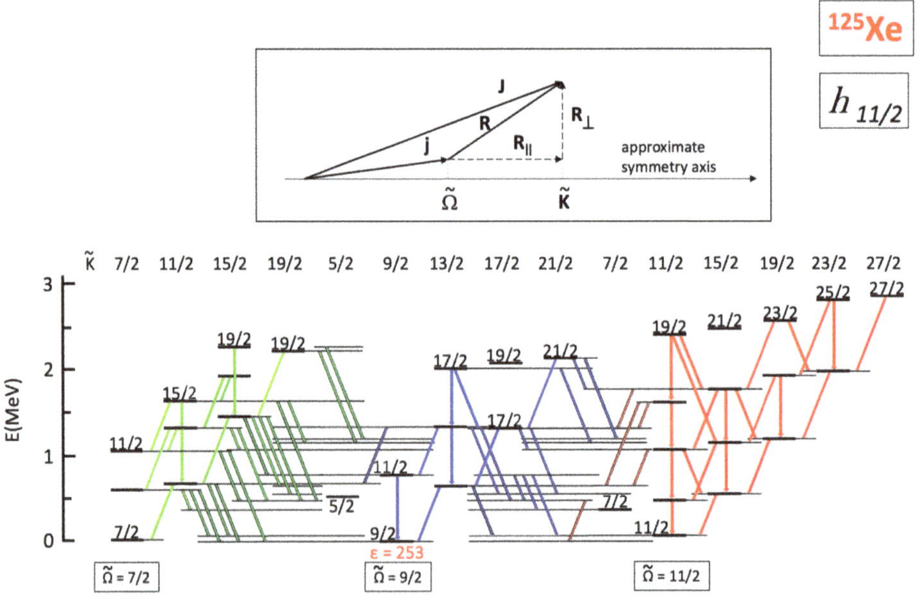

Figure 3.36. States in ^{125}Xe organized into a view developed by J Gizon, A Gizon and J Meyer-ter-Vehn [24]. See text for details. Note that in this figure, the 'base' state with spin 9/2 is displaced to the left, whereas in figure 3.35 it is displaced to the right. The data are taken from ENSDF. Details are discussed in the text. Reprinted from [25] CC-BY 4.0.

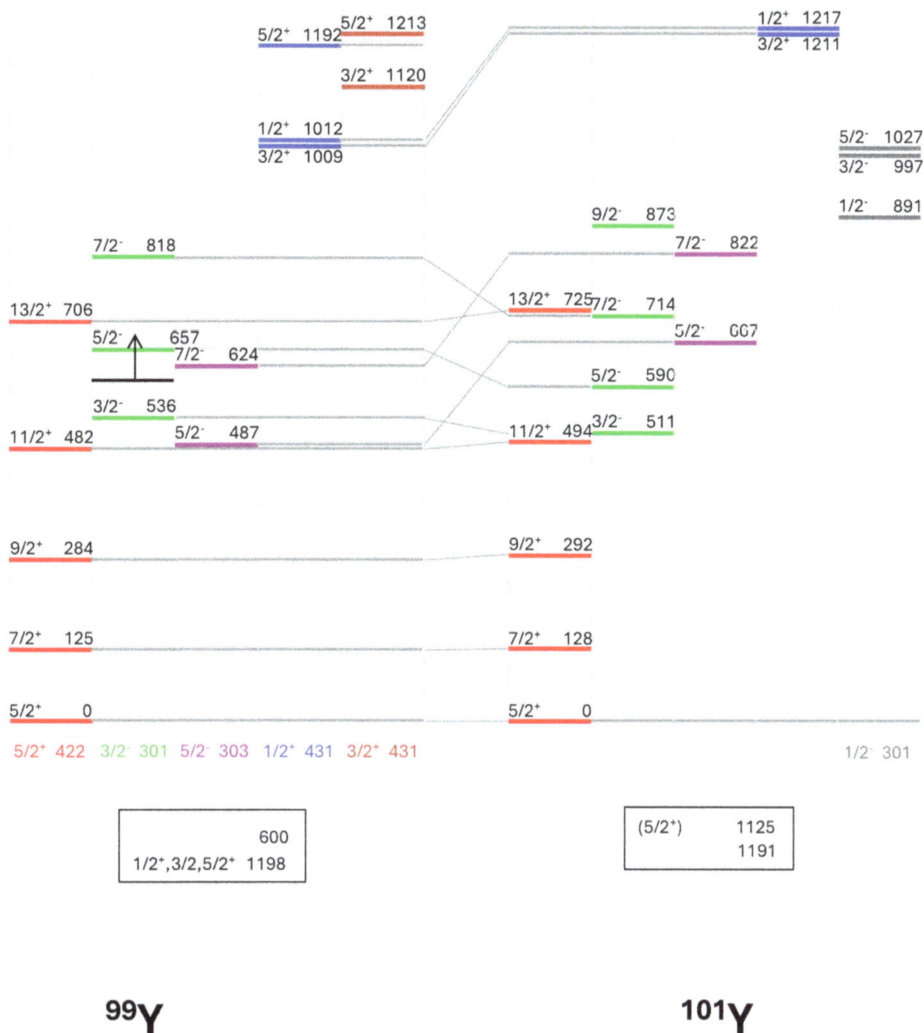

Figure 3.37. Nilsson bands in odd-mass Y isotopes. Horizontal bars with upwards pointing arrows indicate excitation energies above which states are omitted. Some specific details of the lowest-energy states which are omitted are given in the boxes below the level schemes; these are discussed in the text. The data are taken from ENSDF.

uncertainty exist with respect to identification of Nilsson bands in deformed odd-mass nuclei: (a) a level is established but there is a lack of spectroscopic detail to permit a band assignment; (b) gamma-ray spectra contain lines for which assignment as a transition in a level scheme has not been made. There is also the possibility that nucleon transfer reaction spectroscopy reports peaks that may be due to levels in the nucleus under study but may be due to a target contaminant.

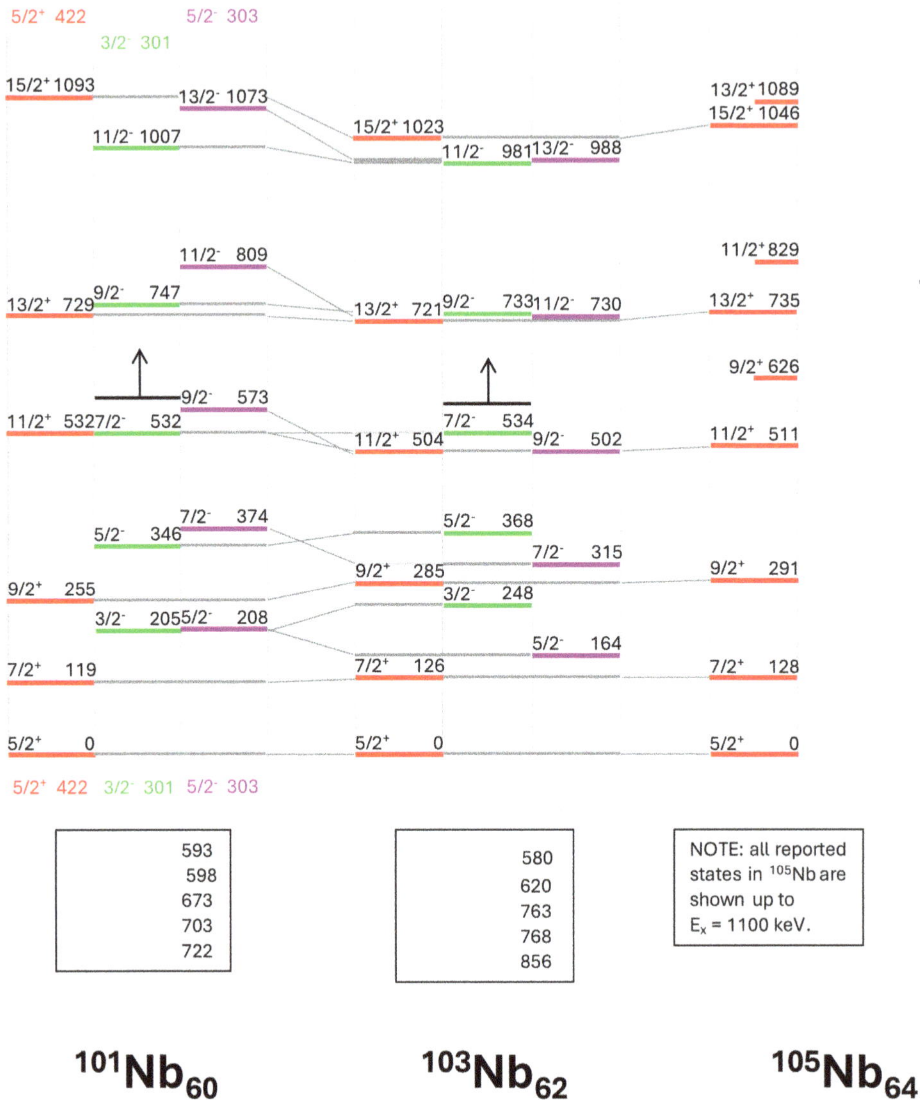

Figure 3.38. Nilsson bands in odd-mass Nb isotopes. Horizontal bars with upwards pointing arrows indicate excitation energies above which states are omitted. Some specific details of the lowest-energy states which are omitted are given in the boxes below the level schemes; these are discussed in the text. The data are taken from ENSDF.

3-15 With reference to figures 3.37–3.39 and 3.40, using data in ENSDF, explore extensions of these systematic views for:
   (a) ^{103}Y; ^{107}Nb; ^{109}Tc;
   (b) $N = 65$, i.e. ^{105}Zr; ^{107}Mo.

3-16 With reference to figure 3.43, using data in ENSDF, attempt similar systematic views for the $N = 61$ and 65 isotones.

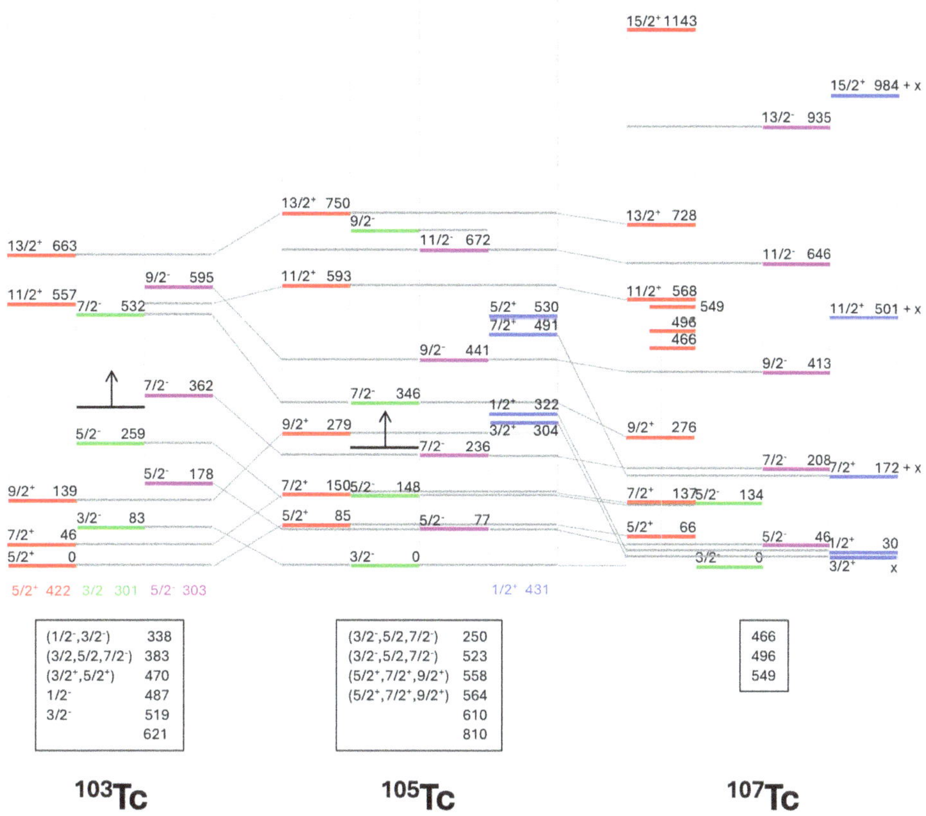

Figure 3.39. Nilsson bands in odd-mass Tc isotopes. Horizontal bars with upwards pointing arrows indicate excitation energies above which states are omitted. Some specific details of the lowest-energy states which are omitted are given in the boxes below the level schemes; these are discussed in the text. The data are taken from ENSDF.

3.4.5 $Z > 28$, $N < 50$ region odd-mass nuclei

The identification of Nilsson states in the $Z \sim 37$, $N \sim 41$ region: figures 3.44 and 3.45 provide systematic views of Nilsson bands in odd-mass Rb and Y isotopes; figures 3.46 and 3.47 provide a similar view in odd-mass $N = 41$ and 39 isotones. A Nilsson diagram for these nuclei is shown in figure 3.48. Note that the Nilsson diagram in figure 3.48 serves equally well for odd-neutron and odd-proton nuclei in this mass region. Note further that the $N = 39$ isotone systematics can be compared with the yttrium ($Z = 39$) systematics.

 3-17 With reference to figures 3.44 and 3.45, using data in ENSDF, explore extensions of these systematic views for:

  (a) ^{81}Rb;

  (b) the bromine ($Z = 35$) isotopes, ^{73,75,77,79}Br.

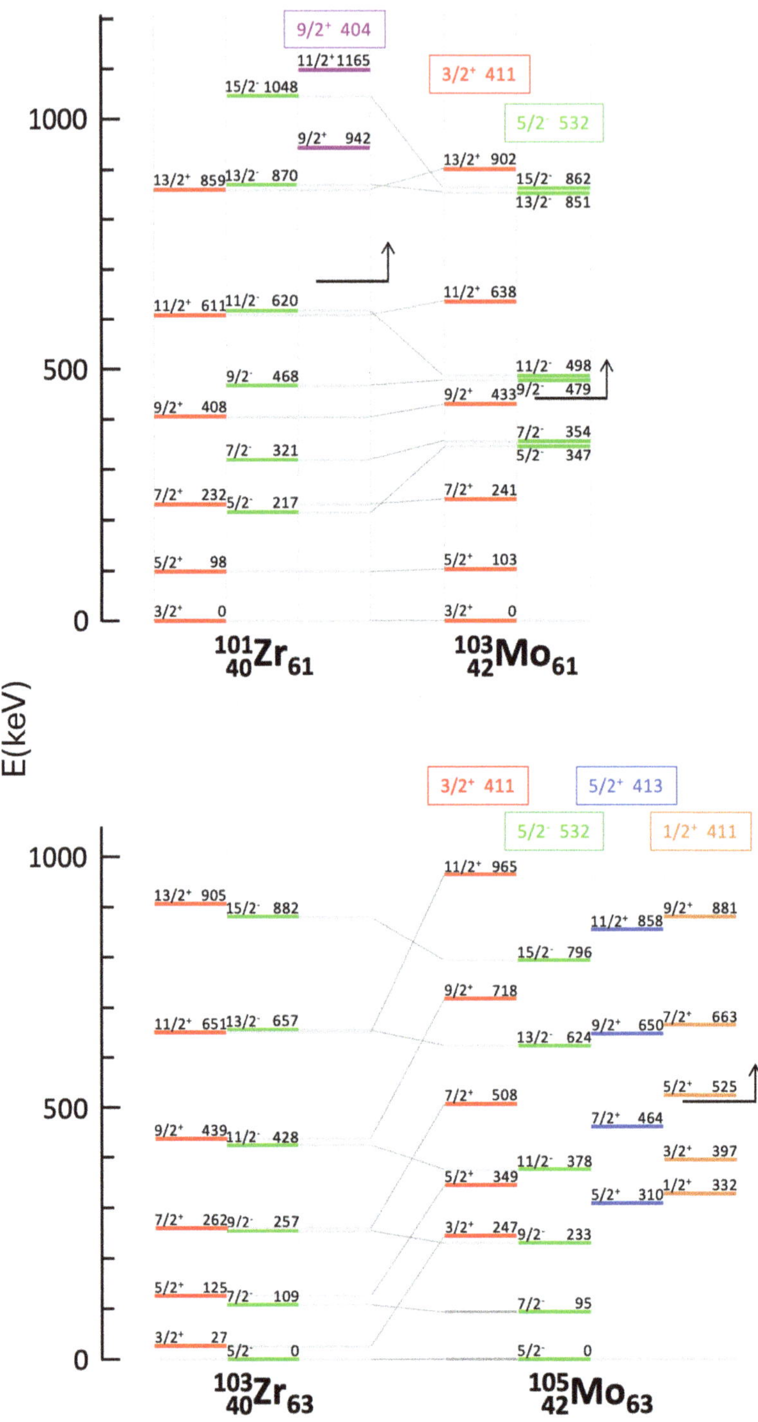

Figure 3.40. Nilsson bands in $N = 61$, 63 isotones. Horizontal bars with upwards pointing arrows indicate excitation energies above which states are omitted. The data are taken from ENSDF, and for ^{103}Zr from [26].

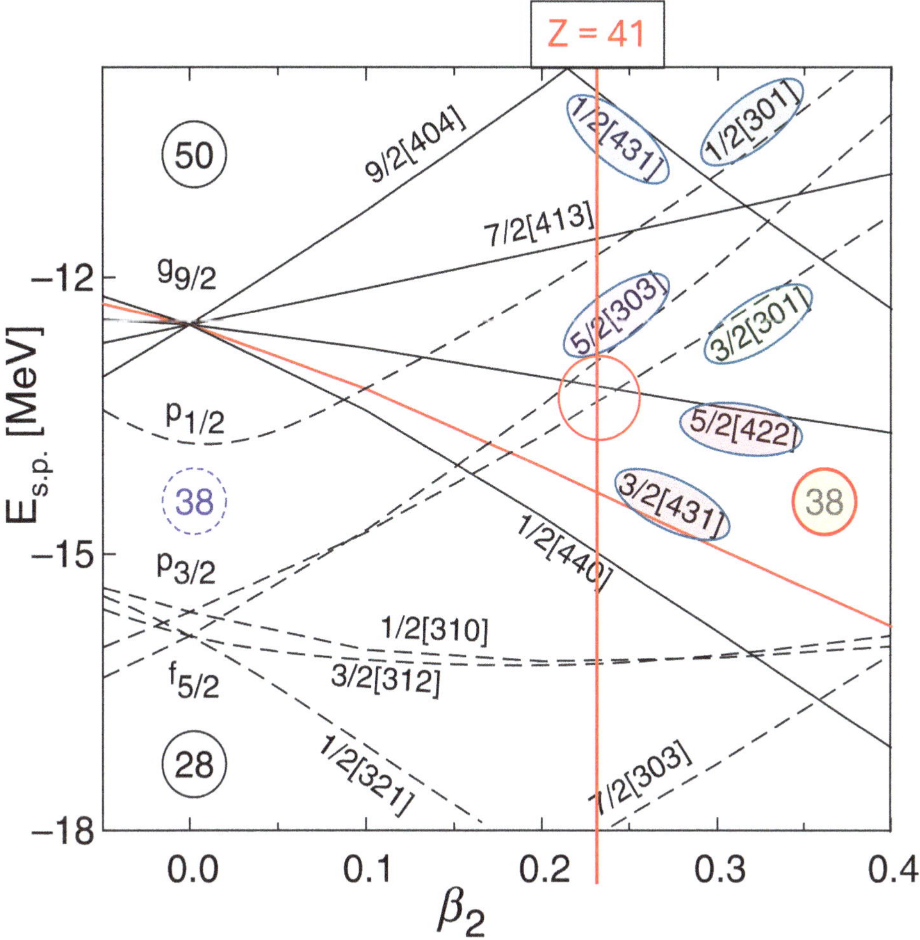

Figure 3.41. Nilsson diagram for odd-proton nuclei in the $Z \sim 40$, $N \sim 60$ region. Reprinted under CC-BY 3.0 license from [27]. Copyright (2015) by the American Physical Society.

3-18 With reference to figure 3.46, using data in ENSDF, explore extensions of these systematic views for:
    (a) the $N = 43$ isotones.
    (b) ^{75}Se.

3-19 Explore, using data from ENSDF, the degree to which the $N = 39$ isotone systematics resemble the $Z = 39$ (yttrium) systematics.

3.4.6 $A \sim 25$ region odd-mass nuclei

The identification of Nilsson states in the $A = 25$ nuclei, ^{25}Mg and ^{25}Al is shown in figure 3.49. A Nilsson diagram for these nuclei is shown in figure 3.50. Note that the Nilsson diagram in figure 3.50 serves equally well for odd-neutron and odd-proton nuclei in this mass region. Indeed, ^{25}Mg/^{25}Al form a so-called mirror pair with

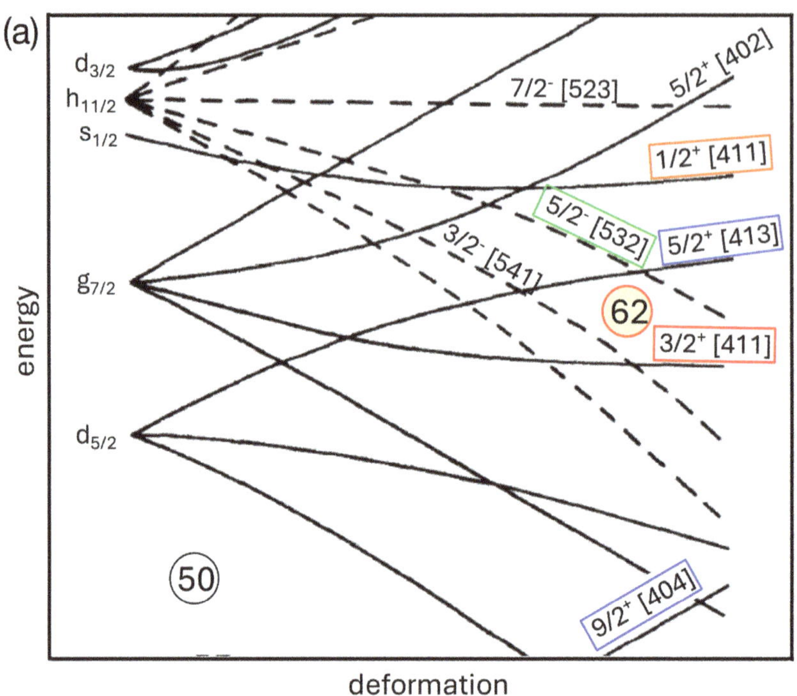

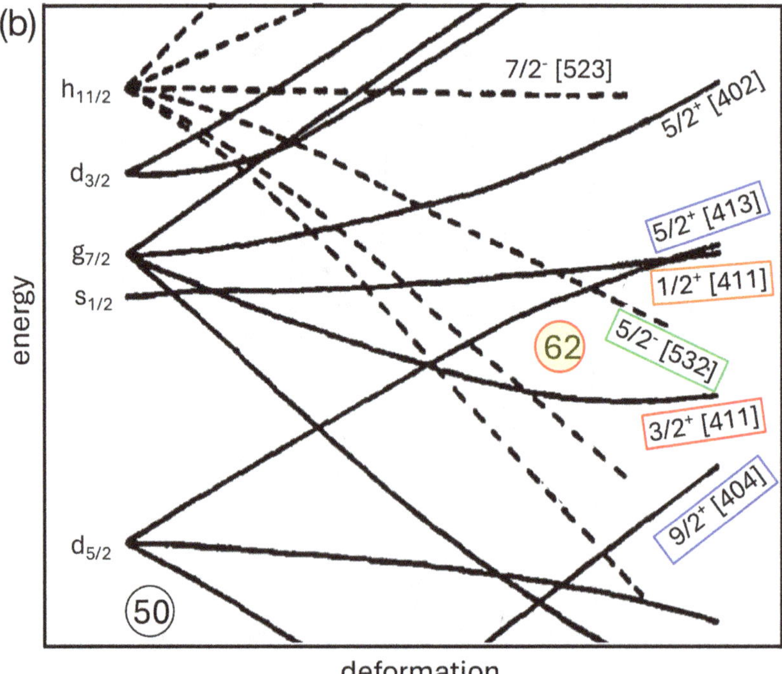

Figure 3.42. Two Nilsson diagrams for odd-neutron nuclei in the $Z \sim 40$, $N \sim 60$ region: (a) Reprinted from [28], copyright (1980) with the permission of Elsevier; (b) Reprinted from [29], copyright (1975) with the

permission of Elsevier. The primary difference between the two Nilsson diagrams shown involves the location of the shell model configurations, especially the ordering of the $g_{7/2}$ and $s_{1/2}$ configurations. Note: there is strong mixing between the configurations stemming from the $d_{5/2}$ and $g_{7/2}$ shell model parent configurations and so the labelling, which uses the asymptotic quantum numbers, must be used with caution.

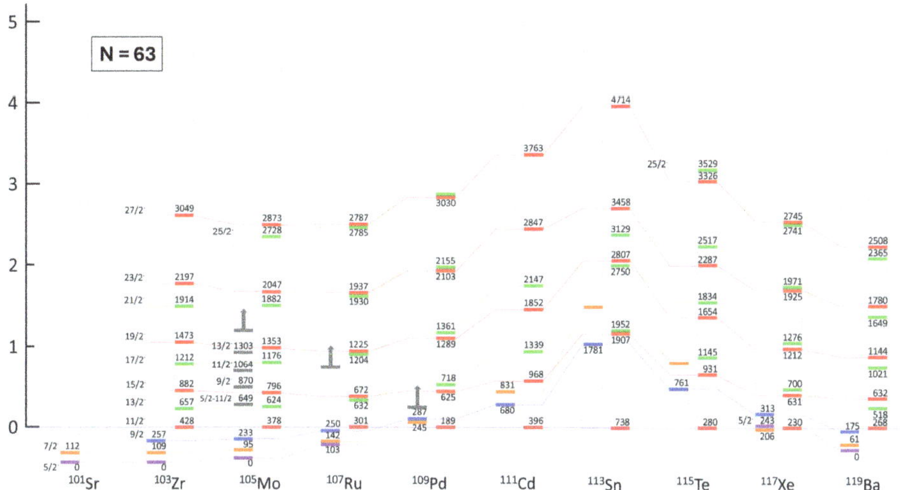

Figure 3.43. Systematics of selected negative-parity states for the $N = 63$ isotones, spanning two open-shell regions. Reprinted from [25] CC-BY 4.0.

respect to isospin symmetry. The identification of Nilsson states in nuclei neighbouring ^{25}Mg/^{25}Al is limited and the exercises below offer some exploration.

3-20 Using data in ENSDF, explore the identification of Nilsson bands in the $A = 23$ and 27 odd-mass nuclei.

3.4.7 Odd-mass nuclei in regions of shape coexistence

In regions of shape coexistence, where strong deformation coexists with weak deformation, Nilsson states can be expected to appear. Examples are depicted in: figure 1.33 for ^{41}Ca, in figure 2.8 for In isotopes, in figure 2.22 for the $N = 49$ isotones and in figure 2.32 for the $N = 81$ isotones. Such views provide new venues for exploring and testing the Nilsson model.

Examples of the systematic appearance of Nilsson configurations, as intruder states in the extensive region of shape coexistence centred on $A \sim 185$, are presented in figures 3.51 and 3.52. The region of deformation centred on $N \sim 105$ is observed to extend into the $N = 99$ isotones, as shown in figure 3.53.

Generally, Nilsson configurations in odd-mass nuclei in regions of shape coexistence can reveal finer details of the shapes involved, e.g. prolate versus oblate. This

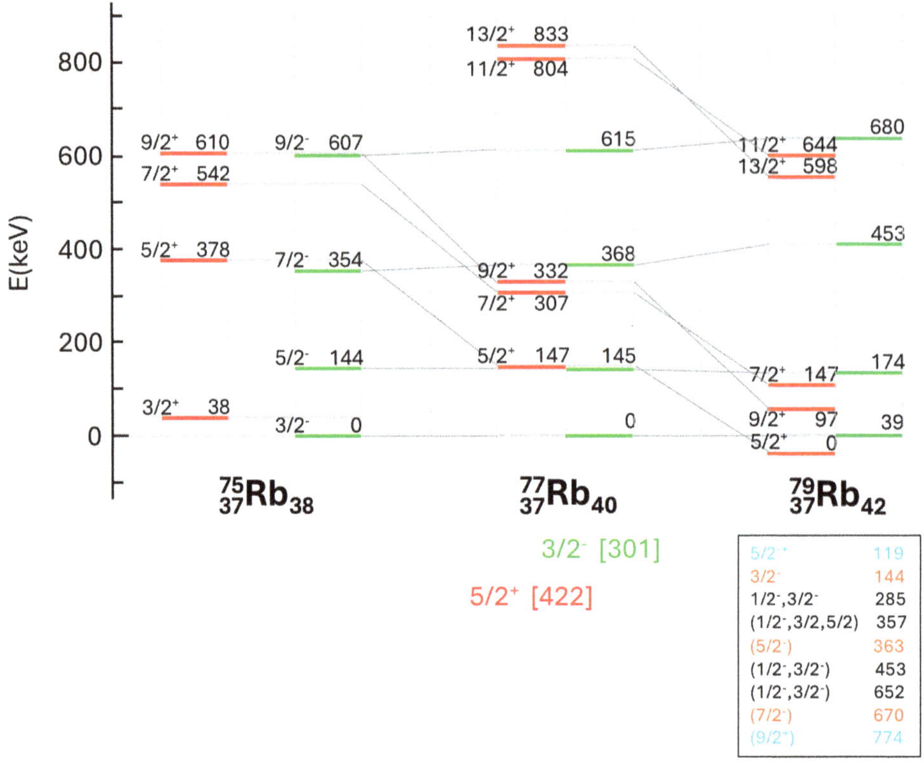

Figure 3.44. Systematics of Nilsson bands identified in the odd-mass rubidium isotopes with $N = 38$, 40 and 42. The data are taken from ENSDF.

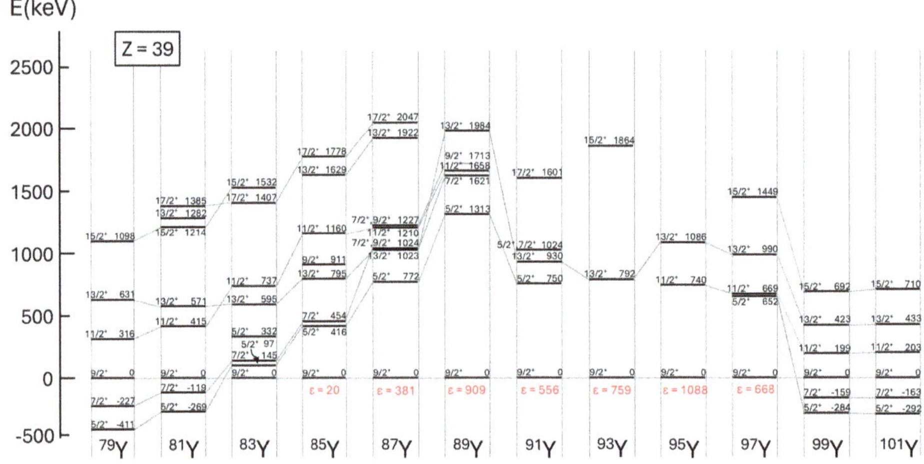

Figure 3.45. Systematics of selected positive-parity states for the yttrium ($Z = 39$) isotones, spanning two open-shell regions. Reprinted from [25] CC-BY 4.0.

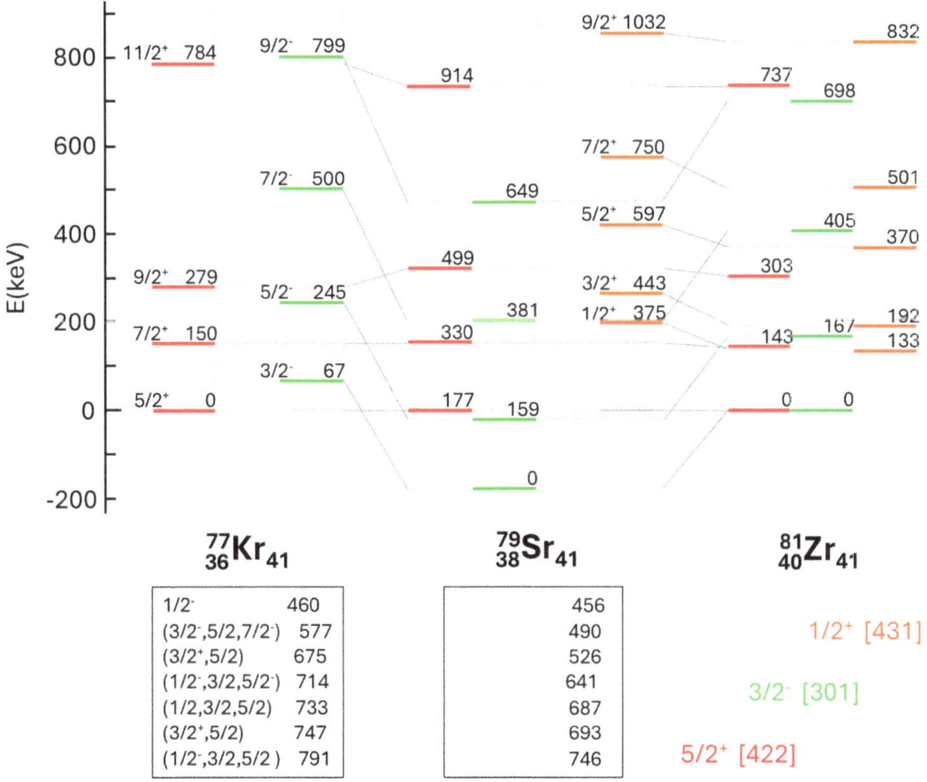

Figure 3.46. Systematics of Nilsson bands identified in the odd-mass $N = 41$ isotones with $Z = 36, 38$ and 40. The data are taken from ENSDF. See the caption to figure 3.44 for relevant commentary.

is illustrated in [8], e.g. for the antimony, thallium and bismuth isotopes. A notable and unexpected example of strong prolate deformation is illustrated in figure 3.55 for the very neutron-deficient gold isotopes.

3-21 In figure 2.32, what is the candidate Nilsson configuration for the intruder state structure?

3-22 In figure 2.32, why can it be inferred that the intruder band has oblate deformation?

3-23 Figures 3.54 and 3.55 suggest that the neutron-deficient gold isotopes ($Z = 79$) possess the $11/2^-[505]$ Nilsson configuration: locate this structure in a Nilsson diagram and estimate the approximate deformation involved.

3-24 By inspection of figure 3.52, for ^{185}Hg and ^{187}Hg, what Nilsson configurations can be expected?

3.4.8 Nilsson model (parameters) viewed from data

From the perspective of the Authors, there is a widely adopted practice of 'imposing' model views on data, as opposed to deducing model views from data. The

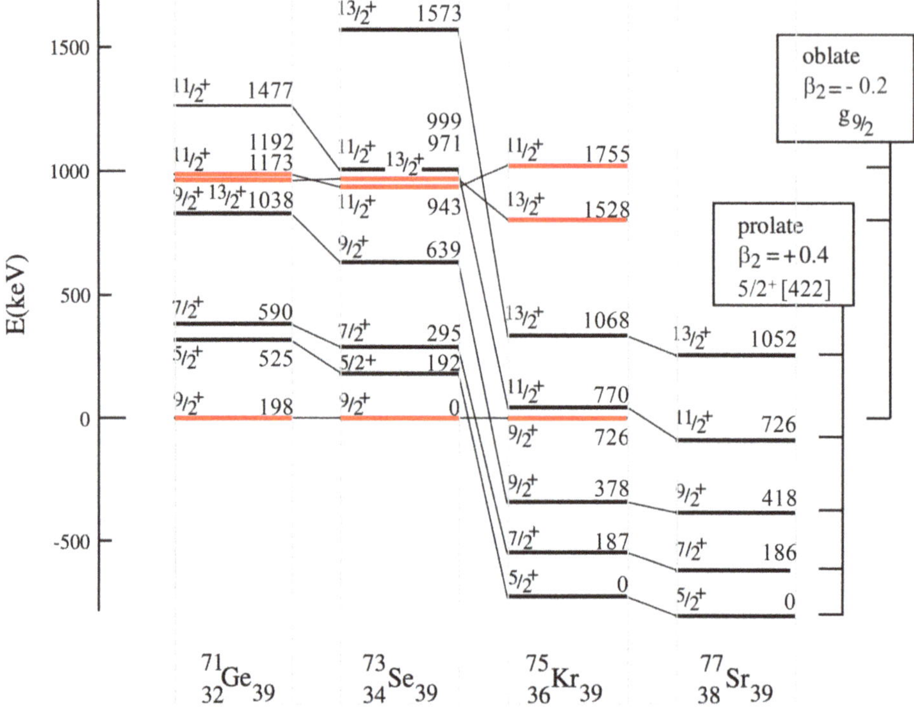

Figure 3.47. Systematics of positive-parity states in the odd-mass $N = 39$ isotones with $Z = 32$–38. The data are taken from ENSDF. This view shows the evidence for coexisting prolate and oblate collective structure. Reprinted with permission from [30]. Copyright (2011) by the American Physical Society.

foundational models of nuclear structure—the shell model, the rotor model and the pairing model were deduced; but there has been a lack of such an approach since these foundational model views were established.

The foundational models have proven to be very good, and so to be fair it is recognized that most model building has been directed at refining these foundational models. But a model view can be 'promoted' to the point that it has been stretched beyond the justification for its use based on data. Many models have moved to a limbo of inactivity as a result, they probably will not see salvation.

The Nilsson model was imposed at inception due to lack of data; but the modern data view would likely result in the Nilsson model view being deduced. However, in the details a data-based view is lacking. Herein, we explore some perspectives that may help to advance a shift to a data-based view of refining the Nilsson model. Figure 3.56 provides a schematic view of the hierarchy of energy factors underlying a Nilsson state. The various energy terms indicated at the bottom of the figure, the effects of which are not shown, are noted here and directions are provided to details given elsewhere. The origins of mixing effects acting on the oscillator shell quantum number, N are subtle; generally N mixing effects can be ignored. Examples of monopole shifts are illustrated in figures 1.16 and 1.29 using data; they must be

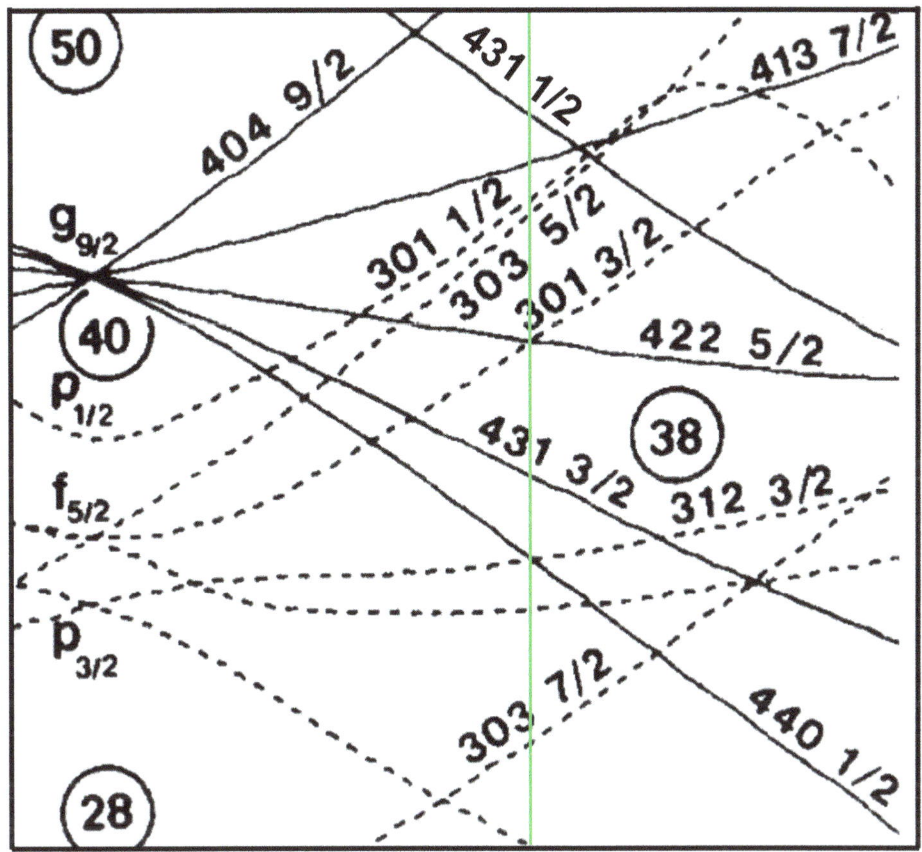

Figure 3.48. Nilsson diagram for odd-proton and odd-neutron nuclei in the $Z \sim 37$, $N \sim 41$ region. Reprinted from [31], copyright (1985) with the permission of Elsevier.

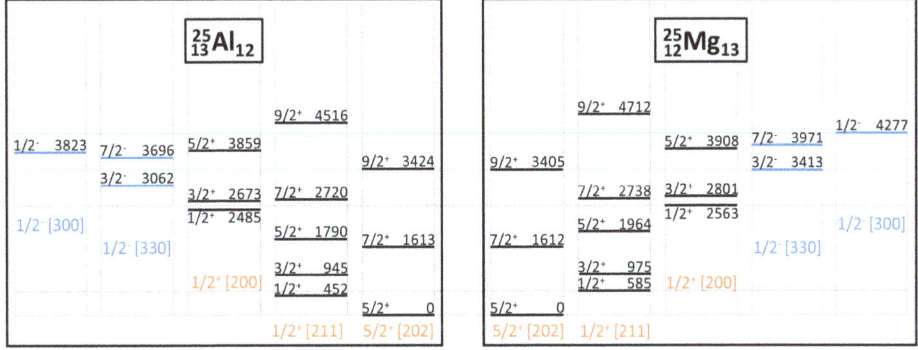

Figure 3.49. Nilsson bands in the isobar (mirror) pair ^{25}Mg/^{25}Al. The data are taken from ENSDF.

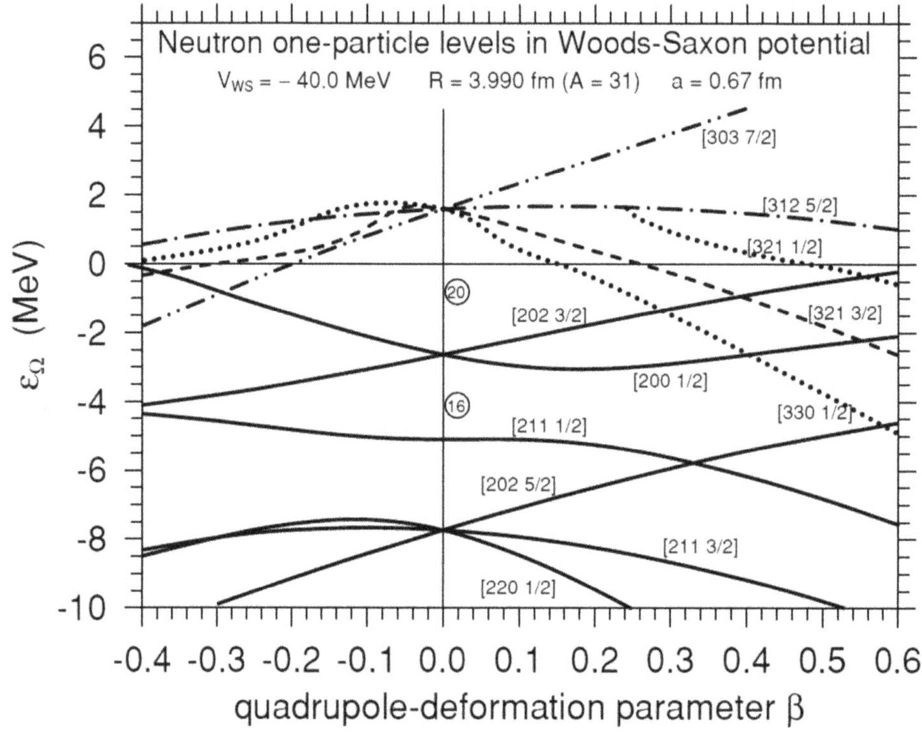

Figure 3.50. Nilsson diagram for odd-proton and odd-neutron nuclei in the $A \sim 25$ region. Reprinted with permission from [32]. Copyright (2007) by the American Physical Society.

Figure 3.51. Systematics of rotational bands built on the 7/2⁻[514] Nilsson configuration in the $N = 105$ isotones. The band-head excitation energy, ε (keV), is given in each nucleus. The data are taken from ENSDF and [33].

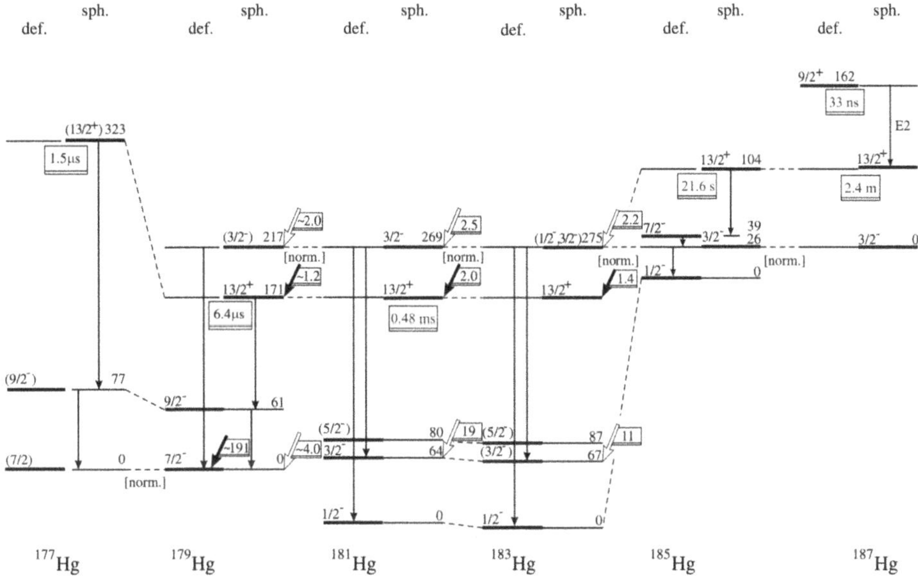

Figure 3.52. Systematics of the low-lying states in the odd-Hg isotopes near $N \sim 104$. Selected γ-ray transitions are shown as vertical arrows: in $^{177-183}$Hg, observed following α decay; in ^{185,187}Hg observed associated with decay of isomers. α-decay feedings (from Pb parent nuclei) are shown as diagonal arrows: solid for decays of the odd-Pb $1i_{13/2}$ isomers, open for the odd-Pb $3p_{3/2}$ ground states. α-decay hindrance factors, normalized to the neighbouring doubly-even ground-states-to-ground-state transitions, are shown in boxes attached to the arrows. States are grouped as spherical and deformed. Reprinted with permission from [30]. Copyright (2011) by the American Physical Society.

handled on a case-by-case basis through proton–neutron model interactions—these lie beyond the scope of the present treatment of data.

The parameterization of the Nilsson model Hamiltonian, equation (3.1) is defined by $\hbar\omega_0 = 41A^{-1/3}$ MeV and the relationship

$$C\mathbf{l} \cdot \mathbf{s} + Dl^2 = -\frac{\kappa(2\mathbf{l} \cdot \mathbf{s} + \mu\mathbf{l} \cdot \mathbf{l})}{2\hbar\omega_0}, \tag{3.10}$$

where typical values of κ are 0.06, and μ are 0.6 (protons) and 0.4 (neutrons). Beyond this basic parameterization, there exists an extensive literature of alternative formulations of the Nilsson model. These are expressed at the level of the shell model, i.e. the terms in equation (3.10); a few more details are presented in chapter 1. These variations are of little practical use because they are 'imposed' views that lack attention to details of data. One major variation in Nilsson potentials is the choice between an infinitely deep harmonic oscillator potential and the Woods–Saxon potential, which has a finite depth. For the specialist, we refer to [35]; but we note that they find no significant differences between the two potentials for Nilsson states in the context of the focus herein.

Figure 3.53. Nilsson bands observed in the $N = 59$ isotones, ^{97}Sr and ^{99}Zr. The ground state structures are weakly deformed. Reprinted with permission from [30]. Copyright (2011) by the American Physical Society.

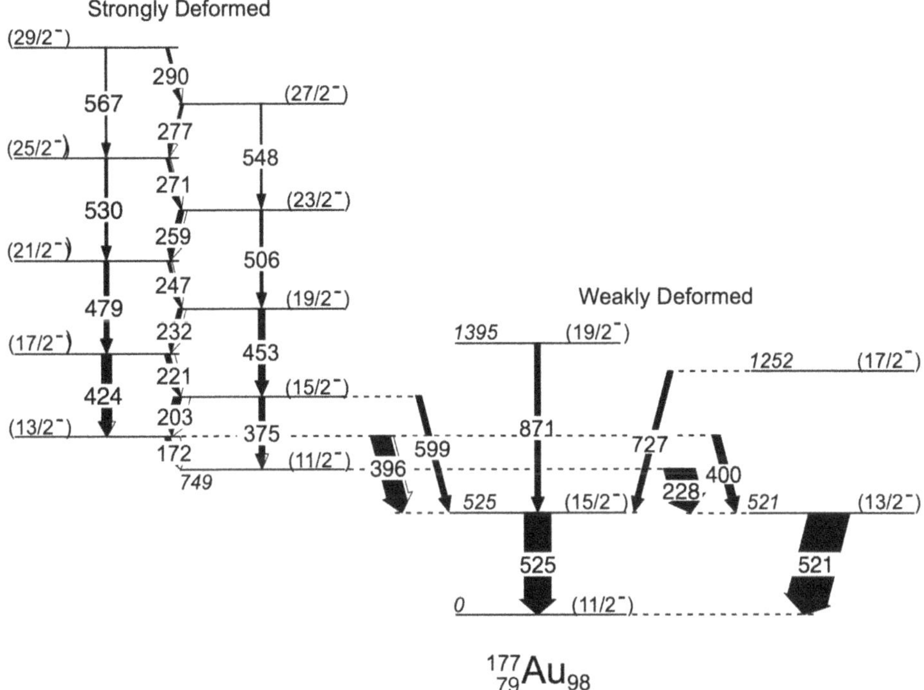

Figure 3.54. Observation of a $K = 11/2$ band in ^{179}Au. The structure coexists with weakly deformed states built on a structure that stems from the $h_{11/2}$ configuration. Reprinted with permission from [34]. Copyright (2017) by the American Physical Society.

A leading question is: 'how quantitative are the energies in Nilsson model diagrams?' Essentially, this reduces to the energy splitting observed in a deformed mean field for a given deformation. Figures 3.57(a) and (b) provide a useful answer to the question. The inset in figure 3.57(a) provides a 'worksheet' that indicates Nilsson diagrams may be able to predict energy splittings of Ω multiplets with $\sim$100 keV reliability. But caution is needed as this is within a single-j multiplet and does not test uncertainties in the implied energy spacing between j subshells for mass regions where nuclei are deformed and a direct view of j subshell spacings is unavailable.

3-25 For the nucleus ^{167}Er, figures 3.57(a) and 3.19 provide complementary views, in particular of the Nilsson bands stemming from the $i_{13/2}$ shell model configuration. For the $5/2^+$[642], $7/2^+$[633] and $9/2^+$[624] Nilsson states using figure 3.19, what can be deduced regarding the energy splitting of the $j = 13/2$ multiplet in a deformed mean-field?

A few points regarding the dependence of the deformation energy on quantum numbers is warranted to further pursue estimates of the reliability of energies in Nilsson diagrams. The deformation energy is given by (cf equation (3.7))

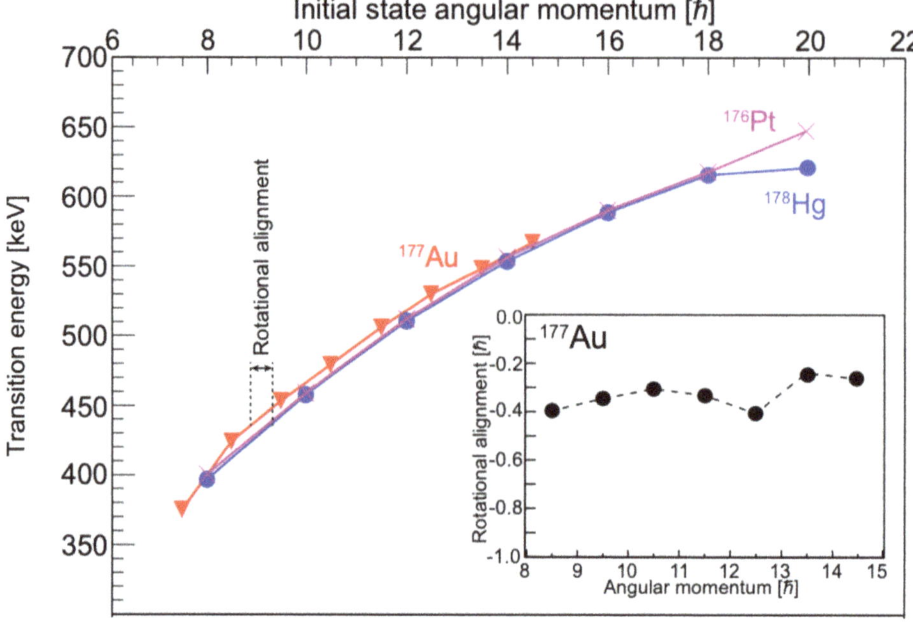

Figure 3.55. A rotation alignment plot for the $K = 11/2$ band in ^{179}Au. The neighbouring even–even cores to ^{179}Au, ^{176}Pt and ^{178}Hg, are both shown as possible cores to which the ^{179}Au structure is coupled. The near collinearity of the plots is consistent with the high-Ω character of the strongly deformed band. Reprinted with permission from [34]. Copyright (2017) by the American Physical Society.

$$E_{\mathrm{def}} = \frac{2}{3}\varepsilon\left(N + \frac{3}{2}\right)\hbar\omega_0\left[\frac{3\Omega^2}{4j(j+1)} - \frac{1}{4}\right]. \tag{3.11}$$

Note that the Ω^2 term is 'attenuated' by a factor $j(j+1)$, i.e. high-j multiplets do not split as widely as low-j multiplets. Thus, the view provided by figures 3.57(a) and (b) will be harder to discern for Nilsson states stemming from low-j shell model parent configurations, quite apart from the fact that there is j mixing which will make such inspections difficult to impossible.

3-26 Plot experimental Dy, Er deformations on a Nilsson diagram for odd-Z rare earth nuclei and attempt a quantified view of Ho Nilsson states (cf figure 3.8).

3-27 Using equation (3.11), estimate the maximum slopes for the highest-Ω orbitals of each j value. Compare with the patterns exhibited in figures 3.4 and 3.13.

3-28 Based on data presented in figures 3.58(a)–(c) and the model view of hexadecapole deformation presented in figure 3.59, can the presence of hexadecapole deformation be deduced from such data?

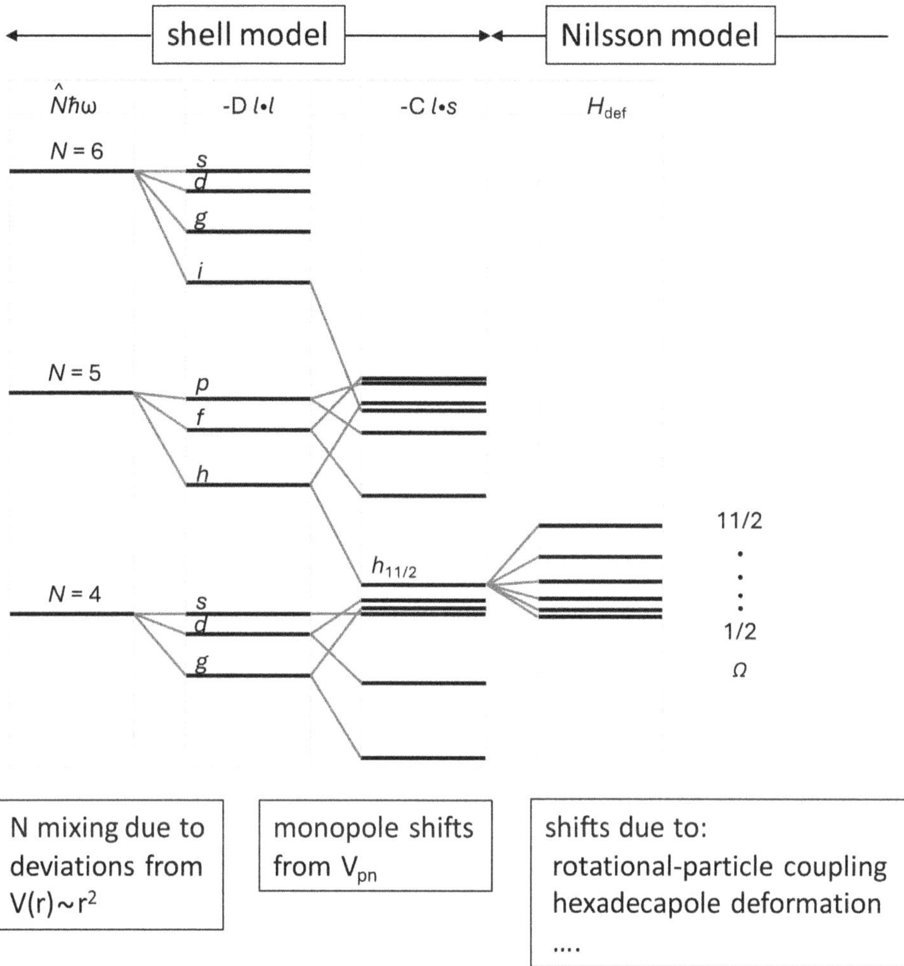

Figure 3.56. Schematic view of the major energy contributions to Nilsson states. On the left-hand side are the shell model contributions due to the harmonic oscillator potential, the model well-flattening term and the spin–orbit coupling term. On the right-hand side is the contribution due to the deformation of the oscillator potential. The quantum numbers are $Nlj\Omega$. The contributions to energies from mixing, interaction shifts, rotation-particle coupling and higher-order mean-field effects are discussed in the text. The relative magnitudes of the energy scales used are not in proportion to the physical scales and are only used to make the figure easier to digest.

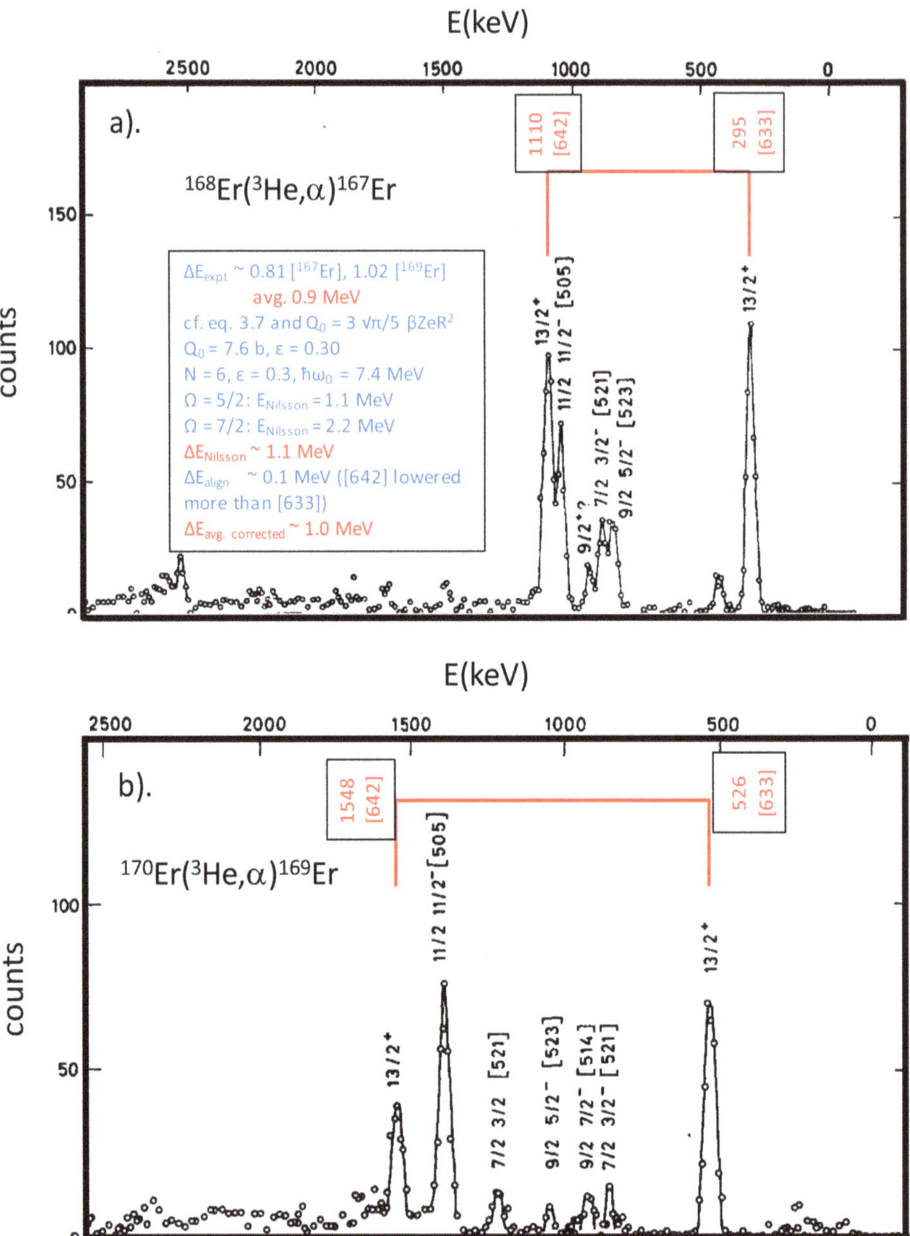

Figure 3.57. Spectra of helions (^{4}He) following the one-neutron removal reactions (a) ^{168}Er(^{3}He, α)^{167}Er and (b) ^{170}Er(^{3}He, α)^{169}Er. These reactions favour the population of high spin states in the final nucleus, in particular there is enhanced population of Nilsson bands stemming from high-j shell model subshells, notably the $I = 13/2$ members of the Nilsson bands associated with the $i_{13/2}$ subshell. Note that there are complementary data for ^{167}Er presented in figure 3.19. The boxed information in panel (a) is a 'worksheet' that sketches details of an estimate for the energy splitting between the $5/2^+$[642] and $7/2^+$[633] Nilsson bands in ^{167}Er. The energies of the states in the final nuclei are indicated in red (in keV) together with the quantum numbers that identify the states. Reprinted from [36], copyright (1972) with the permission of Elsevier.

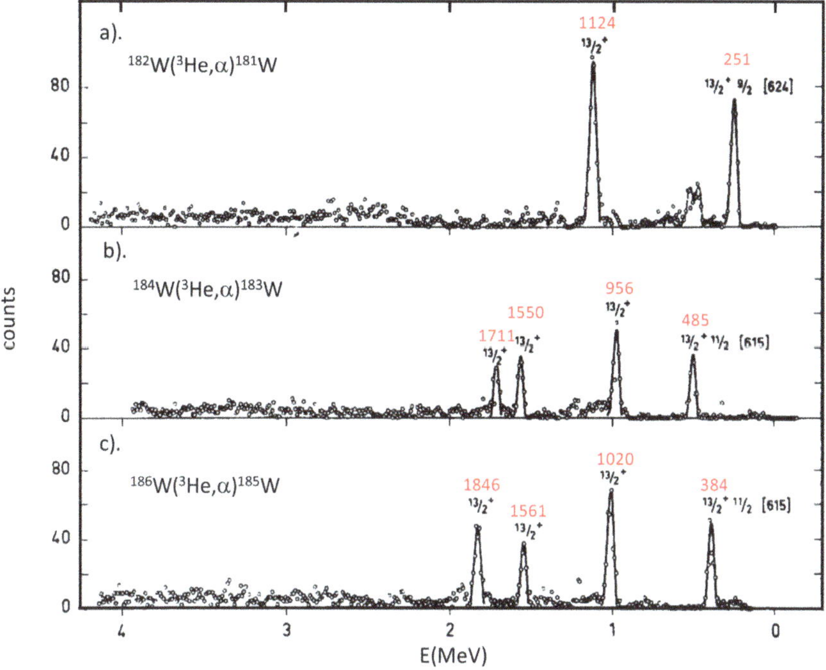

Figure 3.58. Spectra of helions (^{4}He) following the one-neutron removal reactions (a) ^{182}W(^{3}He,α)^{181}W, (b) ^{184}W(^{3}He, α)^{183}W and (c) ^{186}W(^{3}He, α)^{185}W. See the caption to figures 3.57(a) and (b) for other details. Reprinted from [37], copyright (1973) with the permission of Elsevier.

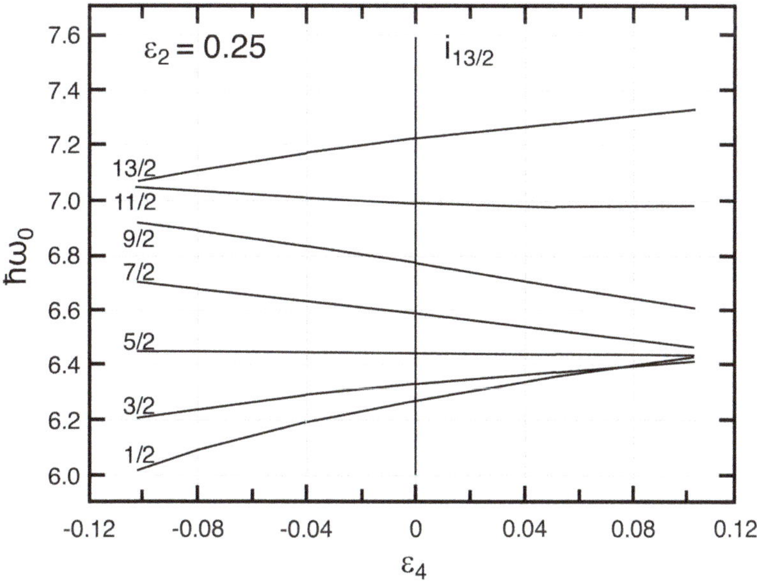

Figure 3.59. The influence of hexadecapole deformation, ε_4, on energies of Nilsson states stemming from the i$_{13/2}$ shell model configuration. The parameters used were $\varepsilon_2 = 0.25$, $\kappa = 0.637$, $\mu = 0.42$. For the rare earth region, $\hbar\omega_0 \sim 7.8$ MeV. Reprinted from [38], copyright (1974) with the permission of Elsevier.

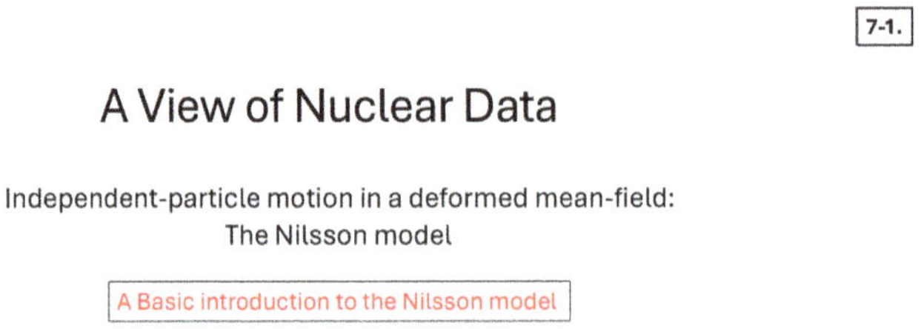

7-1.

A View of Nuclear Data

Independent-particle motion in a deformed mean-field:
The Nilsson model

A Basic introduction to the Nilsson model

Figure 3.60. Video tutorial: A basic introduction to the Nilsson model. Video available at https://doi.org/10.1088/978-0-7503-5648-0.

3.5 Video-based tutorial

3-29 For those reading the e-book, a video-based tutorial providing a basic introduction to the Nilsson model is found in figure 3.60.

References

[1] Jenkins D G and Wood J L 2021 *Nuclear Data: A primer* (Bristol: IOP Publishing)

[2] Løvhøiden G, Burke D G, Hirning C R, Flynn E R and Sunier J W 1978 Levels in ^{165}Ho populated in the ^{166}Er(t,α) reaction *Nucl. Phys.* A **303** 1

[3] Løvhøiden G, Burke D G, Flynn E R and Sunier J W 1979 Levels in ^{167,169}Ho populated in the (t,α) reaction *Nucl. Phys.* A **315** 90

[4] Iwanicki J, Zielińska M, Czosnyka T, Choiński J, Napiorkowski P, Loewe M, Würkner M and Srebrny J 1993 Electromagnetic properties of ^{165}Ho inferred from Coulomb excitation *J. Phys.* G **29** 743

[5] Bunker M E and Reich C W 1971 A survey of nonrotational states of deformed odd-A nuclei (150 < A < 190) *Rev. Mod. Phys.* **43** 348

[6] Jain A K, Sheline R K, Sood P C and Jain K 1990 Intrinsic states of deformed odd-A nuclei in the mass regions (151 $\leqslant$ 193) and ($A \geqslant$ 221 *Rev. Mod. Phys.* **62** 393

[7] Chasman R R, Ahmad I, Friedman A M and Erskine J R 1977 Survey of single-particle states in the mass region A > 228 *Rev. Mod. Phys.* **49** 833

[8] Jenkins D G and Wood J L 2023 *Nuclear Data: A Collective Motion View* (Bristol: IOP Publishing)

[9] Lamm I-L 1969 Shell-model calculations of deformed nuclei *Nucl. Phys.* A **125** 504–30

[10] Tarara R W and Browne C P 1979 Level structure of ^{175}Yb and ^{177}Yb via the ^{174,176}Yb(d,p)^{175,177}Yb and ^{176}Yb(d,t)^{175}Yb reaction *Phys. Rev.* C **19** 674

[11] Ŝtêrba F, Ŝtêrbová J and Rozkoŝ M 1973 Energy levels of ^{167}Er populated by the (d,d$'$) reaction *Czech. J. Phys.* B **23** 601–14

[12] Rasmussen J O 1959 Alpha-decay barrier penetrabilities with an exponential nuclear potential: even-even nuclei *Phys. Rev.* **113** 1593

[13] Casten R F, Kane W R, Erskine J R, Friedman A M and Gale D S 1976 States in ^{243}Pu from the (n,γ), (d,p) and (d,t) reactions *Phys. Rev.* C **14** 912

[14] Ahmad I, Friedman A M, Chasman R R and Yates S W 1977 Proton orbital 1/2$^-$[521] and the stability of superheavy elements *Phys. Rev. Lett.* **39** 12

[15] Hessberger F P *et al* 2005 Energy systematics of low-lying Nilsson levels in odd-mass einsteinium isotopes *Eur. Phys. J.* A **26** 233–9

[16] Chatillon A *et al* 2006 Spectroscopy and single-particle structure of the odd-Z heavy elements ^{255}Lr, ^{251}Md and ^{247}Es *Eur. Phys. J.* A **30** 397–411

[17] Streicher B *et al* 2010 Alpha-gamma decay studies of ^{261}Sg and ^{257}Rf *Eur. Phys. J.* A **45** 275–86

[18] Breitenbach J B, Wood J L, Jarrio M, Braga R A, Kormicki J and Semmes P B 1995 The decay of ^{133}Pm and the structure of ^{133}Nd *Nucl. Phys.* A **592** 194–210

[19] Griffioen R D and Sheline R K 1974 Level structure of ^{129}Ba from the ^{130}Ba(d,t)^{129}Ba reaction *Phys. Rev.* C **10** 624

[20] Bucurescu D *et al* 1998 Single particle excitations in ^{129}Ba from a (d,t) reaction study *Nucl. Phys.* A **630** 643–60

[21] Gizon A *et al* 1996 Level structures of $^{131,\,129}$Ce observed in beta decay *Nucl. Phys.* A **605** 301–33

[22] Meyer-ter-Vehn J 1975 Collective model description of transitional odd-A nuclei: (i) the triaxial-rotor-plus-particle model *Nucl. Phys.* A **249** 111–40

[23] Meyer-ter-Vehn J 1975 Collective model description of transitional odd-A nuclei: (ii). comparison with unique parity states of nuclei in the $A = 135$ and $A = 190$ mass regions *Nucl. Phys.* A **249** 141–65

[24] Gizon J, Gizon A and Meyer-ter-Vehn J 1977 Triaxial shape and onset of high-spin bands in ^{129}Ba *Nucl. Phys.* A **277** 464–76

[25] Stuchbery A E and Wood J L 2022 To shell model, or not to shell model, that is the question *Physics* **4** 697–773

[26] Urban W, Pinson J A, Tzaca-Urban T, Wolska E, Genevey J, Simpson G S, Smith A G, Durrell J L, Varley B and Ahmad I 2009 Mapping neutron levels in the A $\sim$100 region: the $\nu 3/2^+$ [411] band in ^{103}Zr *Phys. Rev.* C **79** 067301

[27] Sotty C *et al* 2015 ^{97}Rb: the cornerstone of the region of deformation around A $\sim$100 *Phys. Rev. Lett.* **115** 172501

[28] Simms P C, Rickey F A and Popli R K 1980 Symmetric rotor interpretation of transitional nuclei *Nucl. Phys.* A **347** 205–29

[29] Rekstad J 1975 Rotational description of some odd-A nuclei in the mass-100 region *Nucl. Phys.* A **247** 7–20

[30] Heyde K and Wood J L 2011 Shape coexistence in atomic nuclei *Rev. Mod. Phys.* **83** 1467

[31] Nazarewicz W, Dudek J, Bengtsson R, Bengtsson T and Ragnarsson I 1985 Microscopic study of the high-spin behaviour in selected $A = 80$ nuclei *Nucl. Phys.* A **435** 397–447

[32] Hamamoto I 2007 Nilsson diagrams for light neutron-rich nuclei with weakly-bound neutrons *Phys. Rev.* C **76** 054319

[33] Zhang W Q *et al* 2022 First observation of a shape isomer and a low-lying strongly-coupled prolate band in neutron-deficient semi-magic ^{187}Pb *Phys. Lett.* B **829** 137129

[34] Venhart M *et al* 2017 De-excitation of the strongly coupled band in ^{177}Au and implications for core intruder configurations in the light Hg isotopes *Phys. Rev.* C **95** 061302

[35] Bengtsson R, Dudek J, Nazarewicz W and Olanders P 1989 A systematic comparison between the Nilsson and Woods-Saxon deformed shell model potentials *Phys. Scr.* **39** 196–220

[36] Løvhøiden G, Tjøm P O and Edvardson P 1972 Study of positive-parity states in ^{165}Er and ^{167}Er and ^{169}Er with the (^{3}He,α) reaction *Nucl. Phys.* A **194** 463

[37] Kleinheinz P, Casten R F and Nilsson B 1973 Hexadecapole deformation in the wolfram nuclei *Nucl. Phys.* A **203** 539

[38] Stephens F S, Kleinheinz P, Sheline R K and Simon R S 1974 Backbending and rotation alignment *Nucl. Phys.* A **222** 235–51

IOP Publishing

Nuclear Data
An independent-particle motion view
David Jenkins and John L Wood

Chapter 4

How are Nilsson configurations affected by rotations?

The influence of rotational degrees of freedom on Nilsson model configurations is introduced.

Concepts: rotation-particle coupling term, Coriolis effects, rotational decoupling, rotation alignment, signature splitting, transfer reactions, spectroscopic factors, magnetic moments.

Learning outcomes: The effect of rotations on the manifestation of Nilsson states in nuclei is shown to be important as a first-order correction to understanding an independent-particle view of odd-mass deformed nuclei.

The separation of independent-particle motion in a nuclear deformed mean field from rotational degrees of freedom of the nucleus must be handled with care. There is an extensive literature that explores this property of nuclei, with models that extend to considerable complexity with intricate parameter fitting. Herein, we adopt the simplest possible view for handling this aspect of nuclear structure.

4.1 The rotation-particle coupling term

The Hamiltonian for an independent particle in a deformed nucleus is usually expressed as

$$H = H_{\text{particle}} + H_{\text{coll}}, \tag{4.1}$$

where H_{particle} is described using the Nilsson model, equation (3.1), and H_{coll} is usually described using a rotational model. Thus, H_{coll} is commonly introduced as

$$H_{\text{coll}} = \mathbf{R}^2/2\mathcal{J} \tag{4.2}$$

where $\mathbf{R}$ is the core angular momentum and $\mathcal{J}$ is the moment of inertia for a symmetric top.

doi:10.1088/978-0-7503-5648-0ch4   4-1

All nucleons possess spin, $s = 1/2$ and may carry angular momentum, l, whence $j = l + s$. Therefore, $\mathbf{R}$ is not equal to the total spin of the nucleus, $\mathbf{I}$. We express the relationship between them as

$$\mathbf{I} = \mathbf{R} + \mathbf{j}, \tag{4.3}$$

whence

$$\mathbf{R} = \mathbf{I} - \mathbf{j} \tag{4.4}$$

and

$$\begin{aligned} \mathbf{R} \cdot \mathbf{R} &= (\mathbf{I} - \mathbf{j}) \cdot (\mathbf{I} - \mathbf{j}) \\ &= \mathbf{I} \cdot \mathbf{I} - 2\mathbf{I} \cdot \mathbf{j} + \mathbf{j} \cdot \mathbf{j}. \end{aligned} \tag{4.5}$$

This leads to the decomposition of the Hamiltonian, equation (4.2) into

$$H_{\text{coll}} = \frac{\mathbf{I}^2}{2\mathscr{J}} - \frac{\mathbf{I} \cdot \mathbf{j}}{\mathscr{J}} + \frac{j^2}{2\mathscr{J}}. \tag{4.6}$$

This is a widely used decomposition of collective model Hamiltonians, for nuclei with an unpaired nucleon, limited to rotational collective behaviour.

The first term in equation (4.6) leads to the familiar expression for rotational energies,

$$E_{\text{rot}} = \frac{I(I + 1)\hbar^2}{2\mathscr{J}}, \tag{4.7}$$

the second term, $-\mathbf{I} \cdot \mathbf{j}/\mathscr{J}$ is the rotation-particle coupling or RPC term, sometimes called the 'Coriolis' term, the third term is sometimes called the 'recoil' term. Caution is needed with respect to this terminology. In many treatments of the Nilsson model, these terms are ignored. Sometimes the influence of the RPC term is handled with respect to its producing $\Delta\Omega = \pm 1$ mixing between Nilsson configurations of the same parity; this is handled in appendix C. With only a few exceptions, the recoil term is neglected, sometimes with the rationale that it is 'intrinsic' to the structure of the nucleus and therefore does not impact rotational energy patterns. Herein, we consult data to obtain clarification of this terminology and the importance of these Hamiltonian terms.

4.2 Bands and rotational effects

We present data for Nilsson bands as plots of gamma-ray transition energy versus spin of the state from which the transition originates. These are observables for the nucleus, with no model dependence to obscure interpretation. In anticipation of complexity due to multiple Nilsson bands in each nucleus, and further complexity due to band mixing, we first focus on odd-proton nuclei in the lanthanide region. Odd-neutron nuclei in the lanthanide region have a higher density of Nilsson bands; odd-mass actinide nuclei have a much higher density of Nilsson bands. Other regions of deformation are of very limited extent, and it is more difficult to discern global trends in data patterns.

Plots of gamma-ray transition energy versus spin for bands in ^{175}Lu are shown in figures 4.1(a)–(d). The results are surprising: for these Nilsson bands the plots are collinear with ground-state bands of neighbouring even–even 'core' nuclei; for those cases that are not collinear, they are near parallel. As we will show in the following, deviations from these two types of behaviour are very rare and usually appear to be minor deviations from a basic pattern of identical rotational 'profiles'.

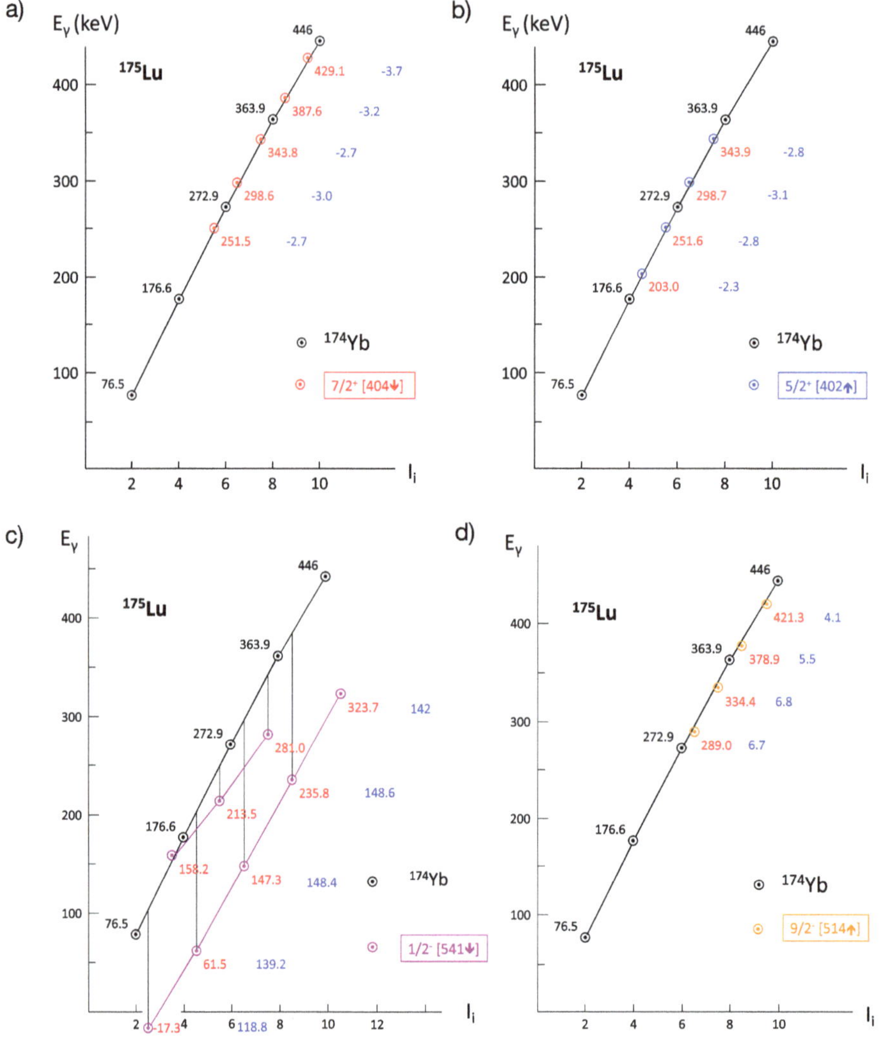

Figure 4.1. Plots of γ-ray energy versus spin of the initial state, I_i for transitions in the ^{175}Lu bands, compared to the neighbouring even–even nucleus, ^{174}Yb. The γ-ray energies in keV are given in red and the interpolated energy differences are given in blue. Recognizing that transition energy is an energy difference for the rotor energy relationship, equation (4.7), it is evident that the odd-mass profiles for such plots are parallel to the profile for the even–even 'core' nucleus, then the term $\mathbf{I} \cdot \mathbf{j}$ has a simple constant contribution to the transition energy and the term j^2 does not change with I.

To interpret the patterns manifested in figures 4.1(a)–(d), first it is important to note that a γ-ray transition energy is a differential quantity with respect to excitation energies in a nucleus. Therefore, comparison should be made with the derivative of equation (4.6). Thus, the j^2 term disappears, the $\mathbf{I} \cdot \mathbf{j}$ term becomes a constant, and the $\mathbf{I}^2$ term contributes a linear term in $\mathbf{I}$. This is essentially what is seen in the data plots, sometimes with (the derivative of) the $\mathbf{I} \cdot \mathbf{j}$ term contributing zero to the observed energies. However, the physics implications of this are far from trivial with respect to the conceptualization of how nuclei 'rotate'. We explore some of the ramifications of this below.

It is important to have a visual perspective of rotational-particle coupling. Figure 4.2 depicts a standard view of a nucleon coupled to a prolate deformed core. The system has an axis of rotational symmetry and a plane of reflection symmetry through the centre of mass at right angles to this symmetry axis. The rotational degree of freedom is often termed the 'symmetric top'. States of the system are described, note $K = \Omega$, by

$$|IMK; \Omega\pi\rangle = \frac{1}{\sqrt{2}}\{|IM, +K\rangle|+\Omega\pi\rangle + (-1)^{I+K}|IM, -K\rangle|-\Omega\pi\rangle\} \tag{4.8}$$

where the Nilsson state component, $|\Omega\pi\rangle$ is expressed as, cf equation (3.8),

$$|\Omega\pi\rangle = \sum_{jl} C_{jl}|Njl\Omega\pi\rangle, \tag{4.9}$$

and recall $\Omega = j_3$.

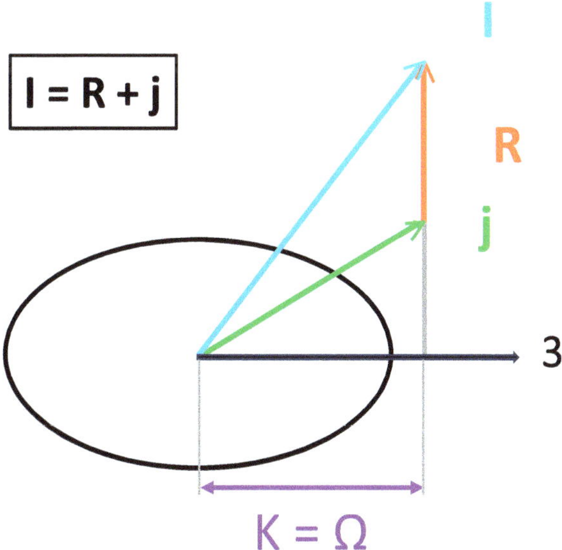

Figure 4.2. Schematic view of the symmetric top model for an odd-mass nucleus. The contributions to the total spin, I (blue) are the spin of the unpaired nucleon, j (green) and the collective rotational angular momentum, R (orange). The vector R is at right angles to the 3-axis, the symmetry axis of the model symmetric top (which cannot rotate about the symmetry axis). The projection of I, K and the projection of j, Ω are equal in this model. Because the unpaired nucleon is in a deformed mean field, j is not a good quantum number, i.e. it is mixed (not shown), but $j_3 = \Omega$ is a good quantum number, as is K.

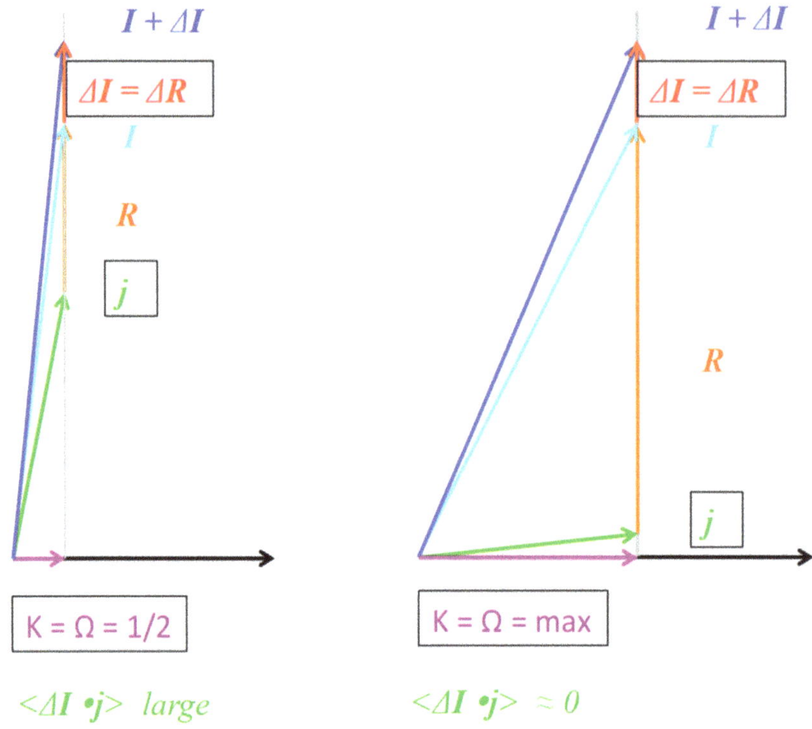

Figure 4.3. A semi-classical explanation of the effect of the $\mathbf{I} \cdot \mathbf{j}$ term on rotational energies in nuclei. The details of the explanation are given in the text. Here, we note that the plots presented in figures 4.1(a)–(d) involve energy differences and they must be matched to vectorial differences of the $\mathbf{I} \cdot \mathbf{j}$ term: recall, a small increment in the direction of a vector is at right angles to the vector, hence the large difference between the alignments shown on the left and that shown on the right.

Figure 4.3 provides a view of the $\mathbf{I} \cdot \mathbf{j}$, RPC term and the key to understanding the patterns of behaviour in figures 4.1(a)–(d). The underlying dynamics are revealed by viewing the information in terms of the quantity $\Delta\mathbf{I} \cdot \mathbf{j}$: this approaches a small value for the Nilsson states with highest Ω values; conversely, $\Omega = 1/2$ Nilsson states possess the largest $\Delta\mathbf{I} \cdot \mathbf{j}$ values and will exhibit the largest alignment effects. Such patterns are shown in figures 4.4(a)–(c). We turn to the quantitative features of the alignment effects below.

The alignment effects increase with decreasing deformation. Decreasing deformation will be associated with decreasing moments of inertia and thus the RPC term will increase in magnitude. Such patterns are shown in figures 4.5(a)–(f). The consequence of this view of rotational energy contributions to the energies of Nilsson states is shown in figures 4.6(a)–(c).

4.3 Signature splitting

A direct effect of rotation-particle coupling is so-called signature splitting: the states in rotational bands with quantum number K split into spin sequences

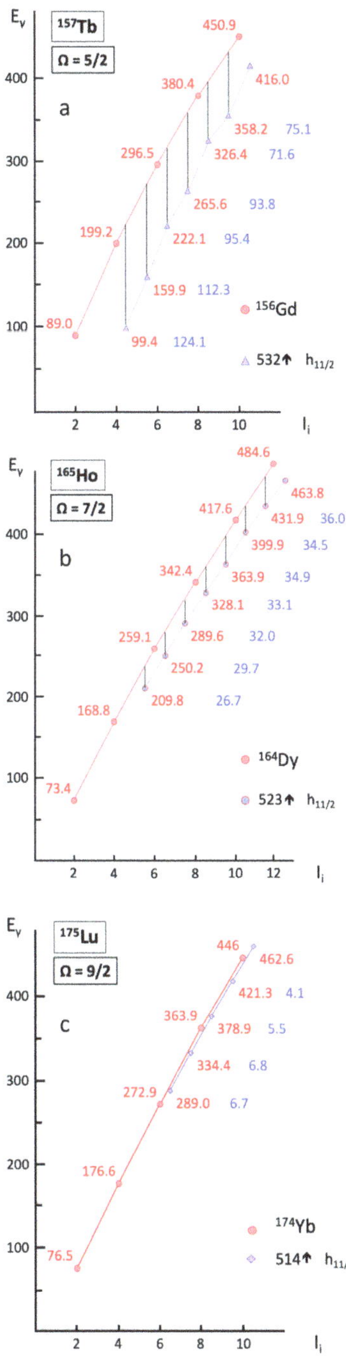

Figure 4.4. Alignment effects associated with Nilsson configurations stemming from the $1h_{11/2}$ spherical shell model configuration. The Nilsson bands (a) 5/2⁻[532] in ^{157}Tb, (b) 7/2⁻[523] in ^{165}Ho, (c) 9/2⁻[514] in ^{175}Lu are shown. Note the progressive decrease in alignment with increase in Ω. The data are taken from ENSDF.

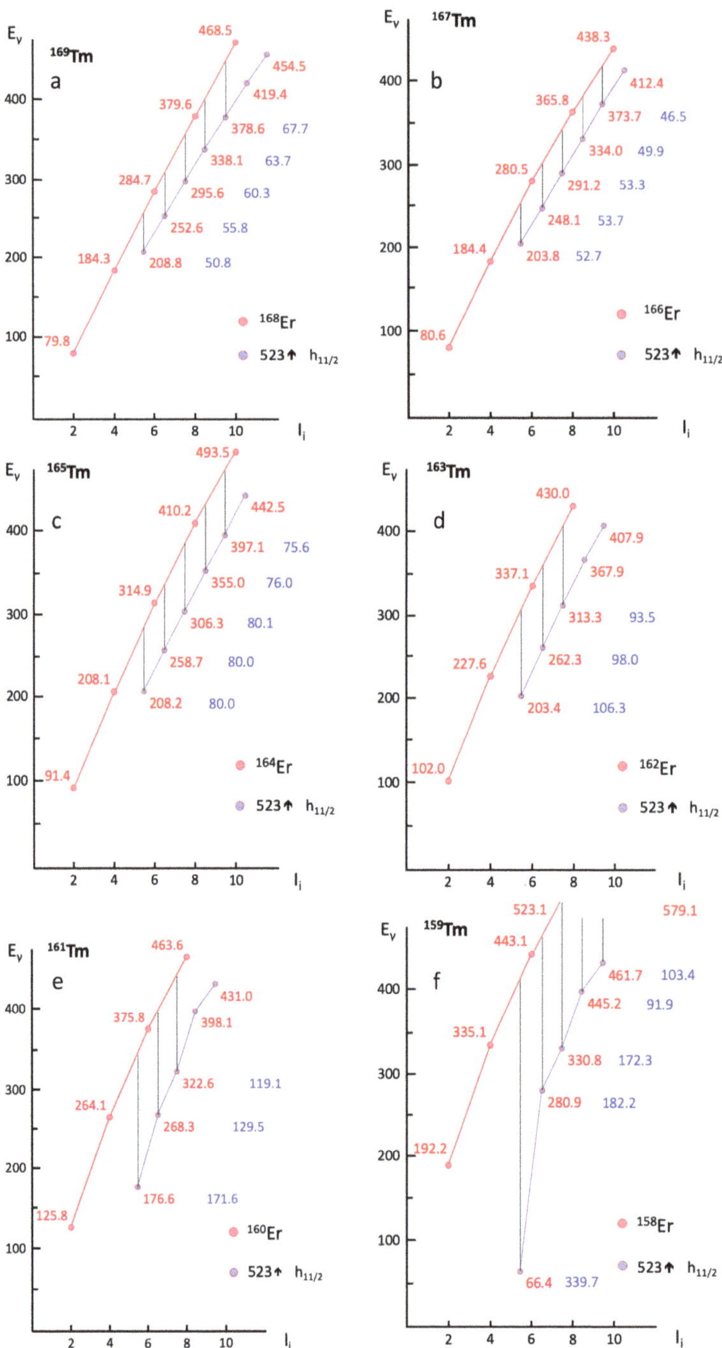

Figure 4.5. Alignment effects associated with the Nilsson configuration $7/2^-[523]$ stemming from the $1h_{11/2}$ spherical shell model configuration: the thulium isotopes, (a)–(f) masses 169, 167, $\cdots$, 159, are shown. Note the progressive increase in alignment with decreasing deformation. The data are taken from ENSDF.

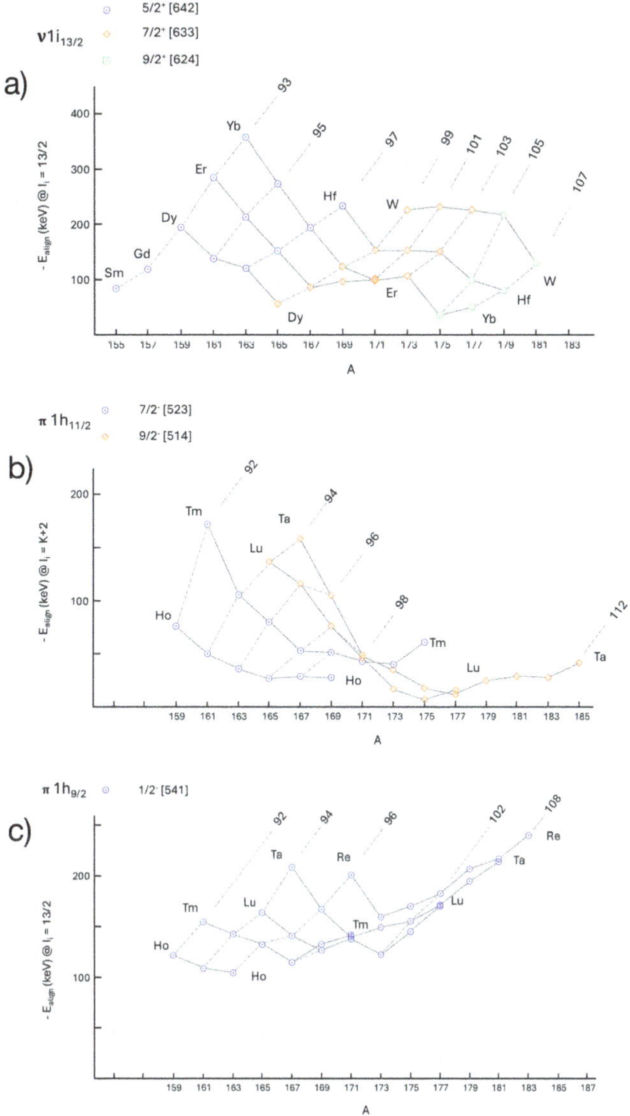

Figure 4.6. Plot of 'alignment' energies in keV at specified spins (indicated) for bands built on Nilsson configurations from (a) the 1i$_{13/2}$ configuration; (b) the 1h$_{11/2}$ configuration; (c) the 1h$_{9/2}$ configuration. See the text for further details.

K, $K + 2$, $K + 4$, ... and $K + 1$, $K + 3$, ... which are displaced in energy with respect to each other.

The most dramatic cases of signature splitting involve $K = 1/2$ bands. The energy patterns of $K = 1/2$ bands are given by the simple formula

$$E = A[I(I + 1) + a(-1)^{I+1/2}(I + 1/2)], \tag{4.10}$$

where A and a are parameters; a is called the 'decoupling' parameter. Figures 4.7(a) and (b) show examples of $K = 1/2$ bands. Figure 4.8 provides a systematics

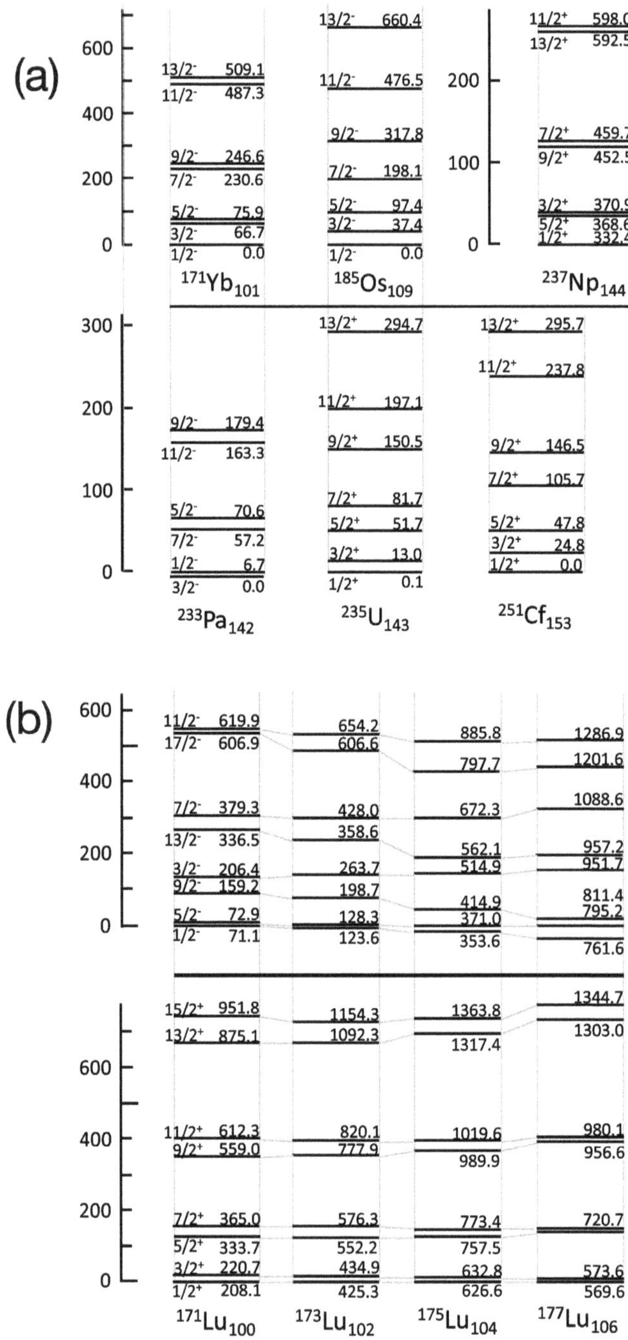

Figure 4.7. (a) Examples of energy patterns, cf equation (4.10), for $K = 1/2$ bands which possess various values of the decoupling parameter, a. The data are taken from ENSDF. (b) Examples of energy patterns, cf equation (4.10), for bands associated with the Nilsson configurations $1/2^-[541]$ (upper panel) and $1/2^+[411]$ (lower panel). The parameter a exhibits values characteristic of the associated Nilsson configuration: more details are given in the text. The data are taken from ENSDF.

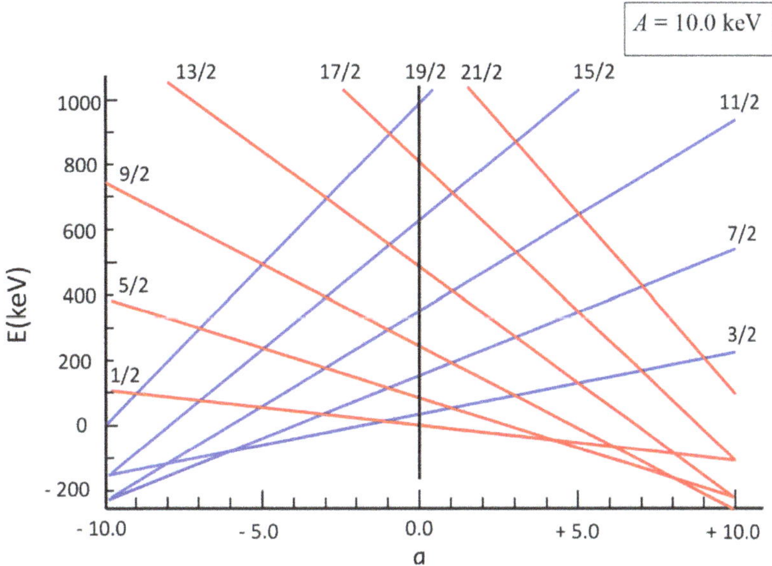

Figure 4.8. Depiction of spin sequences that are associated with equation (4.10) for different values of the parameter, a. The parameter A is set to 10.0 keV.

perspective of spin sequences in $K = 1/2$ bands for selected values of a. The values for the decoupling parameter are determined by the RPC term in association with specific Nilsson bands and are derived in appendix C.

Bands with $K = 1/2$ exhibit a simple and distinctive pattern for plots of gamma-ray transition energy versus spin, as shown in figure 4.9. Note that the RPC self-coupling causes the unfavoured spin sequence to lie above the even–even reference core. This results in alignment energies with negative values where, recall, the RPC term is a negative energy term. One observes alignment plots for some nuclei with small negative alignment energies: these can be interpreted as a consequence of band mixing. Band mixing is handled in appendix C.

Signature splitting and implied decoupling is observed in many rotational bands, not just $K = 1/2$ bands. A sequence of progressively increasing signature splitting patterns is illustrated in figure 4.10. These plots use the simple higher-order correction to the basic $I(I + 1)$ spin dependence of energies, viz.

$$E = AI(I + 1) + B[I(I + 1)]^2, \tag{4.11}$$

whence

$$\Delta E_{I, I-1} = E_I - E_{I-1} = 2AI + 4BI^3, \tag{4.12}$$

and

$$\frac{\Delta E_{I, I-1}}{2I} = A + B(2I^2), \tag{4.13}$$

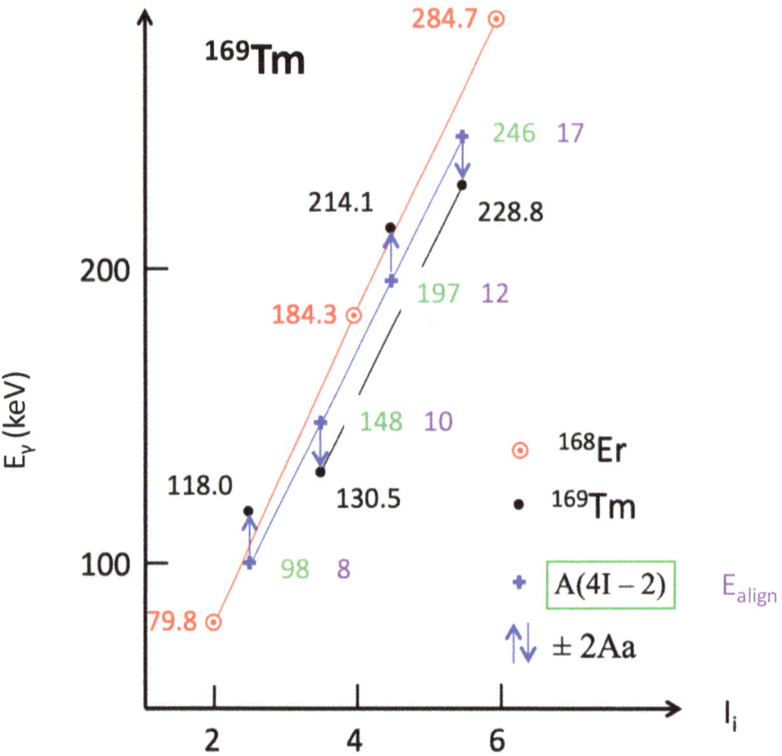

Figure 4.9. Effect of the **I · j** term on a $K = 1/2$ band, illustrated for the $1/2^+[411]$ band in ^{169}Tm. The result is a so-called 'signature splitting'. The quantum mechanical treatment of the **I · j** term leads to the equation (cf equation (4.10))— $E_I = E_0 + A[I(I + 1) + (-1)^{I+1/2}(I + 1/2)a]$. For $\Delta I = 2$ transitions, the first term gives $A(4I - 2)$ shown for $A \sim 12.3$ keV; the second gives $\pm 2Aa$, shown for $a \sim -0.77$; with $E_0 = -18$ keV. The ^{169}Tm data points are shown (energies in keV, black) in comparison with the $A - 1$ 'core', ^{168}Er (energies in keV, red). The data are taken from ENSDF. Further details are discussed in the text.

i.e. a linear relationship should result for a plot of $\Delta E_{I,\,I-1}/2I$ versus $2I^2$. The curvature in the plots in figures 4.9 reveals that there are deeper-lying factors in the rotational motion of nuclei. The staggering in the plots is connected to the RPC term and is addressed in appendix C; here we note that the staggering can be handled phenomenologically by a combination of the terms in equations (4.10) and (4.11).

4.4 Spectroscopic factors

Single-nucleon transfer reaction spectroscopy reveals distinctive patterns for population of rotational band members associated with specific Nilsson model configurations. This was encountered in chapter 3, figure 3.9 and table 3.1. Underlying these patterns are the C_{jl} coefficients expressed in equation (3.8). Figure 4.11 illustrates the essential process whereby a transferred nucleon added to or subtracted from a target to yield the nucleus of interest must conserve angular momentum via population of specific members of the rotational band built on a Nilsson configuration. The population of states in a rotational band built on a Nilsson configuration can be

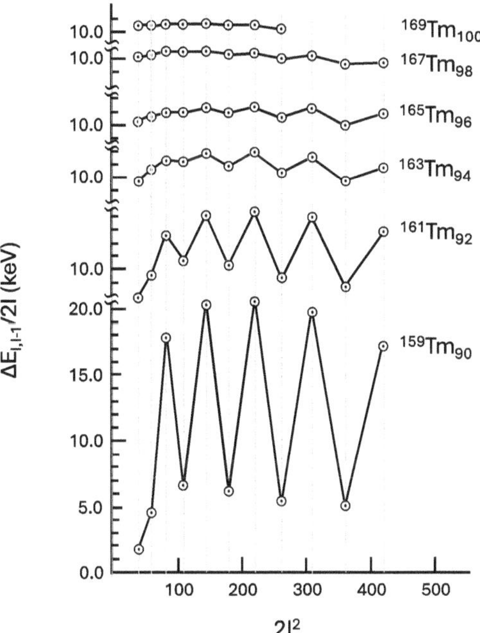

Figure 4.10. Plots of $\Delta E_{I,\,I-1}/2I$ (in keV) versus $2I^2$, cf equation (4.13), for the bands built on the Nilsson configuration 7/2⁻[523], stemming from the $1h_{11/2}$ spherical shell model configuration, in the odd-mass Tm isotopes. Note the increasing 'staggering' with the approach to $N = 90$. These are the same nuclei as depicted in figures 4.5(a)–(f). The data are taken from ENSDF. Details are discussed in the text.

viewed as an observable manifestation (a 'fingerprint') of the transferred j value of an added (or subtracted) nucleon. Figures 4.12(a) and (b) provide a view of the so-called fingerprinting of a given Nilsson band by one-nucleon transfer.

Single-nucleon transfer reactions can be conducted either with the target adjacent to the nucleus of interest or with the projectile adjacent to the nucleus of interest. The latter process is referred to as 'inverse kinematics'; conservation of linear momentum and angular momentum leads to different population of members of a Nilsson band, compared to 'normal kinematics'. Further, target material requires special handling if it is hydrogen, deuterium, tritium, or helium. These issues are beyond the scope of the present treatment; details can be found in reported studies using radioactive beams.

One-nucleon transfer reaction strengths connecting an even–even nucleus to an odd-mass nucleus of interest are quantified using the following equations. The differential cross-section for addition/removal of a nucleon from an even–even nucleus is given by

$$\frac{d\sigma}{d\Omega} = 2NS_j \left(\frac{d\sigma}{d\Omega}\right)_{DWBA},\qquad(4.14)$$

where N is a normalization factor, S_j is the spectroscopic factor and $(\frac{d\sigma}{d\Omega})_{DWBA}$ is the distorted-wave Born approximation, DWBA cross-section for the one-nucleon transfer process. The factor of two comes from the two-fold degeneracy of Nilsson states.

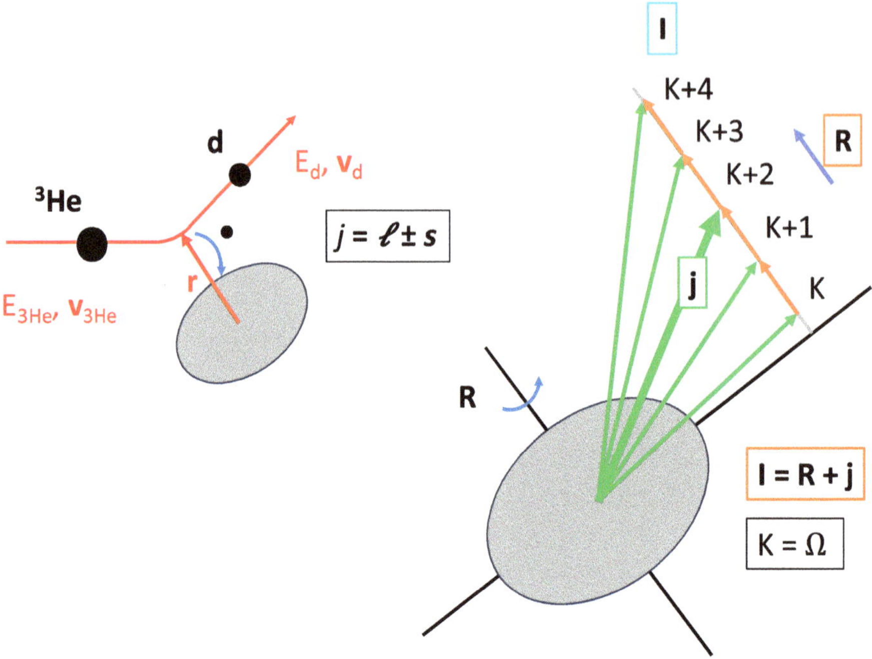

Figure 4.11. Depiction of proton addition in a (^{3}He,d) transfer reaction leading to an odd-proton deformed nucleus (even–even target nucleus with ground-state spin zero). The added proton has a definite total spin, j in the laboratory frame, which must be coupled to rotational angular momentum of the core, R to populate a rotational band member with a specified spin, I. In the body frame, j is not a good quantum number: a given Nilsson configuration is a linear combination of spherical basis components $|Njl\Omega\pi\rangle$, cf equation (3.8), the C_{jl} coefficients. It is through these components that the added nucleon populates various band members with differing probabilities, called a transfer reaction 'fingerprint'. The process is quantified by knowing the beam energy, which implies a beam-particle velocity, and the energies and angular distributions of ejectile-particles. The transferred angular momentum depends on the ^{3}He ion velocity, impact parameter r, and angle of the ejectile deuteron. A polarized beam can distinguish between $j = l + 1/2$ and $j = l - 1/2$ state population, i.e. $\Omega = \Lambda \pm \Sigma$ for the configuration $\Omega^\pi |Nn_z\Lambda\Sigma\rangle$. A video-based tutorial explaining the concept of single-nucleon transfer reactions in deformed nuclei is provided in figure 4.18.

The nuclear structure information is contained in the spectroscopic factor which is expressed as

$$S_j = \left(\sum_i a_i P_i C_{jl}^i \right)^2, \tag{4.15}$$

where a_i is the Coriolis mixing amplitude, P_i is the amplitude for state emptiness, $P_i = U_i$ (nucleon addition) or occupancy, $P_i = V_i$ (nucleon removal), $U_i^2 + V_i^2 = 1$ and C_{jl}^i is the amplitude in equation (3.8), specific to the Nilsson configuration i. Coriolis mixing amplitudes and state occupancy/emptiness amplitudes are discussed shortly. The DWBA is beyond the scope of the present discussion but can be viewed as calculable by standard transfer reaction theory. The normalization factor, N is needed to allow for such issues as kinematic factors for target–projectile

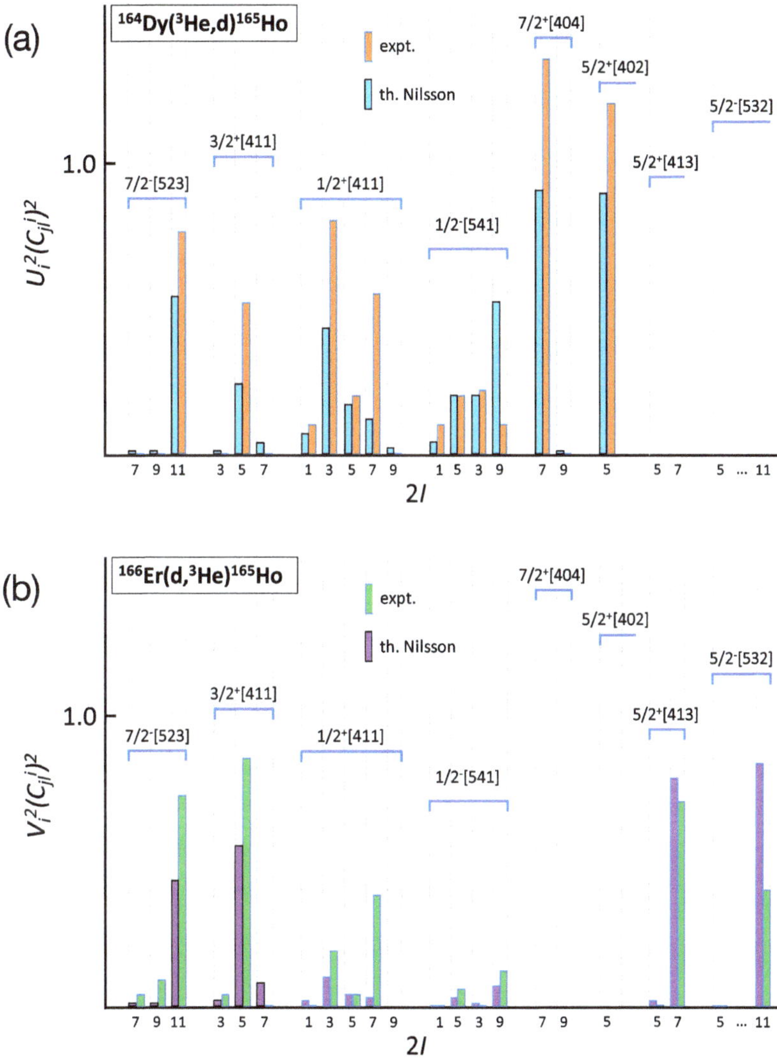

Figure 4.12. Depiction of one-nucleon transfer reaction 'fingerprints'. (a) $U_i^2 C_{jl}^{i\,2}$ values deduced for the ^{164}Dy(^{3}He,d)^{165}Ho one-proton addition reaction. (b) $V_i^2 C_{jl}^{i\,2}$ values deduced for the ^{166}Er(d,^{3}He)^{165}Ho one-proton removal reaction. Both depictions are compared to Nilsson model estimates based on C_{jl}^i coefficients combined with orbital emptiness, U_i^2 and occupancy, V_i^2, where $U_i^2 + V_i^2 = 1$. For a 'sharp' Fermi surface addition reactions can only populate states above the Fermi surface and removal reactions can only populate states below the Fermi surface. Pairing correlations result in diffuse occupancies with the result that both addition and removal reactions can populate states near the Fermi surface, but states far away in energy from the Fermi surface behave as pure particle or pure hole states. Thus, the Nilsson configurations 7/2$^+$[404] and 5/2$^+$[402] behave as (nearly pure) particle states and the Nilsson configurations 5/2$^+$[413] and 5/2$^-$[532] behave as (nearly pure) hole states. Band members are labelled by 2I values. The bands are ordered by increasing excitation energy from the left, cf figure 3.10. Note the dominant population of a given band is often the state with spin equal to the j values of the spherical shell model parent configuration, e.g. the 7/2$^-$[523] Nilsson configuration stems from the 1h$_{11/2}$ shell model configuration, hence the dominance of the $I = 11/2$ band member, cf figure 3.4. The data are taken from [1] for (d,^{3}He) and [2] for (^{3}He,d).

combinations. Note that the sum is over products of amplitudes, not products of squares of amplitudes: this can lead to large interference effects from Coriolis mixing.

With multiple spectroscopic probes of a given nucleus, transfer reaction spectra can usually be matched to excited states with precise energies and, where available, spin-parity information. This can be augmented using angular distributions of ejectile nuclei from transfer reactions or by comparing fingerprints for different projectile induced transfer reactions: such data are beyond the scope of the present discussion. However, above an excitation energy of 500–1000 keV incompleteness in spectroscopic detail eventually makes assignment of a Nilsson configuration to a specific excited state (nearly) impossible.

4.5 Exercises

The exercises explore details of Nilsson states in association with rotational degrees of freedom for (mostly) nuclei other than in the rare earth/lanthanide region. Some exploration of less-deformed nuclei is made, noting that an approach to less-deformed nuclei from the perspective of strongly deformed nuclei necessitates close attention to rotational degrees of freedom.

4.5.1 Rotational effects in regions of strong deformation

4-1 The largest alignment energies exhibited by Nilsson states are for those that stem from high-j shell model configurations. In the actinide region, the highest j shell model configuration is the $j = 15/2$ orbital, which will give rise to the Nilsson states $1/2^-[770]$, $3/2^-[761]$, $5/2^-[752]$, $7/2^-[743]$, $\cdots$. Using data in ENSDF, make alignment plots for nuclei in which rotational bands built on these Nilsson states are known to high spin.

4-2 Using data in ENSDF, identify all the $K = 1/2$ bands that have been established in the actinide region. (figure 4.7(a) shows a few examples.)

4-3 The Nilsson states with the lowest rotational alignments are those with Ω values closest to the j configuration from which they stem. Using data in ENSDF, identify the Nilsson bands in the actinide region most likely to have the smallest alignments and test this empirically inferred property of rotation alignment energies.

4-4 Using data in ENSDF, make alignments plots for the ground-state bands in: ^{133}Nd, ^{99}Y, ^{75}Kr.

4-5 For Nilsson bands stemming from the $j = 15/2$ shell model configuration in the actinide region, using data in ENSDF, make plots similar to those in figure 4.10.

4-6 For Nilsson bands stemming from the $j = 11/2$ shell model configuration in the odd-mass Ta isotopes, using data in ENSDF, make plots similar to those in figure 4.10. (Note that the plots in figure 4.10 are for $\Omega = 7/2$, whereas the lowest energy negative-parity bands in the Ta nuclei possess $\Omega = 9/2$.)

4.5.2 Rotational effects in regions of decreasing deformation

An open question at the research frontier is the 'fate' of Nilsson configurations in nuclei that are weakly deformed. Here, in the spirit of inspiring the reader to explore issues of active research, we present some clues to this major area of research interest.

In some mass regions, the onset of deformation is 'sudden'. This makes detailed structural interpretations difficult for nuclei with weak deformation. Here we present some views of nuclei in a region of gradual onset of deformation, and invite the reader to be sceptical of the views that we present, and to consider the research potential for in-depth, systematic study of nuclei with weak deformation.

There are three nuclear mass regions where the onset of deformation is gradual, centred on (A, Z, N) $\sim$(105, 45, 60), (125, 54, 71) and (191, 76, 115). In all these regions there is evidence for axially asymmetric quadrupole deformation. We limit the details presented for models that describe a particle coupled to a triaxial rotor, and we focus on a data-based view. Already some views are available in figures 3.35 and 3.36. We focus on the (A, Z, N) $\sim$(191, 76, 115) region.

A key clue to the breakdown of the view from strongly deformed odd-mass nuclei provided by the Nilsson model plus symmetric top model is the appearance of 'extra' states of low and medium spin. By 'extra' is meant states that are not found in the model space. Two simple candidate sources of additional model degrees of freedom naturally arise: vibrations from the deformed (Bohr–Mottelson) liquid drop collective model and rotations about the unfavoured axis from the asymmetric top model. We approach these model views from a data perspective. Figure 4.13 provides a view of 'inverted' spin sequences of negative-parity states in the odd-mass W and Os isotopes. Figure 4.14 provides a systematic view of inverted spin sequences of positive-parity states in the odd-mass W and Os isotopes (note that these include some of the isotopes in figure 4.13). There is high confidence with respect to the 'inverted' spin sequences, i.e. $1/2^- \to 3/2^- \to 9/2^-$ and $3/2^+ \to 7/2^+ \to 11/2^+$ because the states are populated by thermal neutron capture on the $A - 1$ nuclei which have ground-state spin zero (they are even–even nuclei) and neutron capture is a low-energy process (little or no angular momentum is added to the final nucleus) and the neutron has intrinsic spin 1/2. The band-head states are identifiable as the Nilsson configurations $9/2^-[505]$ and $11/2^+[615]$ cf exercise 3-2. Such inverted spin sequences, involving intrinsic independent-particle structures stemming from high-j shell model orbitals, are a widely observed but little recognized systematic feature of nuclei which are weakly deformed.

Two high-j-based 'hyper' band structures are elucidated in detail in ^{187}Ir and these are shown in figures 4.16 and 4.17. This view was introduced by Jurgen Meyer-ter-Vehn, but it is largely forgotten. Some basic features already appeared in figures 3.35 and 3.36. The extended band structure manifested in the Meyer-ter-Vehn view, which involves a single high-j particle or hole coupled to an axially asymmetric rotor, is epitomized in figures 4.16 and 4.17 (q.v. figure 3.36).

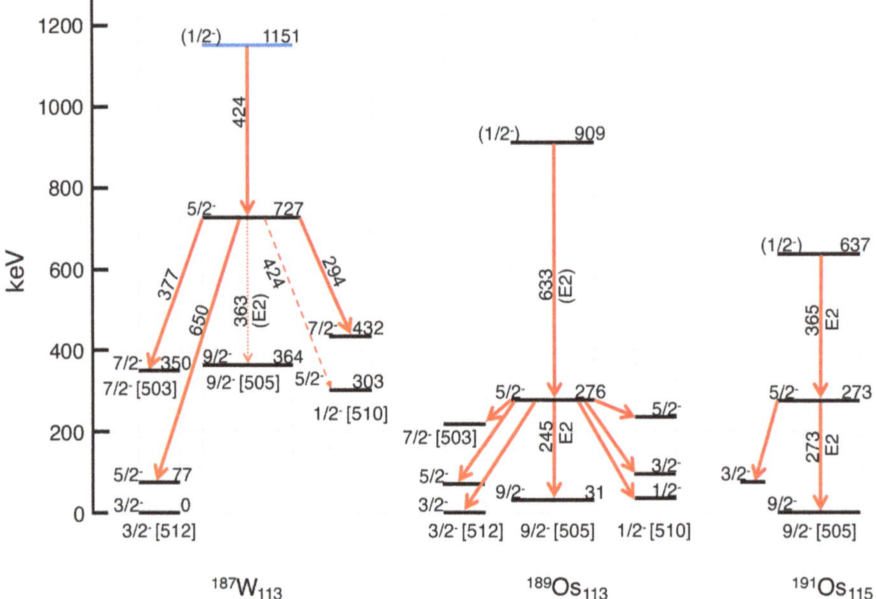

Figure 4.13. Inverted-spin sequences, $1/2^-\rightarrow5/2^-\rightarrow9/2^-$ associated with the Nilsson configuration $9/2^-[505]$, which stems from the shell model $h_{9/2}$ configuration, i.e. $j = 9/2$. Adapted from [3].

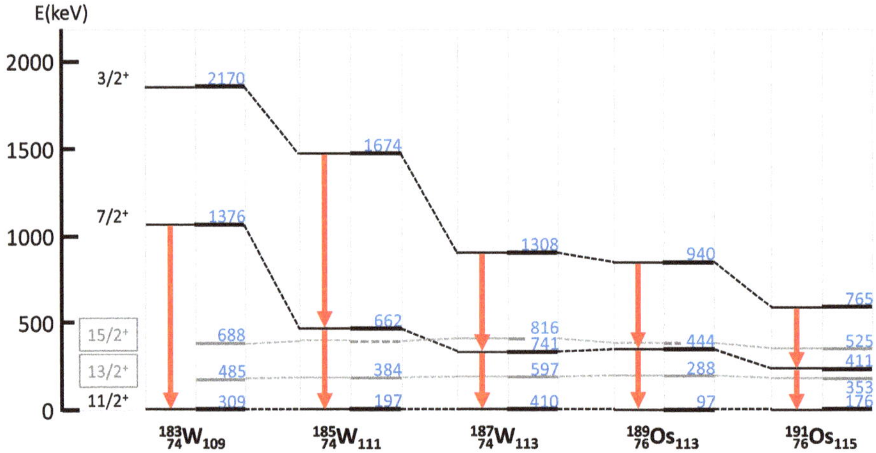

Figure 4.14. Inverted-spin sequences $3/2^+\rightarrow7/2^+\rightarrow11/2^+$ associated with the Nilsson configuration $11/2^+[615]$, which stems from the shell model $i_{13/2}$ configuration, i.e. $j = 13/2$. The figure is based on a more extended view of low-spin positive-parity states in these nuclei appearing in [3]. The $13/2^+$ and $15/2^+$ state energies are taken from ENSDF.

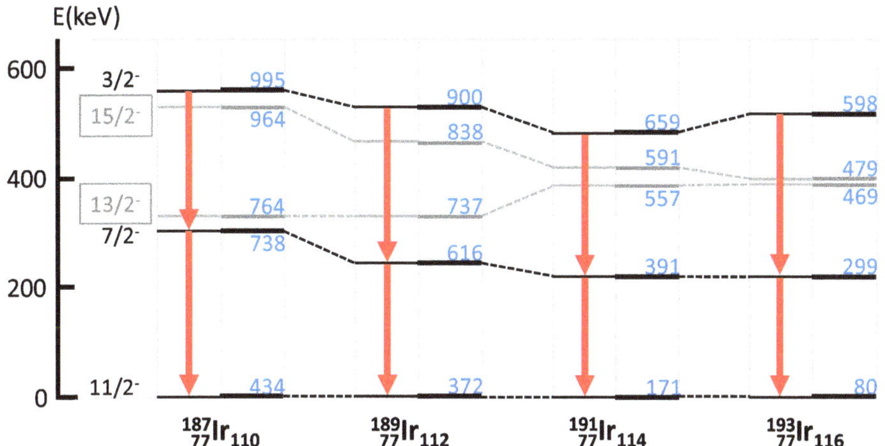

Figure 4.15. Inverted-spin sequences $3/2^-\rightarrow7/2^-\rightarrow11/2^-$ associated with the Nilsson configuration $11/2^-$[505], which stems from the shell model $h_{11/2}$ configuration, i.e. $j = 11/2$. The data are taken from ENSDF.

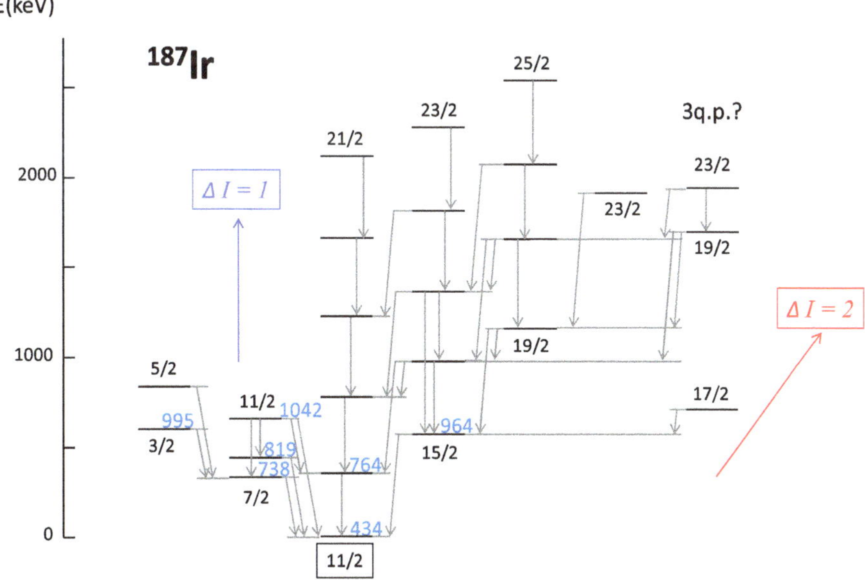

Figure 4.16. Organization of the negative-parity states in ^{187}Ir associated with $j = 11/2$ configuration into a 'hyper-band' pattern due to Meyer-ter-Vehn [4], see also [5]. The vertical columns or 'towers' of states have spin sequence $I, I + 1, I + 2, \cdots$ and the spins of the lowest and highest spins in each tower are indicated. The towers possess, progressively from left to right, tower-base spins 3/2, 7/2, 11/2, 15/2, 19/2. Energies of the lowest few states are given, cf figure 4.15. Vertical spin sequences increase with $\Delta I = 1$, diagonal spin sequences to the 'northeast' increase with $\Delta I = 2$. Possible broken-pair states are indicated as '3qp' (qp = quasiparticle). Details are discussed in the text. Reprinted from [6] CC-BY 4.0.

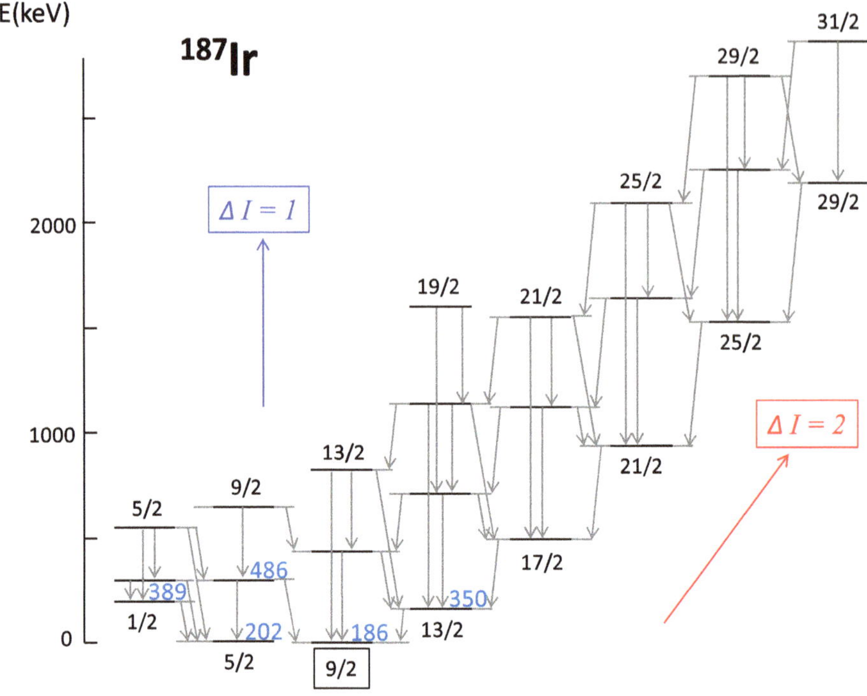

Figure 4.17. Organization of the unique-parity states in ^{187}Ir associated with $j = 9/2$ intruder configuration into a 'hyper-band' pattern due to Meyer-ter-Vehn [4], see also [5]. Other details are given in the caption to figure 4.16. Reprinted from [6] CC-BY 4.0.

We note a few general features that point towards the fate of Nilsson configurations. The 'tower' base spins, noted in the caption to figure 4.16, are ordered by $\cdots$ $j - 4$, $j - 2$, j, $j + 2$, $j + 4$, $\cdots$, where j is the spin of the independent particle stemming from the shell model. The degree(s) of freedom involved would match the so-called 'gamma' vibrations that give rise to second excited 2^+ states in even–even nuclei. In this mass region some of the lowest energy 2_2^+ states are found, e.g. in ^{192}Os $E(2_2^+)$ =489 keV. However, this feature of even–even nuclei, particularly the associated E2 matrix elements [7], is well described by the generalized triaxial rotor, GTRM presented in detail in chapter 2 in [8].

4-7 Using data in ENSDF, make figures similar to figures 4.16 and 4.17 for negative-parity states in ^{189}Ir.

4-8 Using data in ENSDF, make figures similar to figure 3.36 for ^{127}Xe.

4-9 Using data in ENSDF, make a systematic view of the energies of 2_2^+ states in the even–even W, Os and Pt isotopes with $N > 106$.

4-10 Gamma vibrational degrees of freedom would appear in an odd-mass nucleus with $K = \Omega \pm 2$ bands in association with each Nilsson state $\Omega^\pi[Nn_z\Lambda]$. These bands would have the same rotational energy parameter as the 'parent' Nilsson state. Further, two-phonon gamma bands with $K = 0, 4$ would couple to the parent Nilsson state, giving rise to $K = \Omega$,

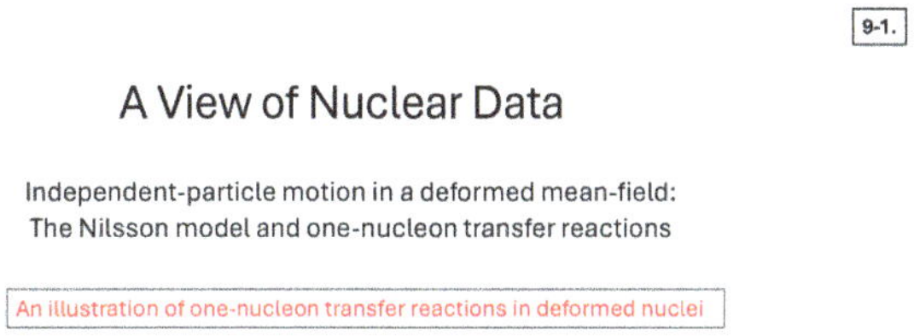

A View of Nuclear Data

Figure 4.18. Video tutorial: Single-nucleon transfer reactions with deformed nuclei. Video available at https://doi.org/10.1088/978-0-7503-5648-0.

$\Omega \pm 4$ bands. Apply this model view to the negative-parity states in ^{187}Ir with $j = 11/2$, i.e. the Nilsson configuration $11/2^-[505]$, using data in ENSDF.

4.6 Video-based tutorials

4-11 For those reading the e-book, a video-based tutorial on single-nucleon transfer reactions with deformed nuclei is found in figure 4.18.

References

[1] Lewis D A and Gray W S 1975 Proton-hole states observed in the ^{166}Er(d,^{3}He)^{165}Ho reaction at 34.5 MeV *Phys. Rev.* C **12** 79

[2] Lewis D A, Broad A S and Gray W S 1974 One-quasi particle states in $^{163,\,165}$Ho observed in the (^{3}He,d) and (α,t) reactions *Phys. Rev.* C **10** 2286

[3] Bondarenko V *et al* 1997 Nuclear levels in ^{187}W *Nucl. Phys.* A **619** 1–48

[4] Meyer ter Vehn J 1975 Collective model description of transitional odd-A nuclei: (ii). comparison with unique parity states of nuclei in the $A = 135$ and $A = 190$ mass regions *Nucl. Phys.* A **249** 141–65

[5] Meyer ter Vehn J 1975 Collective model description of transitional odd-A nuclei: (i) the triaxial-rotor-plus-particle model *Nucl. Phys.* A **249** 111–40

[6] Stuchbery A E and Wood J L 2022 To shell model, or not to shell model, that is the question *Physics* **4** 697–773

[7] Allmond J M, Zaballa R, Oros-Peusquens A M, Kulp W D and Wood J L 2008 Triaxial rotor model description of E2 properties in ^{186,188,190,192}Os *Phys. Rev.* C **78** 014302

[8] Jenkins D G and Wood J L 2023 *Nuclear Data: A Collective Motion View* (Bristol: IOP Publishing)

IOP Publishing

Nuclear Data
An independent-particle motion view
David Jenkins and John L Wood

Chapter 5

How are Nilsson states manifested in even-mass nuclei?

The use of Nilsson model configurations in deformed even-mass nuclei is introduced.
Concepts: broken-pair states, K isomers, backbending, cranking.
Learning outcomes: The Nilsson model is shown to provide a widely applicable framework of organization for broken-pair states in even-mass nuclei.

The use of Nilsson model configurations for classifying excitations in deformed even-mass nuclei is indispensable. For even–even nuclei, this appears above the pairing energy gap which can be as small as $\sim$1 MeV. For odd–odd nuclei, this is mandated already at the level of the ground state. By implication, such structures in even–even nuclei extend to their occurrence in the 'cores' of odd-mass nuclei.

5.1 Broken-pair states in even–even nuclei

The extent to which the concepts of intrinsic excitation and rotations can be used to organize excited-state data for a deformed even–even nucleus is illustrated in figures 5.1 and 5.2(a) and (b). Figure 5.1 shows 70 excited states in ^{168}Er, assigned uniquely to 14 rotational bands: this constitutes a powerful organizing principle for an enormous body of data. The task at hand is to provide an introductory guide to the interpretation of the various bands shown in figures 5.1 and 5.2(a) and (b), i.e. the intrinsic excitations.

Figure 5.3 shows the population of excited states in ^{168}Er by inelastic deuteron scattering. Besides the first few members of the ground-state band, only a few excited states are populated. In particular, the first intrinsic excitation, the head of a band with $K^\pi = 2^+$, is populated (also the 4^+ member of this band). Deformed nuclei widely exhibit first intrinsic excitations with $K^\pi = 2^+$. Second only to ground-state bands, these $K^\pi = 2^+$ bands are moderately strongly populated in Coulomb excitation: this is shown for ^{168}Er in figure 5.4. Remarkably, the only consensus regarding the nature of these $K^\pi = 2^+$ bands is that they involve a quadrupole

doi:10.1088/978-0-7503-5648-0ch5　　　　5-1　　　　

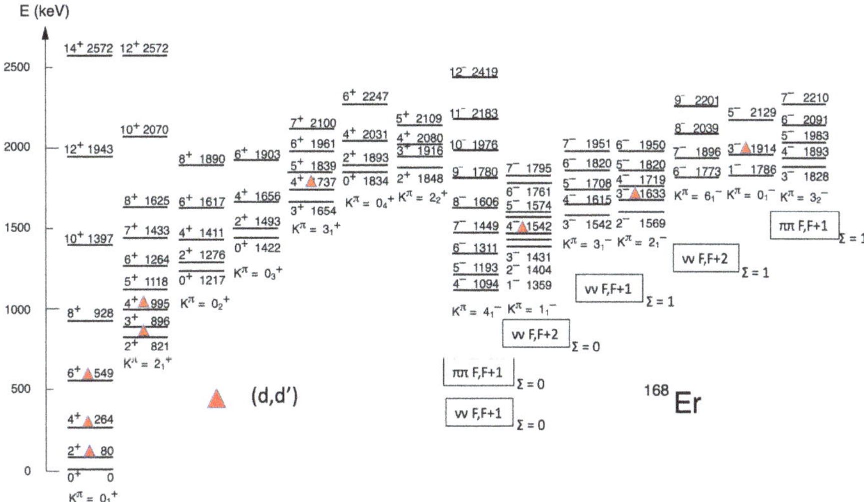

Figure 5.1. Excited states in the strongly deformed even–even nucleus, ^{168}Er organized into rotational bands. The lowest 70 states can be assigned to 14 bands, 7 with positive parity and 7 with negative parity. Reproduced from [1]. Copyright IOP Publishing Ltd. All rights reserved.

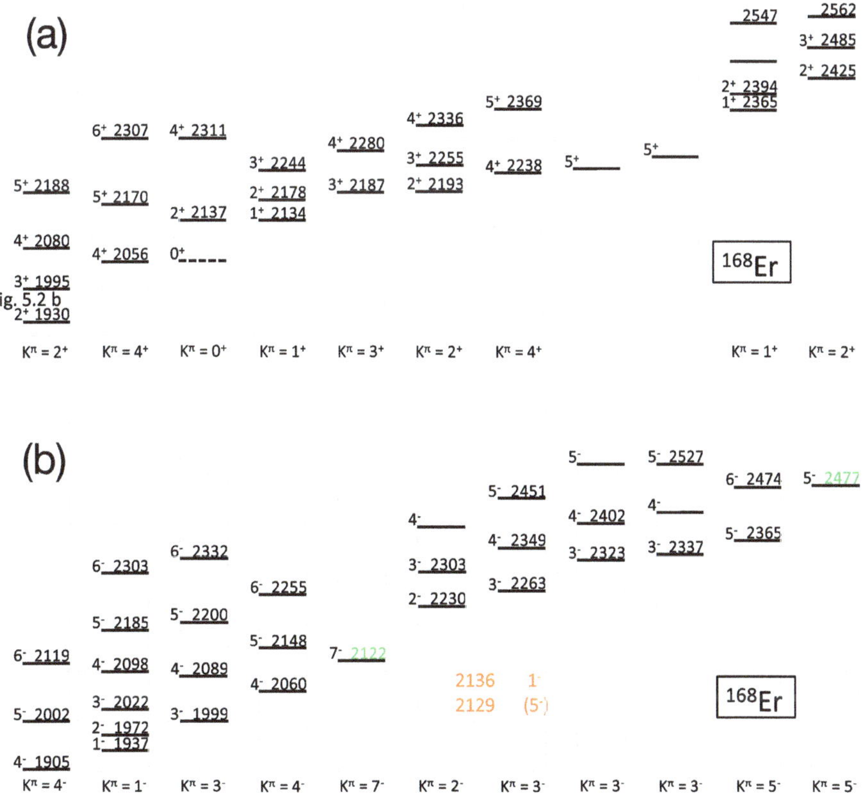

Figure 5.2. (a, b) Excited states in ^{168}Er above an excitation energy of 1900 keV organized into bands for (a) positive-parity states, (b) negative-parity states. The colour coding in the boxed items indicates: green— likely band-head states with K indicated; red—states lacking a K assignment.

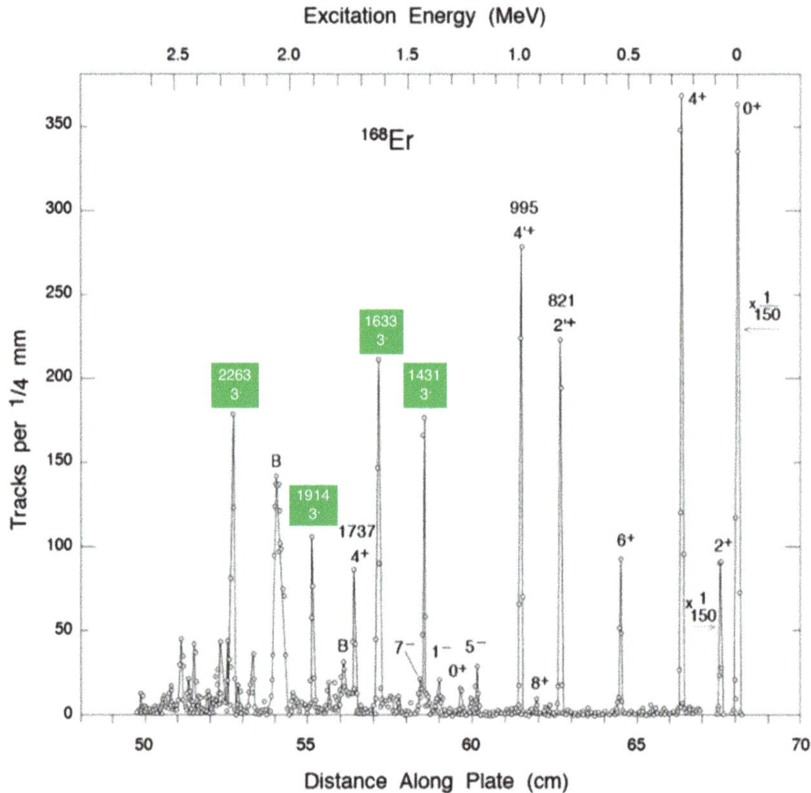

Figure 5.3. Population of excited states in ^{168}Er by inelastic scattering of deuterons. The data are for an incident energy of 12.098 MeV and a scattering angle of 125° with respect to the incident beam. Events are counts per 1/4 mm interval as a function of distance in cm along a photographic plate in the focal plane of a magnetic spectrometer, cf chapter 6, figure 6.13 in [1]. Deuteron lines are labelled by the spins and parities of the corresponding levels (cf figure 5.1). Lines marked with a B are background events resulting from target contaminants. Note the scale reduction factor for the ground state and first excited 2^+ state. The strongly populated states are indicated in figure 5.1 as red triangles. The spin-parities and energy of states are from ENSDF. The original data for this figure are derived from [2]. The figure is reproduced from [3], copyright 2010 World Scientific Publishing Company.

collective degree of freedom; but whether the intrinsic excitation is rotational or vibrational is unclear. For present purposes, we refer to such bands as 'collective' and use the popular name of 'gamma' bands when referring to them.

A further observation with respect to figure 5.3 is that inelastic deuteron scattering also populates states in ^{168}Er with spin-parity 3^-. This is a widely occurring feature observed via inelastic scattering in strongly deformed nuclei, with variable numbers of such negative-parity states observed. They are interpreted as resulting from intrinsic collectivity and are termed collective octupole bands. In ^{168}Er, inspection of the bands shown in figures 5.1 and 5.2(a) and (b) reveals that of the four 3^- states, three are members of bands with band heads $K^\pi(E_x) = 1^-(1359)$, $2^-(1569)$ and $0^-(1786)$, cf figure 5.1(a); the fourth 3^- state at 2263 is assigned to a

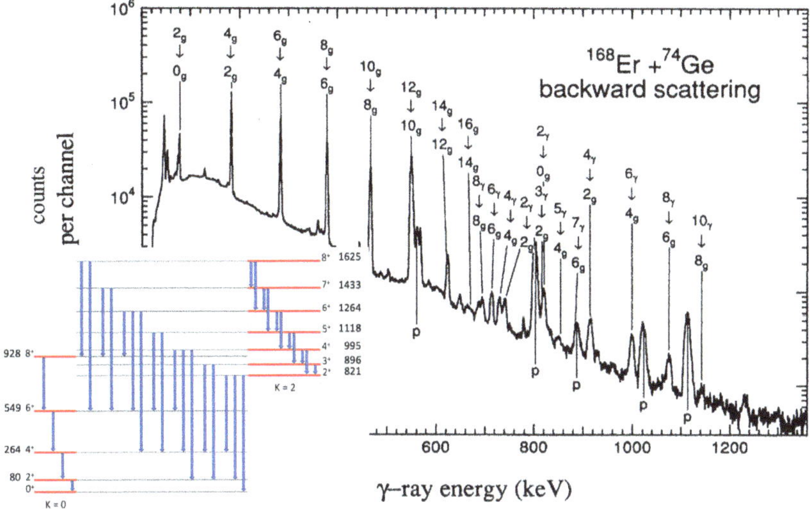

Figure 5.4. Coulomb excitation of states in [168]Er. Peaks labelled 'p' result from Coulomb excitation of [74]Ge. The inset can be used as a guide for identifying strong γ-ray lines in the spectrum. Other data are taken from ENSDF. Reprinted with permission from [4]. Copyright (1995) by the American Physical Society.

$K^\pi = 3^-$ band with this state as the band head, cf figure 5.2(b). However, we note that there are other bands ($K^\pi = 3^-$ at 1542 and 1828), many bands in figure 5.2(b) that possess 3^- states and these states are not strongly populated by inelastic scattering of deuterons.

A final observation with respect to figure 5.3 is that a state with spin-parity 4^+ at 1737 keV is observed to be significantly populated by inelastic deuteron scattering and it is a member of a $K^\pi = 3^+$ band (cf figure 5.1(a)). A much more subtle point is that the 4^+ member of the gamma band (at 995 keV) is more strongly populated than the 2^+ band member. Taken together, these observations support the presence of hexadecapole collectivity, manifested with $K^\pi = 2^+$ and 3^+.

Many other bands, involving what must be intrinsic excitation, are evident in figures 5.1 and 5.2(a) and (b). It is observed that selections of these bands are populated in one-proton and one-neutron transfer reactions. The spectrum of alpha particles produced in the ^{169}Tm(t, α)^{168}Er (one-proton removal) reaction is shown in figure 5.5. This reaction reveals the role of so-called 'broken-pair' excitations: they can be identified as specific Nilsson configurations involving the odd-mass target nucleus ground state. The population of more than one member of a rotational band by one-nucleon transfer is discussed in chapter 4. The assignment of Nilsson configurations to the various bands reveals a number of detailed features. First, the bands populated all involve the configuration $1/2^+[411]$ in $\pi\pi$ broken-pair states because this is the ground-state of the ^{169}Tm target nucleus. A more subtle feature is the evidence for a residual interaction between the members of each broken pair: the $\Sigma = 0$ coupling lies lower in energy than the $\Sigma = 1$ coupling. This applies to both proton broken-pair states and to neutron broken-pair states. While this is a detail

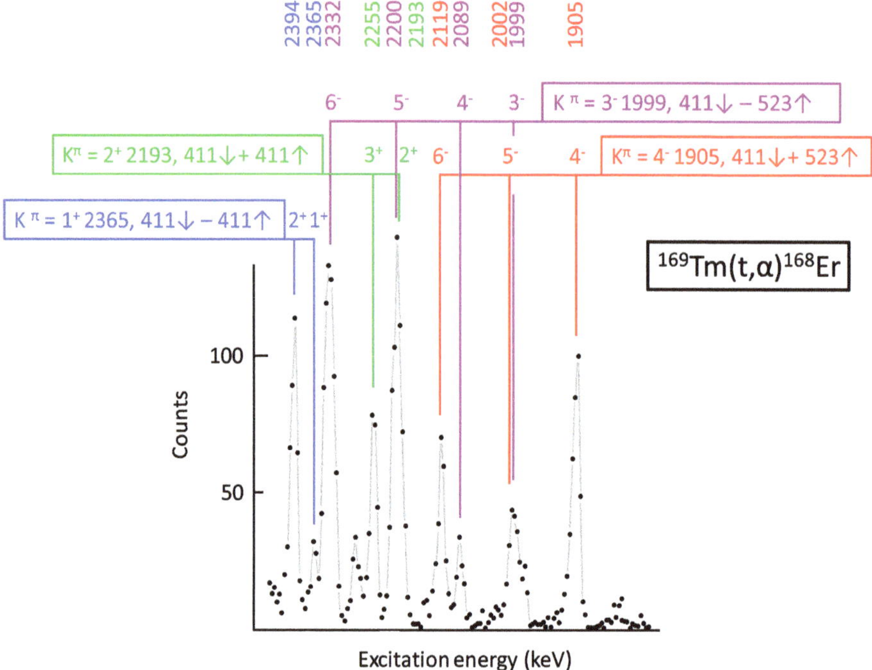

Figure 5.5. Spectrum of alpha particles (helions) following the proton removal reaction ^{169}Tm(t, α)^{168}Er. The Nilsson configuration of the target nucleus is $1/2^+$[411] which confines population of broken-pair states in ^{168}Er to the configurations $1/2^+$[411] $\pm$ $7/2^-$[523], $1/2^+$[411] $\pm$ $3/2^+$[411], $\cdots$. The spectrum is based on data appearing in [5].

that lies beyond the present level of interest, it is worthy of note that such properties of the nuclear many-body problem can be probed via detailed spectroscopic study. Further details of broken-pair states are explored in the exercises.

5.2 K isomers

One of the most remarkable manifestations of broken-pair states in deformed even–even nuclei is the occurrence of K isomers. The most dramatic example occurs in ^{178}Hf and is shown in figure 5.6. There are many examples of K isomers, and these extend to broken-pair states in odd-mass nuclei wherein we speak of three unpaired nucleons.

The characterizing feature of K isomers is their long half-lives. Selected examples are presented in table 5.2. A useful conceptual quantity is the hindrance factor for the decay of such structures. The Nilsson model provides a detailed interpretation of their structure, and the Nilsson configurations are often easily identified using Nilsson diagrams, as illustrated in figure 5.7. Through the underlying Nilsson configurations, isotopic and isotonic sequences can be tracked, and predictions can be made. Examples of such sequences are shown in figure 5.8.

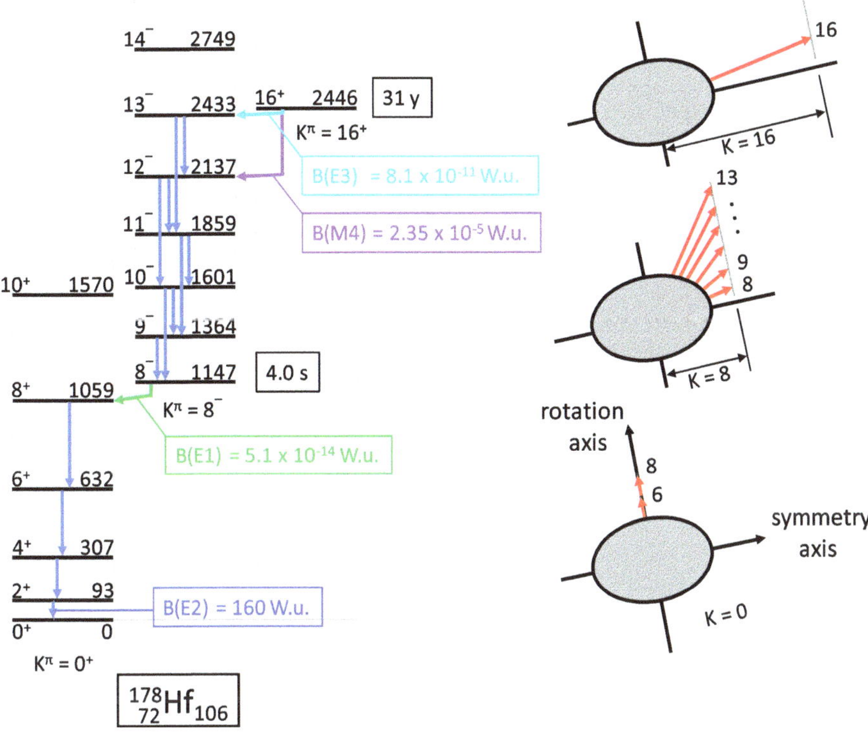

Figure 5.6. The $K = 8$ and $K = 16$ isomers in ^{178}Hf. The cartoons of the spin-angular momenta are drawn to dramatize the vectorial changes that underlie the highly hindered electromagnetic transitions between the different K bands. The strengths of selected transitions are given in Weisskopf units. The data are taken from ENSDF. A video-based tutorial on the topic of K isomers in ^{178}Hf is presented in figure 5.32.

An attractive feature of K isomers is their selective decay paths. This leads to the 'isolation' of subsets of (lower energy) excited states in nuclei. Particularly, sequences of lower K bands may be isolated, as is evident in figure 5.6. Interesting questions are: 'what is the highest K that has been observed?' and 'what is the greatest number of broken pairs observed and in which nucleus?' These questions are explored as exercises.

The reason for the long half-lives of such states is due to the large change in the K quantum number when the spin change is small. Electromagnetic transitions in nuclei are dominated by transitions with low multipolarity. This is understood at the level of the Weisskopf estimates for electromagnetic transitions, shown in table 5.1. For example, the decay of the 8^- state in ^{178}Hf at 1147 keV could, in principle, occur by any transition to a lower state, e.g. by a $\Delta I = 8, \Delta\pi =$ yes transition to the ground state. The Weisskopf estimate for such a transition gives a half-life of $\sim 10^8$ y. The 'compromise' is the decay path shown via an E1 transition, which would normally be fast. However, a spin change of 1 cannot effect a change in K quantum number of 8. One nucleus where such K hindrance plays a major role is ^{180}Ta where the ground state ($I^\pi = 1^+$) has a half-life of 8.154 h; and above a rotational excited

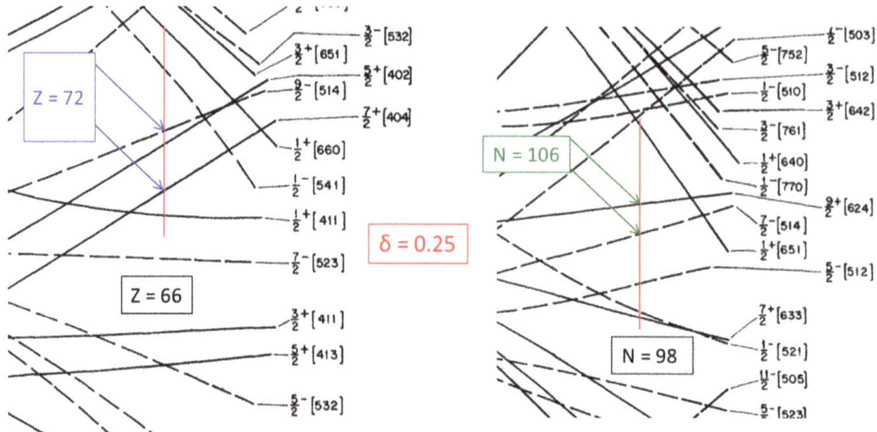

Figure 5.7. Nilsson diagrams for (a) protons and (b) neutrons in the rare earth region, pointing to the high-Ω configurations which can give rise to broken-pair proton and neutron structures which result in high-K bands for $Z \sim 72$ and $N \sim 106$.

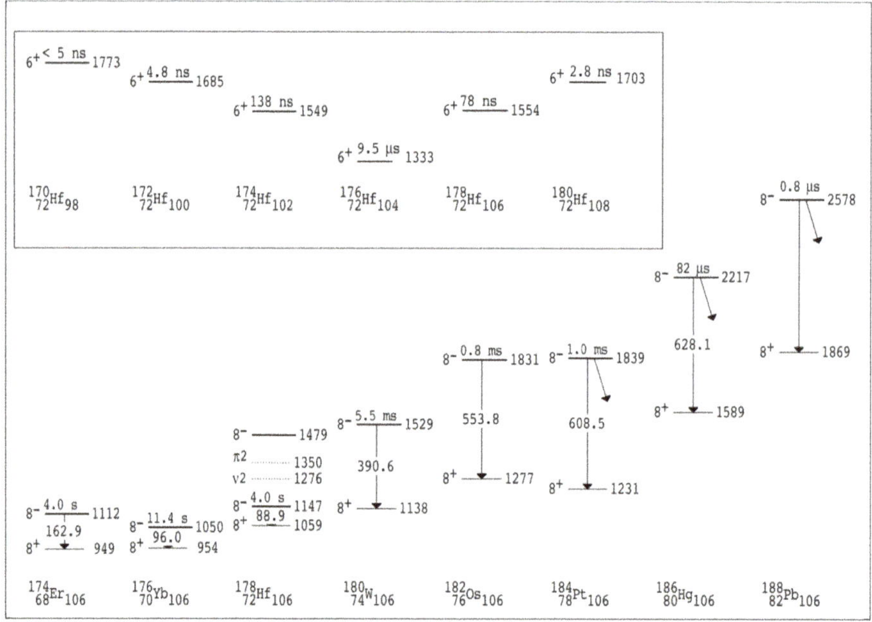

Figure 5.8. K isomer energy systematics: as a function of changing neutron number for $K^{\pi} = 6^{+}$ bands in the even-mass Hf isotopes; as a function of changing proton number for $K^{\pi} = 8^{-}$ bands in the even-mass $N = 106$ isotones. Reprinted from [6], copyright (2015) with the permission of Elsevier.

state ($I^{\pi} = 2^{+}$, 39.54 keV) is a state at 77.2 keV with a half-life of $>2.9 \times 10^{18}$ y [7]. This state has $I^{\pi} = 9^{-}$ and $K = 9$. It could in principle decay by an E7 electromagnetic transition of 37.7 keV. The drama is that the isotope ^{180}Ta should not be found in Nature; but it is, due entirely to K isomerism.

Nuclear Data

Table 5.1. Selected K isomer decays

Isotope	Mλ or Eλ	E_i (keV)	E_γ (keV)	$I_i, K_i^{\pi_i} \to I_f, K_f^{pi_f}$	ΔK	Trans. rate (W.u.)
^{244}Cm	M1	1040	744	$6, 6^+ \to 6, 0^+$	6	5.6×10^{-13}
^{174}Yb		1518	992	$6, 6^+ \to 6, 0^+$	6	7.0×10^{-12}
^{178}Hf		2574	141	$14, 14^- \to 13, 8^-$	6	1.8×10^{-8}
^{174}Yb	E2	1518	992	$6, 6^+ \to 6, 0^+$	6	8.5×10^{-9}
^{244}Cm		1040	744	$6, 6^+ \to 4, 0^+$	6	2.4×10^{-10}
^{174}Yb		1518	1265	$6, 6^+ \to 4, 0^+$	6	8.7×10^{-11}
^{178}Hf	E1	1147	89	$8, 8^- \to 8, 0^+$	8	5.1×10^{-14}
^{180}Hf		1141	58	$8, 8^- \to 8, 0^+$	8	3.5×10^{-17}
^{182}Hf		1173	51	$8, 8^- \to 8, 0^+$	8	1.3×10^{-16}
^{184}Hf		1272	73	$8, 8^- \to 8, 0^+$	8	3.5×10^{-15}
^{180}Hf	M2	1141	501	$8, 8^- \to 6, 0^+$	8	9.8×10^{-15}
^{182}Hf		1173	507	$8, 8^- \to 6, 0^+$	8	4.0×10^{-12}
^{178}Hf	E3	2447	13	$16, 16^+ \to 13, 8^-$	8	8.1×10^{-11}
^{180}Hf		1141	501	$8, 8^- \to 6, 0^+$	8	7.1×10^{-10}
^{178}Hf	M4	2447	310	$16, 16^+ \to 12, 8^-$	8	2.4×10^{-5}
^{178}Hf	E5	2447	587	$16, 16^+ \to 11, 8^-$	8	1.2×10^{-4}

5.3 Backbending

A more subtle manifestation of broken-pair configurations in even–even nuclei is the phenomenon described by the term 'backbending'. Backbending can be due to a more-deformed[1] band 'crossing' the ground-state band, with the occurrence of some band mixing at and near the crossing point. This is illustrated in figures 5.9–5.11. Figure 5.9 shows the two bands depicted as their excitation energies: the crossing is manifest. Figure 5.10 shows an 'enhanced' view of excitation energies by subtracting a rotational 'reference' energy of $9.000I(I + 1)$ keV: the minor departure from two smooth trajectories near the implied crossing point (visible for $I = 14$) suggests mixing and repulsion. More dramatic examples of mixing and repulsion are shown below. Figure 5.11 shows a view of the gamma-ray transition energies of the yrast spin sequence (often only the yrast[2] states = highest-spin-at-lowest energy states) were observed in early work, i.e. the crossing bands were not characterized, and the down-turn in gamma-ray transition energy is the origin of the term 'backbending'. The term 'backbending' no longer has any relevant operational use.

The phenomenon, which from the perspective presented here is rather mundane, was sensational when it was first observed. It was thought to be due to the 'collapse' of pairing correlations in nuclei, the so-called 'Mottelson–Valatin' effect. As a result, an enormous experimental effort was dedicated to the study of backbending and so the database is extensive. Such an effort led to the development of large arrays of

---

[1] But also see the conclusion in section 5.6.5.
[2] The term 'yrast' is taken from the Swedish language and means 'dizziest'.

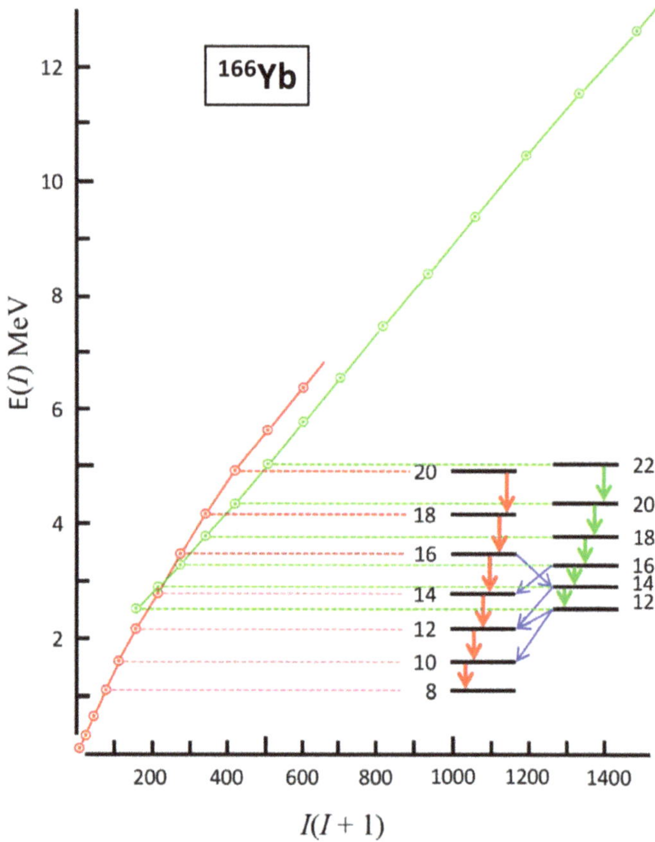

Figure 5.9. Depiction of band crossing involving the ground-state band in ^{166}Yb. The structure of the crossing band is uncertain: it may extend to lower spins. It is discussed further in the text and is portrayed using other details in figures 5.10, 5.11 and 5.13(a)–(d). The data are taken from ENSDF.

detectors, needed to elucidate the high-multiplicity of the gamma-ray cascades involved in the observation of high-spin states. We present here a few more details of backbending, which it appears needs much further study of the structures involved.

The modern view of backbending is that it is due to the crossing of the ground-state band by a broken-pair band. As such, each case can be interpreted using the Nilsson model. However, rotational effects on Nilsson configurations become of major significance in such interpretations. The popular view is that to understand such structures, rotational effects progressively become dominant with increase in rotational frequency of the nucleus. This has led to so-called 'cranking' models. Cranking models involve the constraint on the independent particles in a deformed mean field being forced to rotate through a 'cranking' of the mean-field potential about a fixed axis with variable rotational frequency. Here, we take some preliminary steps into this model view.

At the outset it is important to recognize two distinct *model views* of the dynamics of nuclear behaviour when high spins are involved; the first view is from the $\mathbf{I} \cdot \mathbf{j}$ term

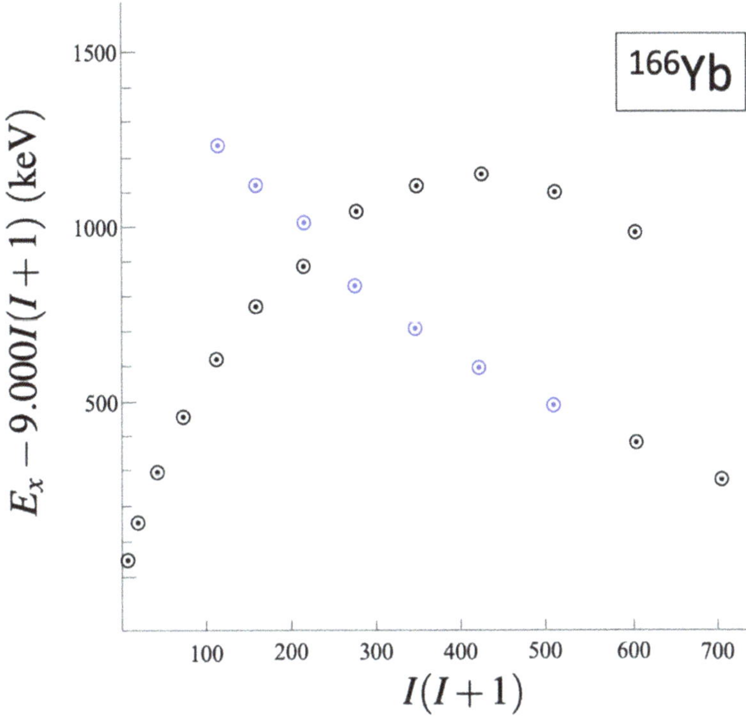

Figure 5.10. Depiction of band crossing involving the ground-state band in ^{166}Yb. The energies are shown with the subtraction of the quantity $9.000I(I + 1)$ keV: this is to permit an expanded energy scale so that finer details of the ground-state band and the crossing band are revealed, notably that at the crossing point there are small deviations from a smooth interpolation indicative of two-state mixing and 'repulsion' at spin 14. Further details are discussed in the text. The data are taken from ENSDF.

arising from the particle-rotor model, the second is of an $\boldsymbol{\omega} \cdot \mathbf{j}$ term, where ω is the imposed cranking model view of nuclear rotation in terms of a rotational frequency about a fixed axis. With adherence to our chosen data-based view of nuclear structure, we first look at data to obtain guidance on the model views involved. Figures 5.12(a)–(c) show 'alignment' energies for Nilsson configurations most likely to dominate high-spin structures at low energy in deformed rare earth nuclei. The appearance of broken-pair states that involve these structures must be considered at the outset of any interpretation of the crossing bands that underly the observation of backbending. Evidently, a broken pair in high-j, low-Ω Nilsson configurations will need close attention. A way to inspect this in some depth is by comparing backbending in a given even–even nucleus with that in neighbouring odd-mass nuclei wherein the unpaired nucleon can 'block' the Nilsson orbital being considered as the cause of the backbending. A brief look at the blocking perspective is given in the following.

Figures 5.13(a)–(d) illustrate an odd-particle blocking view of backbending in the nucleus ^{166}Yb. The idea is that the backbend is 'delayed', i.e. appears at a higher spin, if an unpaired nucleon is in an orbital that is behind the broken-pair alignment.

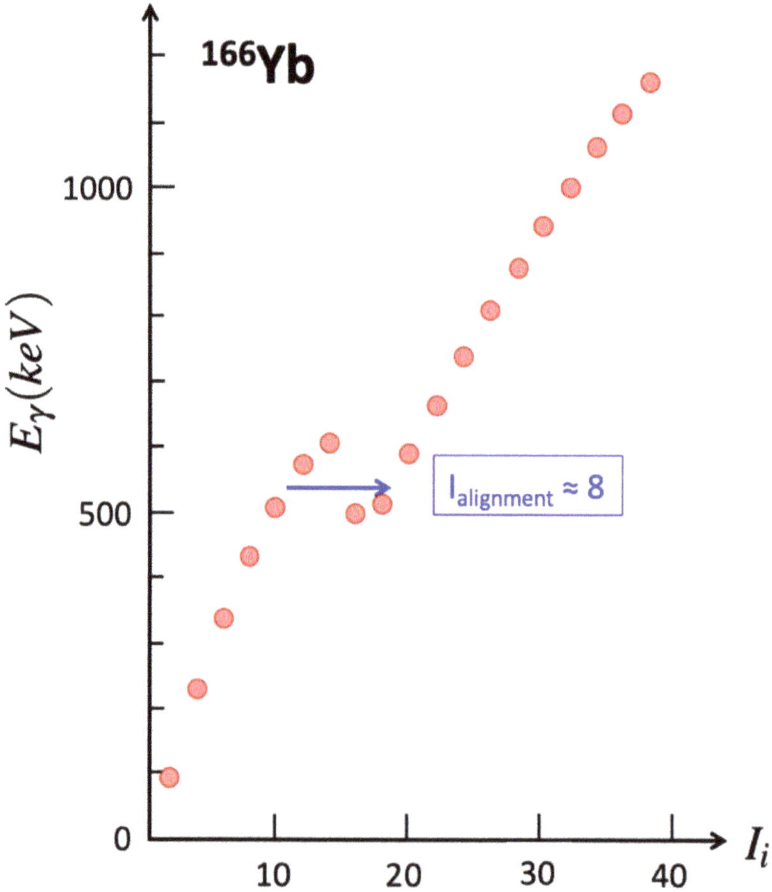

Figure 5.11. Depiction of the yrast band in ^{166}Yb using an E_γ versus I_{initial} plot. This suggests an alignment spin of about $8\hbar$ for the crossing band. Further details are discussed in the text. The data are taken from ENSDF.

The conventional view is that backbending results from a broken pair occupying Nilsson configurations that stem from the $i_{13/2}$ orbital. Figure 5.13(a) is consistent with this view in that the backbending does indeed occur at a higher spin for the $5/2^+[542]$ Nilsson configuration. But this also occurs for the $1/2^-[541]$ Nilsson configuration as shown in figure 5.13(b); however, backbending for the $9/2^-[514]$ and $7/2^+[404]$ Nilsson configurations, cf figures 5.13(c) and (d) closely parallel that in the ^{166}Yb core. The alignment spins are large for the $5/2^+[542]$ and $1/2^-[541]$ Nilsson configurations, and the backbend behaviour of the $1/2^-[541]$ configuration appears to be as important a candidate for alignment as the $5/2^+[542]$ configuration. Further, beyond the backbend, the $5/2^+[542]$ aligned band appears identical to the ^{166}Yb aligned band, so the alignment mechanism cannot be a simple involvement of the $i_{13/2}$ orbital. Evidently, further research into backbending is needed. There are other interesting observations regarding backbending which follow.

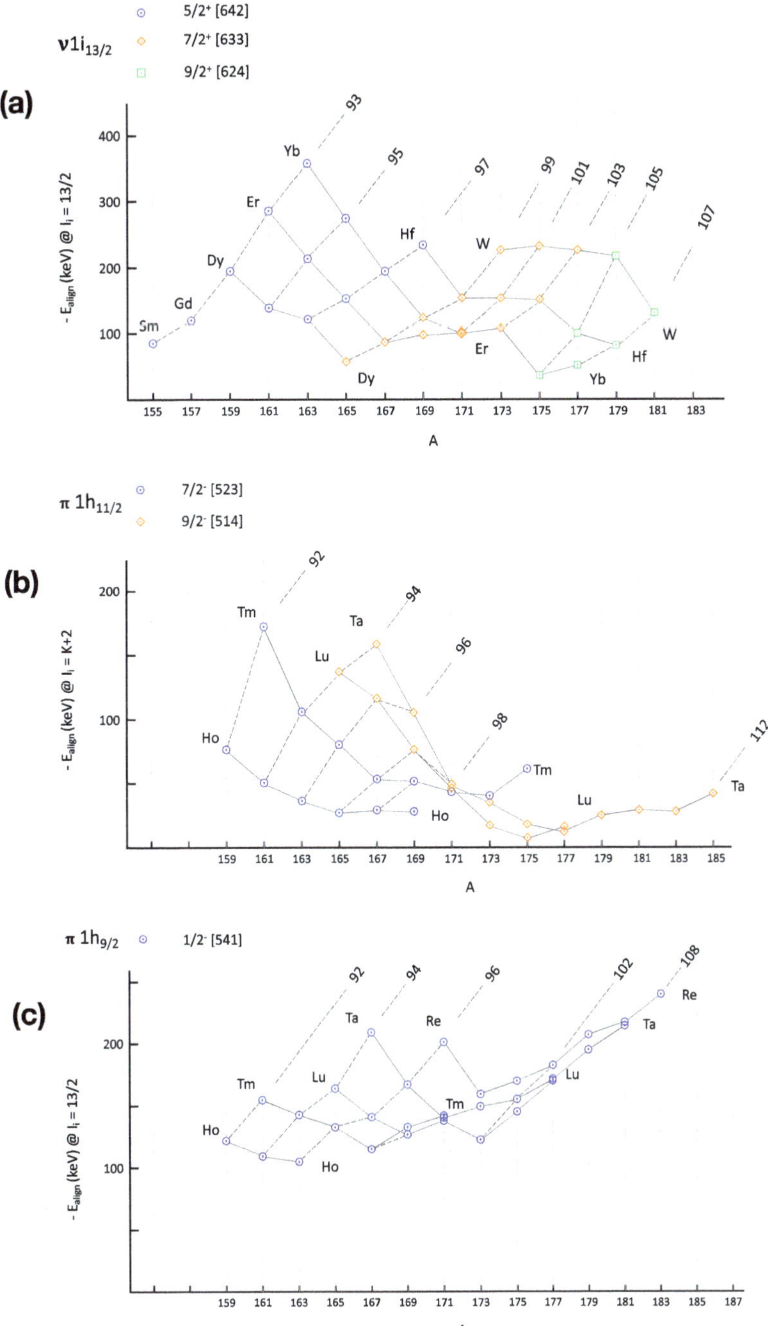

Figure 5.12. Plot of 'alignment' energies in keV at specified spins (indicated) for bands built on Nilsson configurations from: (a) the $1i_{13/2}$ configuration; (b) the $1h_{11/2}$ configuration; (c) the $1h_{9/2}$ configuration. See the text for further details.

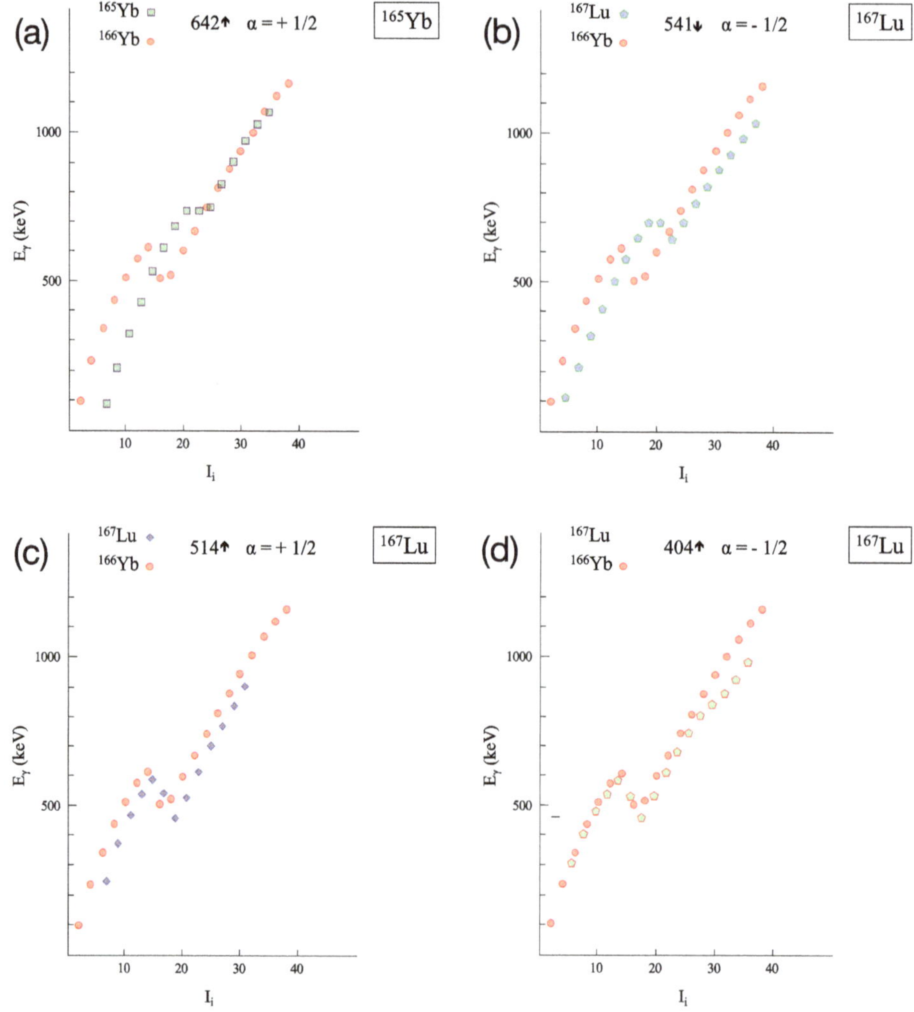

Figure 5.13. Depiction of selected bands in odd-mass nuclei adjacent to ^{166}Yb using E_γ versus I_{initial} plots, with inclusion of the ^{166}Yb yrast band. The 'delayed' backbends for the [642] and [541] configurations suggest that both configurations may be involved in the structure of the crossing band. Further details are discussed in the text. The data are taken from ENSDF.

Another view of backbending can be taken based on the 'sharpness' of the backbend, which translates into the interaction strength between the band structures and the resulting mixing between them at the crossing point. Figures 5.14(a) and (b) illustrate such manifestations of band mixing. In ^{160}Dy the mixing is strong as revealed by the virtual crossing point not being approached in the plot in figure 5.14(a). In ^{162}Dy, on the other hand, the mixing must be very weak as revealed by the proximity of the two data points for $I = 18$. Some discussion of this exists, see, e.g. [8], but more exploration appears to be warranted when such large differences in interaction strength occur in adjacent nuclei.

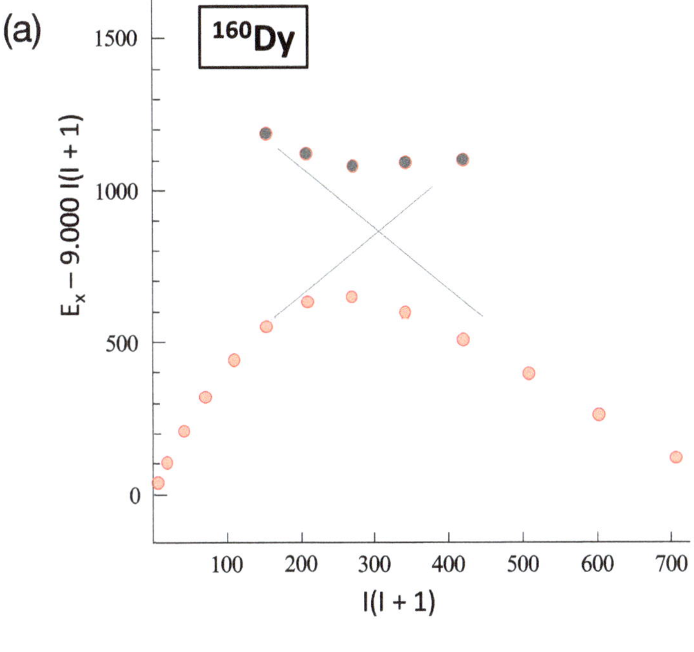

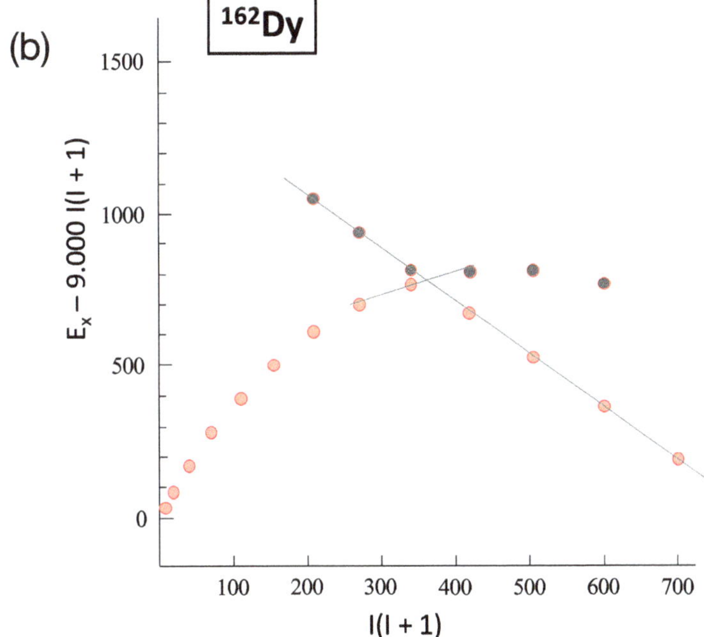

Figure 5.14. Depiction of band crossing involving the ground-state bands in (a) ^{160}Dy and (b) ^{162}Dy. The energies are shown with the subtraction of the quantity $9.000I(I + 1)$ keV. Note the implied very different band interaction energies at their crossing points. Also note the near-linear trend for the crossing band versus $I(I + 1)$ in ^{162}Dy. Further details are discussed in the text. The data are taken from ENSDF.

The view presented in figures 5.14(a) and (b) also reveals some details of the crossing bands; in particular that some crossing bands appear to be rigid, i.e. there is no curvature to their excitation energy profiles. One finds that the literature on backbending is dominated by an 'imposed' view using the cranking model, which presents plots that involve a derived quantity, a rotational 'frequency'. Possibly this view has obscured some of the details such as shown in figures 5.13(a)–(d) and 5.14(a) and (b). The appearance of a rigid crossing band, such as in ^{162}Dy, cf figure 5.14(b), is not expected. It appears that much more work is needed to understand these structural features: some further exploration is given in the exercises. There appears to be interesting scope for research regarding this aspect of nuclear structure.

In conclusion, it appears that some issues relating to the above-posed model views need clarification. There are energy contributions from the $\mathbf{I} \cdot \mathbf{j}$ model term that need further study at lower spins. At very high spins, the $\boldsymbol{\omega} \cdot \mathbf{j}$ (cranking) model term can be expected to be approached asymptotically as a useful view. A unification of the two model views is needed. The database appears to lack critical information for such exploration. For example, where are the odd-spin members of the crossing bands?

5.4 Weak collectivity in deformed even–even nuclei

Excitation of deformed even–even nuclei by inelastic scattering of light ions, as already noted, reveals weakly collective modes in the energy range 0.8–2.5 MeV. There are universally occurring $K = 2$ bands, which are termed gamma bands, that exhibit enhanced E2 transition strength to the ground-state band. There are states with spin-parity 3^- and 4^+ that exhibit enhanced population via inelastic scattering. These excitation modes are discussed in some detail in [1]. An unanswered question is the distribution of K values deduced for the 3^- states: notably that there is a dominance particularly of $K = 2$ over $K = 0$ bands. We identify spectroscopic information that may be used to explore the structure of 3^- states in deformed nuclei and invite the reader to pursue this in the exercises.

Figures 5.15(a)–(d) show the identified K values in the rare earth region for 3^- states. Figures 5.16(a) and (b) shows an interpretation of the observed K values in terms of Nilsson configurations. The focus here is on estimating the excitation energies of these candidate Nilsson configurations using experimentally observed Nilsson state energies in odd-mass nuclei. Figures 5.1(a) and (b), 5.3 and 5.5 provide a basic view of the challenge.

5.5 Odd–odd nuclei

An example of an odd–odd nucleus, ^{166}Ho is shown in figure 5.17. This nucleus is a neighbour of the nuclei mentioned above. Rotational bands can be identified in a manner such as presented for broken-pair states in ^{168}Er. Again, a weak residual interaction is evident within a broken pair, but that results in the $\Sigma = 1$ coupling lying lower in energy than the $\Sigma = 0$ coupling.

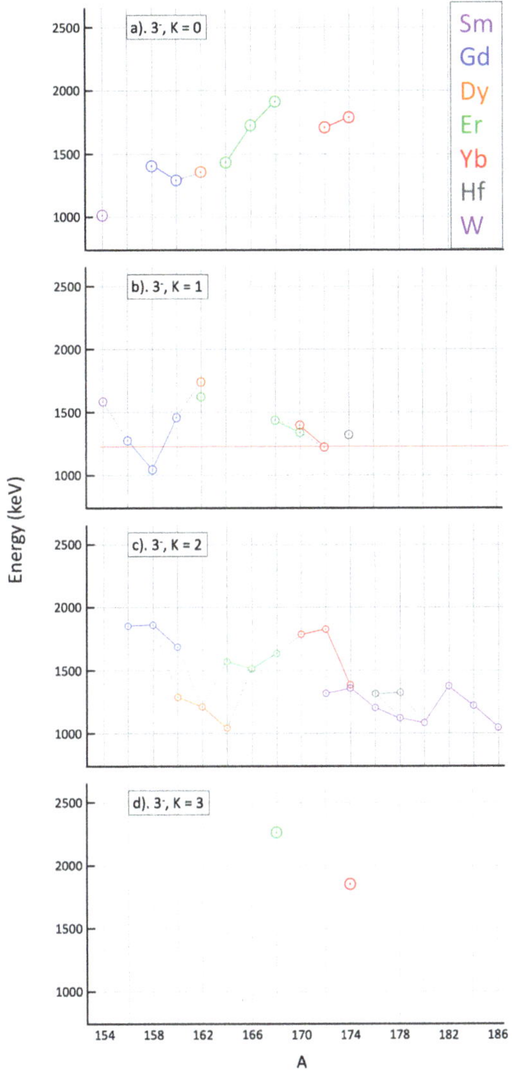

Figure 5.15. Systematics of excitation energies of collective 3^- states for nuclei in the rare earth region according to K quantum number assignments: (a) $K = 0$, (b) $K = 1$, (c) $K = 2$, (d) $K = 3$. Note the suppressed energy zero. Details are discussed in the text.

5.6 Exercises

The exercises expand the view of Nilsson states and their role in the interpretation of states in even-mass nuclei, both broken-pair states in even–even nuclei and states in odd–odd nuclei. Particular attention is given to a deeper look into K isomerism and into backbending. A look at excited $K = 0$ bands and non-rotational collective excitations is made.

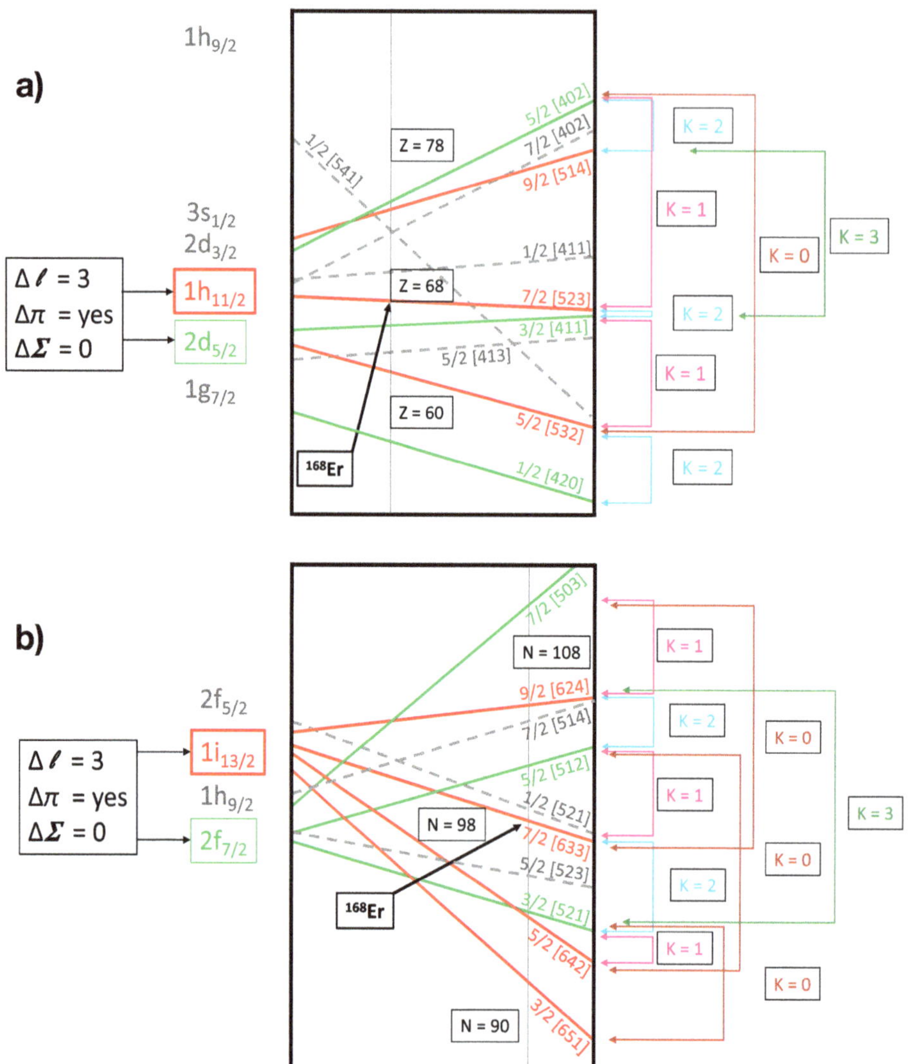

Figure 5.16. Schematic view of Nilsson diagrams for the rare earth region, (a) for protons and (b) for neutrons, illustrating the role of configurations that give rise to candidate collective octupole bands with $K = 0$, 1, 2, 3.

5.6.1 Broken-pair states and the role of pairing correlations

Figures 5.18(a) and (b) present a view of pairing energies in the rare earth/lanthanide region. The quantities ΔS_n and ΔS_p are defined as

$$\Delta S_n = S_n(A, \text{even}) - 1/2\{S_n(A + 1) + S_n(A - 1)\} \tag{5.1}$$

and

$$\Delta S_p = S_p(A, \text{even}) - 1/2\{S_p(A + 1) + S_p(A - 1)\}, \tag{5.2}$$

Figure 5.17. Excited states in the odd–odd nucleus, ^{166}Ho organized into rotational bands. The Nilsson configurations of the bands are deduced from the possible Nilsson states expected for $Z = 67$ and $N = 99$ with combinations $p\Omega^{\pi}[Nn_z\Lambda] \pm n\Omega^{\pi}[Nn_z\Lambda]$. The relative energies of these couplings is beyond the present level of discussion. Reproduced from [1]. Copyright IOP Publishing Ltd. All rights reserved.

where the values of S_n and S_p are taken from [9].

    5-1 Make figures, similar to figures 5.18(a) and (b), for the actinide region using data in [9].

5.6.2 Broken-pair states in ^{168}Er: a case study

Figures 5.1, 5.2(a) and (b) and 5.5 provide a view of broken-pair proton states in ^{168}Er. To complete the picture, figure 5.19 shows the proton spectrum from the one-neutron addition reaction ^{167}Er(d,p)^{168}Er. Table 5.2 summarizes details of some broken-pair configurations in ^{168}Er. An important complement to the (d,p) spectroscopic view is shown in figure 5.20, which presents the low-energy helion (alpha) spectrum from the one-proton removal reaction ^{169}Tm(t,α)^{168}Er (q.v. figure 5.5 for the spectrum above 1900 keV). The conclusion is that the low-energy $K^{\pi} = 4^-$ band is a mixture of broken neutron-pair and broken proton-pair configurations. Figure 5.21 summarizes the negative-parity bands in ^{168}Er with band-head excitations below 2270 keV (q.v. figures 5.1 and 5.2(b)). Further details are discussed below.

    Figure 5.21 provides a comprehensive view of negative-parity states in ^{168}Er. The most important structural factor to note is the energy estimates for breaking a

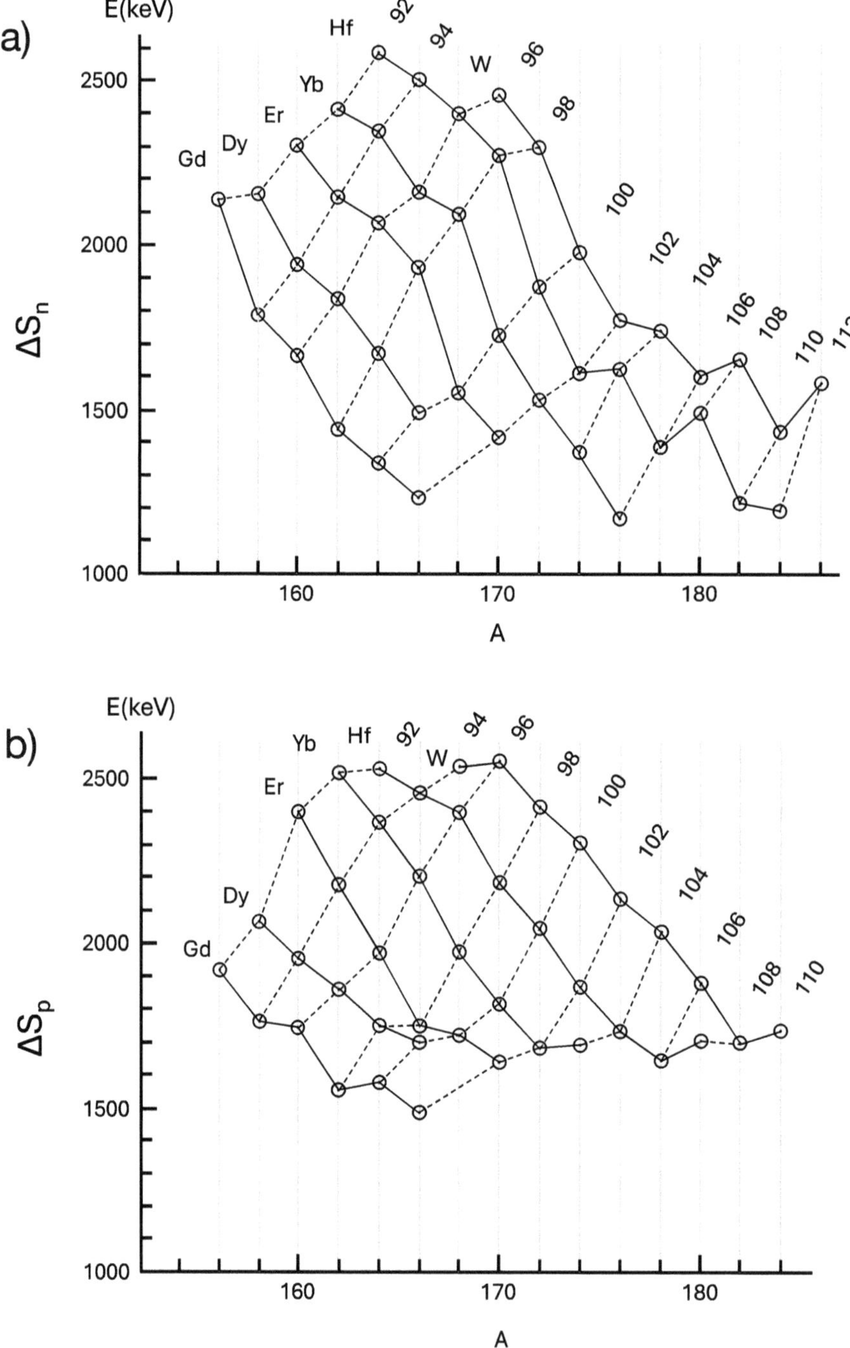

Figure 5.18. (a) Systematics of the quantity ΔS_n for the rare earth region. (b) Systematics of the quantity ΔS_p for the rare earth region. These quantities are defined in the text. They are based on data in [9].

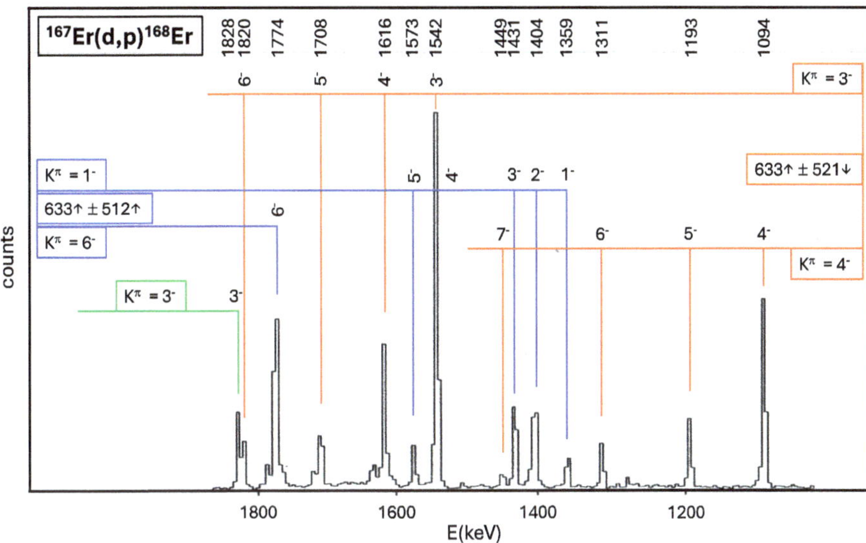

Figure 5.19. Spectrum of protons following the neutron addition reaction ^{167}Er(d,p)^{168}Er. The Nilsson configuration of the target nucleus is $7/2^+[633]$, which confines population of broken-pair states in ^{168}Er to the configurations $7/2^+[633] \pm 1/2^-[521$, $7/2^+[633] \pm 5/2^-[512]$, ⋯. The spectrum is based on one appearing in [10]. Reproduced with permission from [10].

Table 5.2. Configurations in ^{168}Er identified via one-nucleon transfer reactions.

J^π	Configuration	Excitation energy (keV)
3^-	$\pi 1/2^+[411 \downarrow] - \pi 7/2^-[523 \uparrow], \Sigma = 1$	1828 $(+\nu\nu)$
4^-	$\pi 1/2^+[411 \downarrow] + \pi 7/2^-[523 \uparrow], \Sigma = 0$	1905 $(+ 30\% \, \nu\nu)$
1^+	$\pi 1/2^+[411 \downarrow] - \pi 3/2^+[411 \uparrow], \Sigma = 1$	2365
2^+	$\pi 1/2^+[411 \downarrow] + \pi 3/2^+[411 \uparrow], \Sigma = 0$	2193
3^-	$\nu 7/2^+[633 \uparrow] - \nu 1/2^-[521 \downarrow], \Sigma = 1$	1542 $(+\pi\pi; +\nu\nu$ 633-512 band member)
4^-	$\nu 7/2^+[633 \uparrow] + \nu 1/2^-[521 \downarrow], \Sigma = 0$	1094 $(+ 30\% \, \pi\pi)$
1^-	$\nu 7/2^+[633 \uparrow] - \nu 5/2^-[512 \uparrow], \Sigma = 0$	1359 $(+ $ oct.$)$
6^-	$\nu 7/2^+[633 \uparrow] + \nu 5/2^-[512 \uparrow], \Sigma = 1$	1773

neutron pair (1551 keV) or a proton pair (1720 keV), which are estimated from figures 5.18(a) and (b). A secondary factor of importance is that the spin–spin interactions favour $\Sigma = 0$ coupling over $\Sigma = 1$ coupling. This is universal for like nucleons (but is the reverse for the coupling of an unpaired proton with an unpaired neutron, see section 5.5). Thus, there is a reasonable consistency between the observed broken-pair configurations and expectations of the energy cost inferred from pair breaking as quantified in figures 5.18(a) and (b).

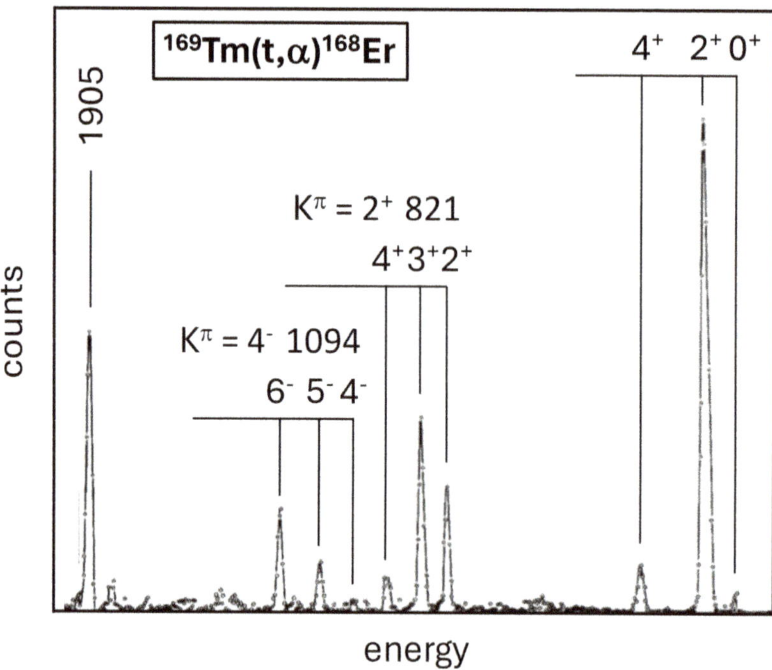

Figure 5.20. Spectrum of alpha particles (helions) following the proton removal reaction ^{169}Tm(t,α)^{168}Er. The Nilsson configuration of the target nucleus is $1/2^+[411]$ which confines population of broken-pair states in ^{168}Er to the configurations $1/2^+[411] \pm 7/2^-[523]$, $1/2^+[411] \pm 3/2^+[411]$, $\cdots$. The spectrum is based on data appearing in [5]. This is the low-energy region of the spectrum, the high-energy region is presented in figure 5.5. The ground state band is on the right-hand side.

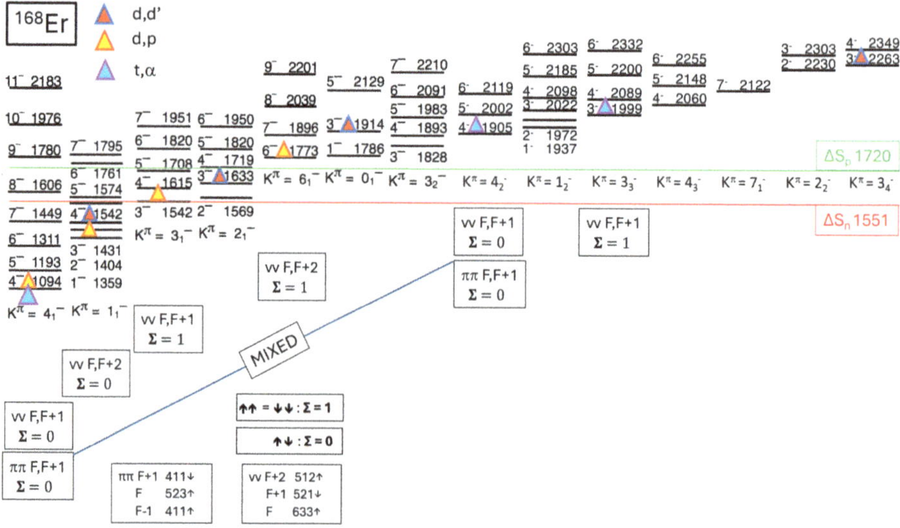

Figure 5.21. The negative parity bands in ^{168}Er identified below 2270 keV. The data are taken from ENSDF. The spectroscopic identifications via (d,d'), (d,p) and (t,α) are indicated by the colour-coded triangles. The Nilsson independent-particle states are indicated as $\cdots F-1, F, F+1, \cdots$ where 'F' designates the ground state configuration of the neighbouring $A-1$ nucleus, respectively for protons and for neutrons. The spin–spin couplings are indicated: this is left for the reader to fully interpret.

Four factors must be considered in estimating excitation energies of broken-pair states in an even–even nucleus:

1. The energy needed to break the pair, q.v. figures 5.18(a) and (b).
2. The single particle excitations, which can be estimated from the spectra of the neighbouring odd-mass nuclei.
3. The spin–spin interaction energy: this must be systematically explored through identification of $\Sigma = 0$ and $\Sigma = 1$ coupling.
4. Possible occurrences of configuration interactions of a specified K^π.

5-2 Using the above four criteria, explore how far it is possible to classify broken-pair negative-parity states in ^{168}Er. Note that configuration interactions may result in collectivity, e.g. octupole collectivity associated with $K^\pi = 3^-$ states.

5-3 Explore this approach applied to the negative parity states in ^{172}Yb using data in ENSDF.

5-4 Using the various data sources provided herein, explore the microscopic view of octupole collectivity for ^{168}Er (cf figure 5.21) conjectured in figure 5.16.

5.6.3 Broken-pair states in ^{234}U: a case study

Figure 5.22 presents a view of the low-energy band structure in ^{234}U. The red triangles indicate states that are populated by inelastic deuteron scattering, as deduced from the spectrum shown in figure 5.23. States in ^{234}U populated by the (d,p) and (d,t) one-neutron addition and removal reactions are shown in figure 5.24.

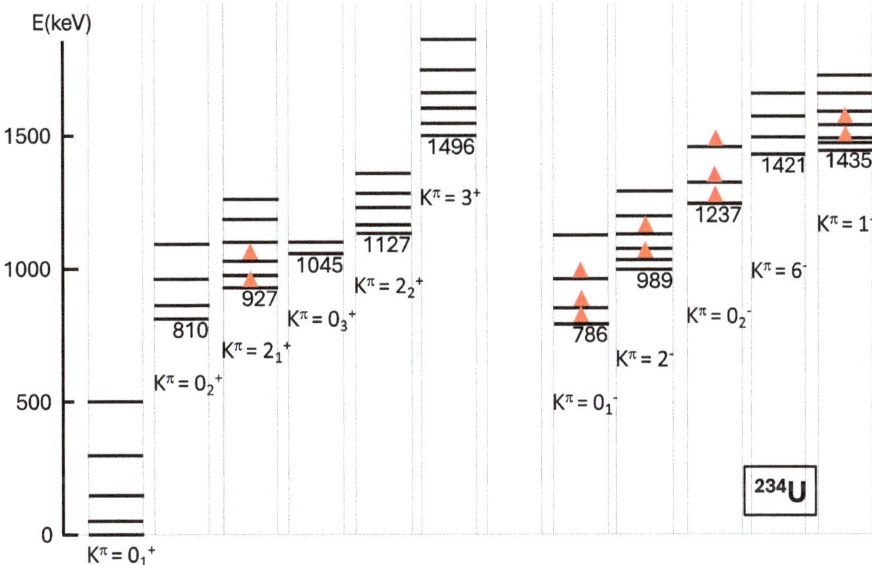

Figure 5.22. Band structure in ^{234}U. The data are taken from ENSDF. The red triangles identify states that are populated by inelastic deuteron scattering, cf figure 5.23.

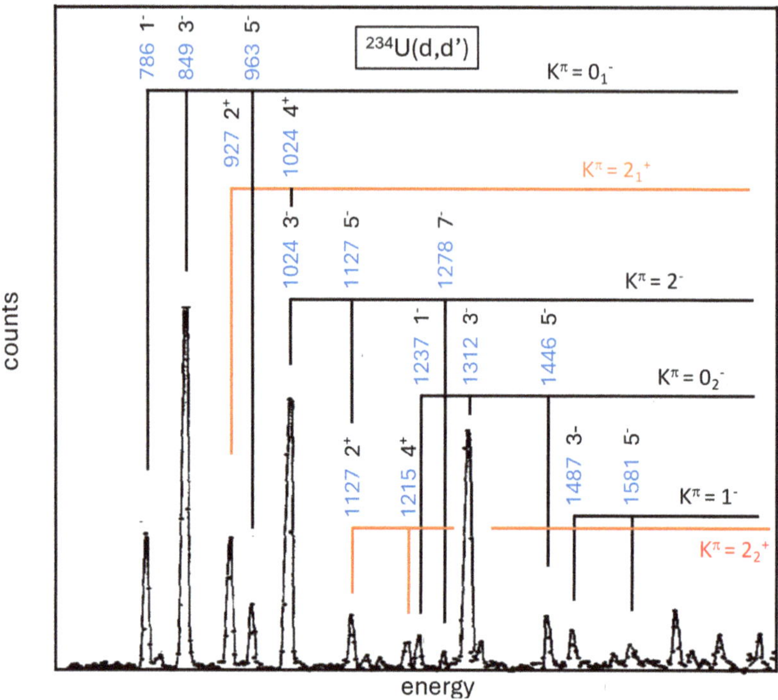

Figure 5.23. States in ^{234}U which are populated in the (d,d') reaction. Reprinted from [11], copyright (1973) with the permission of Elsevier.

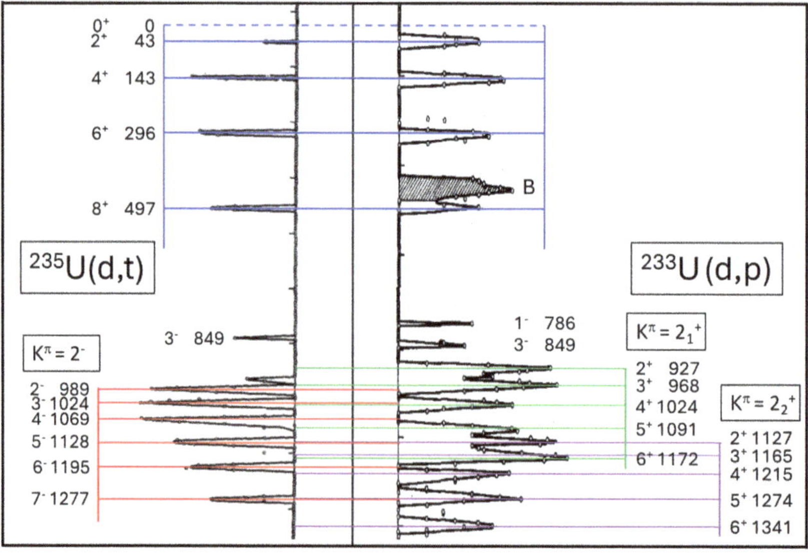

Figure 5.24. States in ^{234}U which are populated in the (d,t) and (d,p) reactions. Reprinted from [12], copyright (1968) with the permission of Elsevier.

5-5 Following the procedures introduced for an interpretation of negative-parity broken-pair states in ^{168}Er, attempt an interpretation of the negative-parity states in ^{234}U.

5-6 Using the various data sources provided herein, explore the microscopic view of octupole collectivity for ^{234}U.

5.6.4 *K* isomers in the lanthanide/rare earth and actinide regions

Figures 5.7(a) and (b) presents a view of Nilsson diagrams annotated to point out the proximity of high-Ω configurations for protons and separately for neutrons in the lanthanide/rare earth region; figures 5.25(a) and (b) provide a similar view for the actinide region. Figure 5.26 shows the systematic occurrence of $K^\pi = 8^-$ isomers in the $N = 150$ isotones.

There are a number of aspects to K isomers that warrant searches for such structures in various mass regions. First, their interpretation is often restricted to only one or a few possible Nilsson broken-pair configurations. Second, their decay modes define sequences of 'daughter' states that often reveal recognizable rotational bands (q.v. figure 5.6). Third, and less commonly, the electromagnetic decay may be so hindered that beta decay is observed: this may populate a K isomer in the beta-decay daughter nucleus. An example is shown in figure 5.27. These points form a 'base' view of K isomer occurrences which can be used to make systematic searches along isotopic and isotonic sequences (q.v. figure 5.8). Some of the lighter mass regions, where Nilsson states are beginning to show a systematic occurrence (cf sections 3.4.3 and 3.4.4), exhibit K isomerism: an example of a systematic occurrence in the $Z > 50$, $N < 82$ region is shown in figure 5.28; an example of an occurrence in the $Z < 50$, $N > 50$ region is shown in figure 5.29.

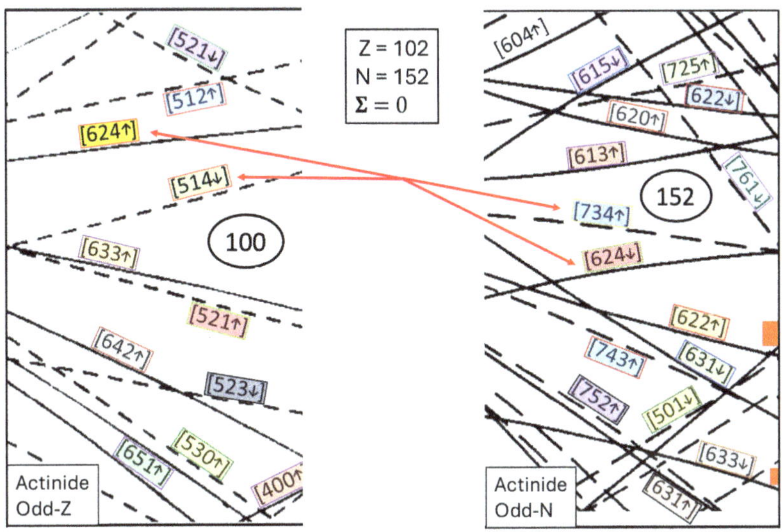

Figure 5.25. (a,b) Nilsson diagrams showing combinations of high Ω that can give rise to K isomers at low excitation energy at and near ^{254}No.

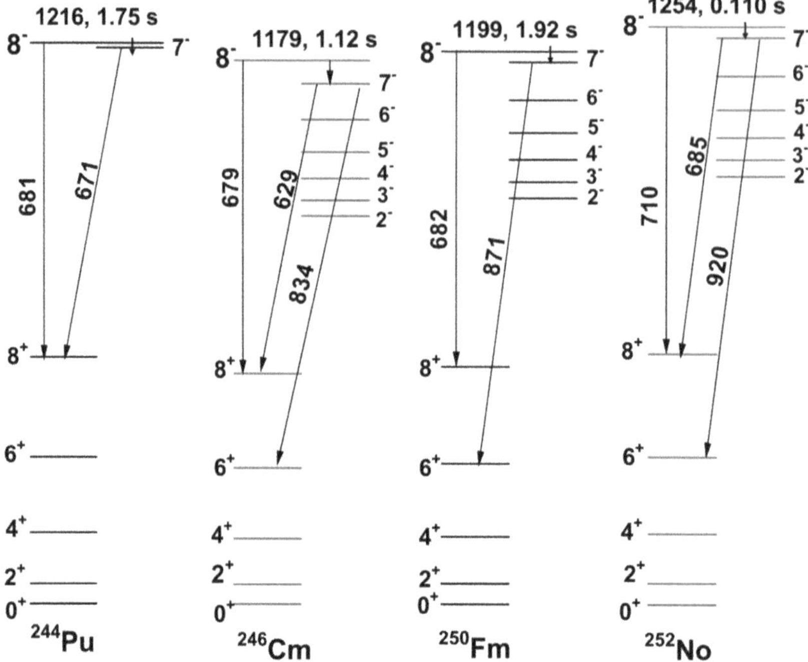

Figure 5.26. $K^\pi = 8^-$ isomers in the $N = 160$ isotones. More detailed information can be found in ENSDF. Reprinted from [13] CC-BY 4.0.

5-7 Figure 5.25(a) suggests that $K^\pi = 8^-$ isomers should occur in the nobelium ($Z = 102$) isotopes. Using data in ENSDF, make a systematic view of candidate states for such an interpretation.

5-8 Figures 5.25(a) and (b) suggest that a $K^\pi = 16^+$ isomer should be observed for $Z \sim 102$, $N \sim 150$. Using data in ENSDF explore what experimental evidence exists for such a structure.

5-9 Using the view provided in figures 5.18(a) and (b) and the results from exercise 5-1, together with the view provided by figures 5.25(a) and (b), explore possibilities for K isomerism searches in the actinide $A \sim 250$ region and compare with data in ENSDF.

Broken-pair states occur in odd-mass nuclei and are called three-quasiparticle states. Examples are involved in the structures shown in figure 5.27. They can be directly related to broken-pair states in neighbouring even–even nuclei as the ground-state configuration of the odd-mass nucleus Ω value combined with the even–even broken-pair configuration K value. The combinations are by addition of Ω to K; recall parities are combined multiplicatively.

5-10 For ^{177}Lu and ^{177}Hf, relate the K isomers shown in figure 5.27 to the K isomerism observed in ^{178}Hf (q.v. figure 5.6).

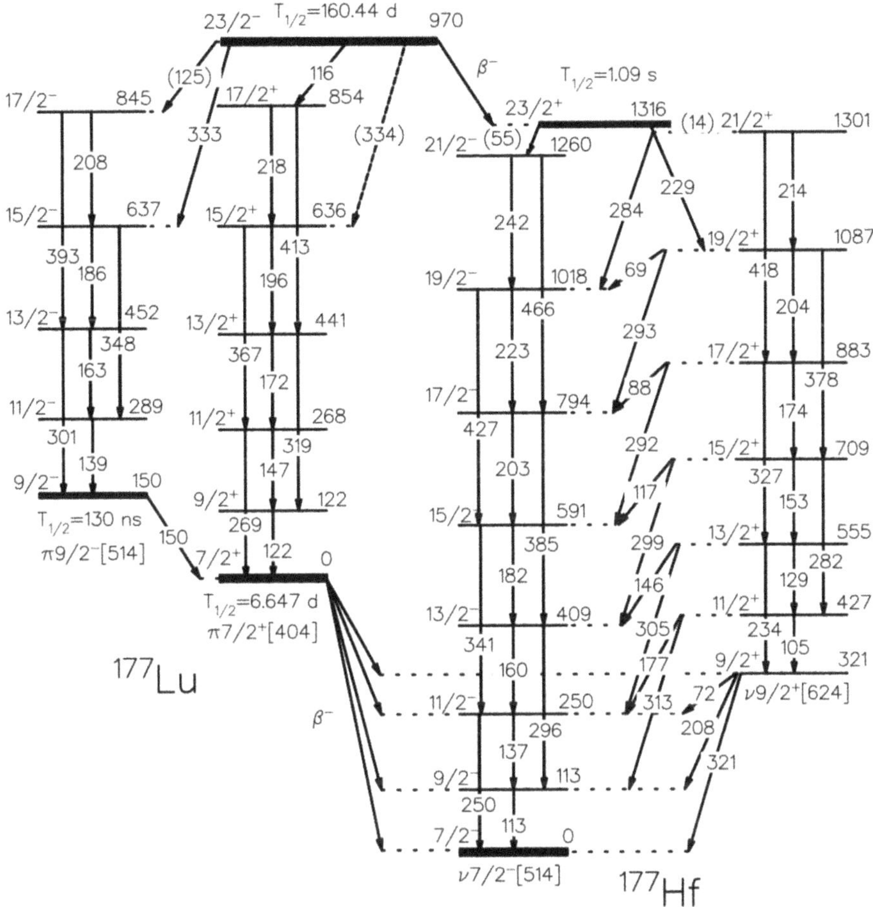

Figure 5.27. Example of beta decay of a high-K isomer. This illustrates the manner in which such decays, albeit rare, provide detailed information on bands. Further details can be obtained from ENSDF. Reprinted with permission from [14]. Copyright (2012) by the American Physical Society.

5-11 Using the approach in exercise 5-8, explore K isomerism in odd-mass nuclei in the $A \sim 250$ region using data in ENSDF [Suggestions: ^{255}Lr, ^{257}No].

5-12 K isomerism is beginning to be explored in the mass regions $A \sim 160$, 190. Using ENSDF make a basic identification of the Nilsson configurations involved. [Suggestions: ^{162}Dy, ^{188}W, ^{165}Er].

5.6.5 Backbending

The key ingredient that underlies backbending is a crossing band with respect to the ground-state band. This can arise in two ways. Shape coexistence of a more deformed, excited band is the simplest mechanism to understand but is less common. The most widely occurring crossing bands arise from configurations with large

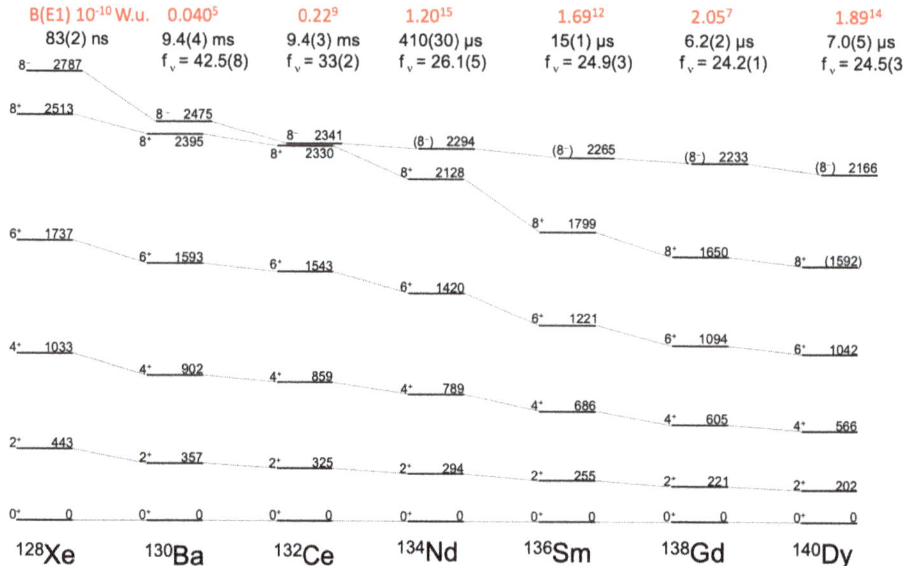

Figure 5.28. $K^\pi = 8^-$ isomers in the $N = 74$ isotones; see reference [15], wherein the more specialized details are defined. Further details can be obtained from ENSDF. Reprinted from [15] CC-BY 4.0.

alignment energies. The alignment energy plots, figures 5.12(a)–(c) identify the configurations most likely to produce backbending effects. Note that alignment energies are deduced from plots of gamma-ray transition energies versus spin of the state from which the gamma-ray transition originates. Recognizing that gamma-ray transition energies are a *differential* view of rotational excitations in such nuclei, this view 'integrates' into expressions for excitation energies of the form

$$E = AI(I + 1) + BI, \tag{5.3}$$

where the first term is the standard rigid-rotor model view of excitations in deformed nuclei and the second term is the alignment energy with B quantifying this energy, and with $B < 0$. The observed behaviour of rotational bands, i.e. that they have (near) identical rotational energy profiles leads to the conclusion that band crossing is almost entirely due to the second (negative) energy term.

A key feature of backbending is mixing. At and near the crossing point, quantum mechanical configuration mixing occurs. This results in varying degrees of 'sharpness' of the backbend. A global view of the rare earth/lanthanide region is presented in figure 5.30. The large difference in band crossing behaviour between ^{160}Dy and ^{162}Dy, shown schematically in figure 5.14, can be seen in figure 5.30 (note the backbend in ^{162}Dy occurs at spin 18). Figure 5.31 provides a larger-scale view of ^{162}Dy which possesses an S band which probably involves shape coexistence as inferred from the profile of the crossing band which is near linear suggesting a rigid rotational dynamics.

The large backbending manifestations in the 'northwest corner' of figure 5.30 are consistent with the large alignment energies that can be expected by extrapolating

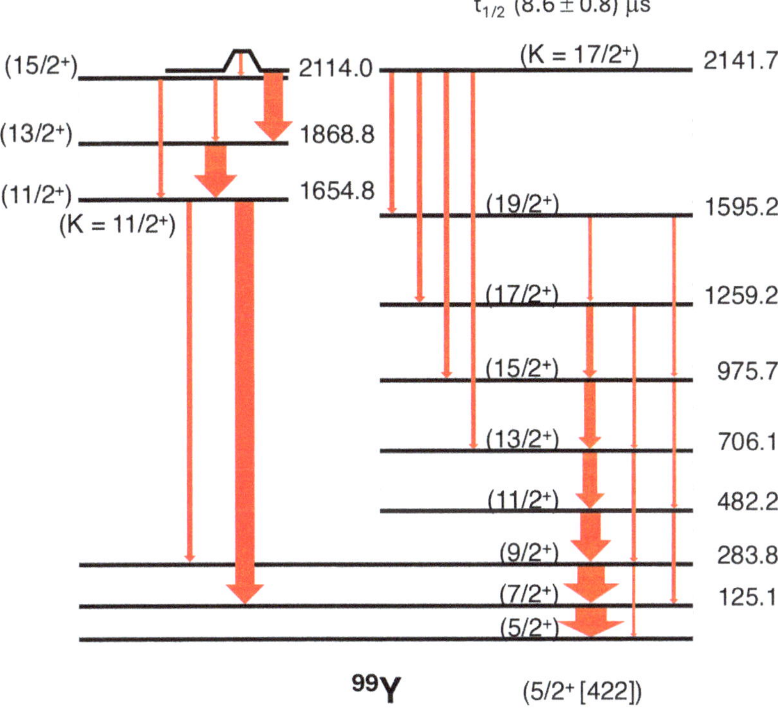

Figure 5.29. A high-K isomer in an $N = 60$ nucleus, ^{99}Y; see reference [16] wherein the more specialized details are defined. Further details can be obtained from ENSDF. Adapted from [16].

trends in figure 5.12. Note that these backbends occur at a low-spin limit of 10–12. Further note that the maximum alignment of a neutron pair is $J = 12$ (for the $i_{13/2}$ configuration) and for a proton pair is $J = 10$ (for the $h_{11/2}$ configuration).

Figure 5.31 shows the S band crossing the ground band with only small deviations from the profiles as interpolated to the crossing point. The S band crossing of the gamma band shows larger deviations, but note that the odd-spin members, as would be expected (there is no mixing), do not show a deviation from a smooth profile. The S band is the $K = 0$ band with band head at 1400 keV. Note that this band has an energy difference of 53 keV for the spin-0 and spin-2 band members, which can be compared with 81 keV for the ground band.

5-13 For the nuclei ^{164}Er and ^{166}Yb, using data from ENSDF, make plots similar to those in figure 5.31.

5-14 Using the plots from exercise 5-11, for the S bands in ^{164}Er and ^{166}Yb, compare their profiles for similarities.

5-15 Using B(E2) data for ^{162}Dy from ENSDF, what can be said about collective strengths in the ground-state band and the S band?

5-16 Explore the actinide region using plots such as in figure 5.31. Use data from ENSDF. [Suggested nuclei for exploration: ^{232}Th, ^{238}U, ^{240,242}Pu.]

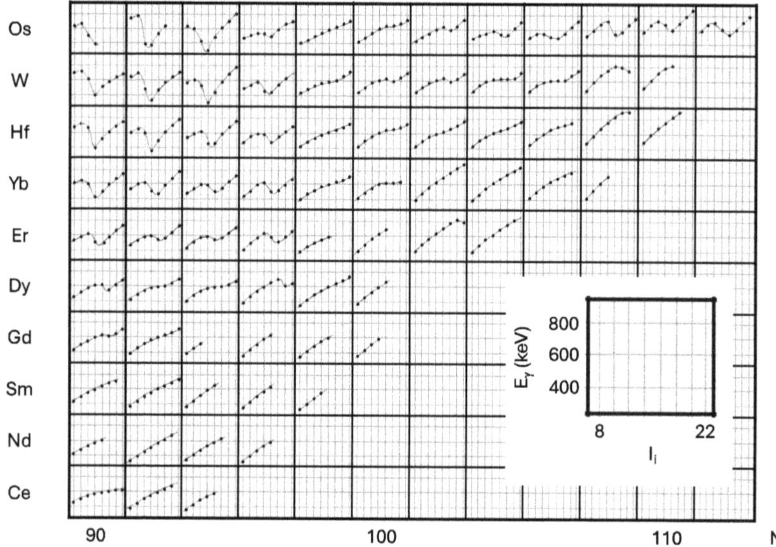

Figure 5.30. A global view of backbending in the rare earth region. The plots are for initial spin of the state from which the plotted gamma-ray transition energy occurs. The inset shows the scales used in the boxes.

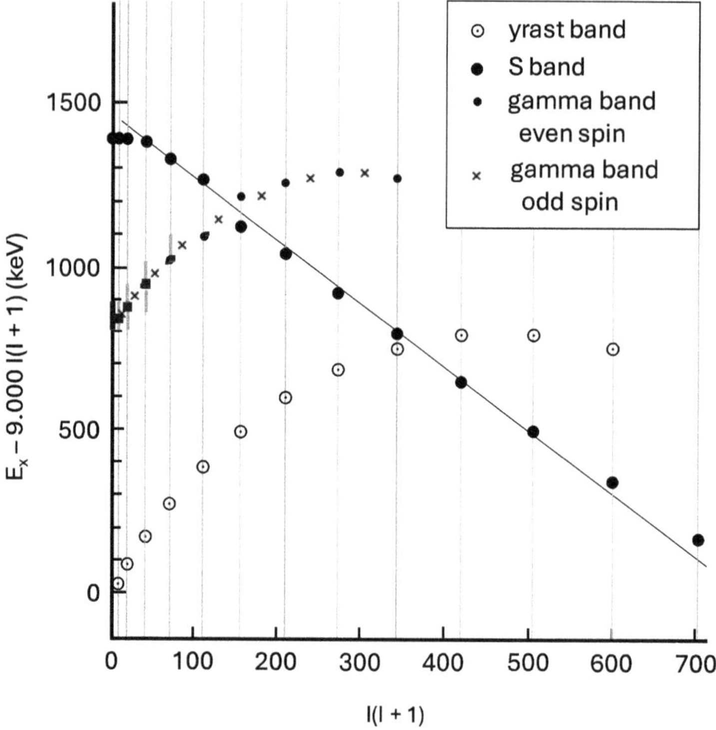

Figure 5.31. A high-resolution view of backbending in the rare earth region: the nucleus is [162]Dy. The plot shows the band crossings involving the ground band, S band and the $K = 2$ gamma band. Details are discussed in the text. The data are taken from ENSDF.

5.6.6 Excited $K = 0$ bands in the rare earth and actinide regions

In strongly deformed nuclei, excited $K = 0$ bands are always associated with excited states possessing spin-parity 0^+. There is no consensus regarding the structure of such states. Probably, such states involve excitations of proton pairs and neutron pairs, i.e. they likely do not involve broken-pair excitations.

There is no discernable pattern to excited 0^+ states in strongly deformed nuclei. Formerly, the first such excited state was called a 'beta vibration', but this does not appear to be a useful view. The status of the beta vibration concept was addressed in [17], and was placed in doubt. A basic issue is that there are many excited 0^+ states and so, with respect to the beta vibrational concept, what degrees of freedom are involved in producing these other states?

Manifestly, excited states in deformed nuclei should possess an explanation in terms of protons and neutrons occupying Nilsson orbitals. Consultation of figure 5.1 shows that excited 0^+ states can occur well below the pairing energy gaps for both protons and neutrons (q.v. figure 5.18): thus, correlations can be expected to be important.

We provide a few basic exercises for the more adventurous reader.

5-17 A simple mechanistic view of excited 0^+ states in deformed nuclei would be to excite an unbroken proton pair or neutron pair from a lower energy occupied Nilsson configuration to a higher energy Nilsson configuration. This would incur $2\times$ the energy difference between such pairs of Nilsson configurations. Using experimentally observed energy differences for Nilsson states in odd-mass deformed nuclei, refute this simple view.

5-18 Using data in ENSDF, for rare earth/lanthanide nuclei with $N > 90$, $Z < 74$, make a catalog of excited 0^+ states with energies <1600 keV.

5-19 Repeat exercise 5-16 for actinide nuclei with $Z > 86$, $N > 138$.

5.6.7 Odd–odd deformed nuclei

Odd–odd nuclei can appear to be the 'unwelcome' visitor to the venue of nuclear structure study—they are complex. The most important aspect to their structure is that we wish to understand the ground-state spins and parities of 25% of all nuclei. The spin-parity of an odd–odd nucleus controls its beta decay in an important manner. In particular, it determines the spin-parities of the states populated in the daughter nucleus.

In deformed nuclei, the ground state spin-parity of an odd–odd nucleus can be conjectured by coupling the ground-state configurations of a neighbouring odd-proton nucleus to that of an odd-neutron nucleus. There is one proviso, such a mechanism usually favours $\Sigma = 1$ coupling. Note that this is the opposite to coupling spins of broken proton-pair or broken neutron-pair configurations in even–even nuclei (q.v. figure 5.21). Figure 5.17, which shows the low-lying states in ^{166}Ho, captures the essential features of a deformed odd–odd nucleus.

5-20 There is the possibility of populating some odd–odd nuclei by one-nucleon transfer on stable odd-mass target nuclei. For example, states in ^{166}Ho can

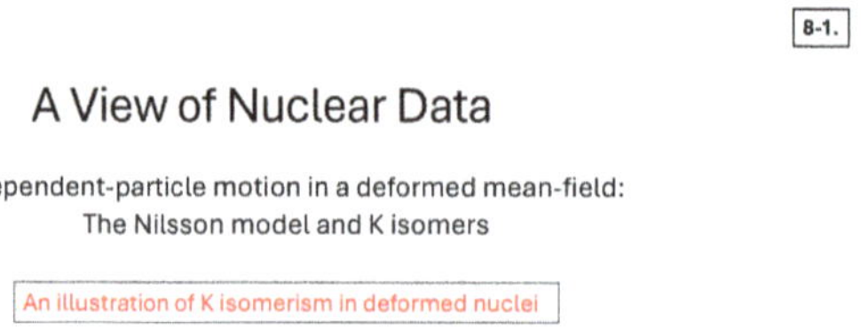

A View of Nuclear Data

Figure 5.32. Video tutorial: K isomers in deformed nuclei. Video available at https://doi.org/10.1088/978-0-7503-5648-0.

be populated by the ^{165}Ho(d,p)^{166}Ho one-neutron addition reaction. They can also be populated by the ^{167}Er(^{3}He,d)^{166}Ho one-proton removal reaction. Identify which rare earth/lanthanide odd–odd nuclei can be studied in this manner.

5-21 For some odd–odd nuclei, K bands have been elucidated up to high spins. Make E$_\gamma$ versus I_i plots, cf figures 4.1, 4.4, and 4.5, for odd–odd nuclei for a selection of cases where high-spin band members have been determined. Then compare the alignment energies with those in the neighbouring odd-mass nuclei to determine if there is 'additivity' for alignment energies.

5.7 Video-based tutorials

5-22 For those reading the e-book, a video-based tutorial on K isomerism in deformed nuclei is found in figure 5.32.

References

[1] Jenkins D G and Wood J L 2021 *Nuclear Data: A primer* (Bristol: IOP Publishing)

[2] Tjøm P O and Elbek B 1968 Collective vibrational states in even erbium nuclei *Nucl. Phys.* A **107** 385

[3] Rowe D J and Wood J L 2010 *Fundamentals of Nuclear Models: Foundational Models* (Singapore: World Scientific)

[4] Oshima M, Morikawa T, Hatsukawa Y, Ichikawa S, Shinohara N, Matsuo M, Kusakari H, Kobayashi N, Sugawara M and Inamura T 1995 Two-phonon γ-vibrational state in ^{168}Er *Phys. Rev.* C **52** 3492

[5] Burke D G, Davidson W F, Cizewski J A, Brown R E and Sunier J W 1985 Two-quasi proton states in ^{168}Er studied by the ^{169}Tm(t,α)^{168}Er reaction *Nucl. Phys.* A **445** 70

[6] Kondev F G, Dracoulis G D and Kibédi T 2015 Configurations and hindered decays of K isomers in deformed nuclei with $A > 100$ *At. Data Nucl. Data tables* **103–104** 50–105

[7] Arnquist I J *et al* 2023 Constraints on the decay of ^{180m}Ta *Phys. Rev. Lett.* **131** 152501

[8] Hagemann G B and Hamamoto I 1992 Interaction strength of two crossing bands *Phys. Rev.* C **46** 838

[9] Wang M, Huang W J, Kondev F G, Audi G and Naimi S 2021 The AME2020 atomic mass evaluation (II). Tables, graphs and references *Chin. Phys.* C **45** 030003

[10] Mayerhofer U, von Egidy T, Hlawatsch G and Lindner H 1996 The nucleus ^{168}Er studied with the (d,p) reaction *Fizika* B **5** 103–12

[11] Boyno J S, Huizenga J R, Elze T W and Bemis C E 1973 Levels of ^{234}U and ^{236}U excited by the (d,d') reaction *Nucl. Phys.* A **209** 125–34

[12] Bjørnholm S, Dubois M J and Elbek B 1968 Energy levels in ^{234}U populated by the ^{233}U(d,p) and^{235}U(d,t) processes *Nucl. Phys.* A **118** 241–60

[13] Hessberger F P 2023 K isomers in transuranium nuclei arXiv:2309.10468 https://arxiv.org/abs/2309.10468

[14] Kondev F G *et al* 2012 M3 and E4 K-forbidden decays of the $K^{\pi} = 23/2^-$ isomer in ^{177}Lu *Phys. Rev.* C **85** 027304

[15] Stuchbery A E and Wood J L 2022 To shell model, or not to shell model, that is thequestion *Physics* **4** 697–773

[16] Meyer R A, Monnand E, Pinston J A, Schussler F, Ragnarsson I, Pfeiffer B, Lawin H, Lhresonneau G, Seo T and Sistemich K 1985 Deformation in odd-mass nuclei near $A \sim 100$: one- andthree-quasiparticle Nilsson states in ^{99}Y *Nucl. Phys.* A **439** 510–34

[17] Jenkins D G and Wood J L 2023 *Nuclear Data: A Collective Motion View* (Bristol: IOP Publishing)

IOP Publishing

Nuclear Data
An independent-particle motion view
David Jenkins and John L Wood

Chapter 6

Epilogue

The message that emerges from the data presented herein is that independent-particle motion is realized in many deformed nuclei, but in only a few nuclei that can be called 'spherical'. Indeed, even the issue of which nuclei possess spherical shapes appears to need detailed investigation.

Independent-particle motion in a spherical mean field appears to be realized in a few nuclei, i.e. those nuclei adjacent to doubly closed shells. Nuclei adjacent to singly closed shells are a logical extension of the view, but data suggest that deformation may be emerging in such nuclei. Pairing correlations in singly closed shell nuclei, which are natural candidates for further study of spherical nuclei, necessitate deferring discussion of such nuclei until pairing models are handled.

Independent-particle motion in a deformed mean field is widely realized in many nuclei. These are the so-called Nilsson states. However, there is a proviso: rotational degrees of freedom must be included in any treatment that aims at full description of these states. This is reliably achieved with the particle-rotation coupling model. The combined description is extraordinarily successful at a near quantitative level.

An issue that appears to need future investigation is the limit to applying the particle-rotation coupling model to data for weakly deformed nuclei. From the limit of structure manifested in strongly deformed nuclei, patterns of behaviour that appear useful for handling the structure of weakly deformed nuclei are suggested. These patterns are usefully described as 'rotation-alignment' effects, and some details are discussed herein. From a spherical perspective, there is a lack of a useful view that describes how the structures of weakly deformed nuclei are organized.

We note that there are many model descriptions of weakly deformed nuclei, but there is no consensus on which model should be used. The present approach adheres closely to a data-based view and intentionally avoids data for nuclei that appear to be deformed, but not strongly deformed. (We note, based on data which have not been discussed here, it appears that axially asymmetric shapes may be important in weakly deformed nuclei, but at present this view lacks clear model-independent support.)

doi:10.1088/978-0-7503-5648-0ch6 6-1

The success of the Nilsson-plus-rotor model in describing data for deformed odd-mass nuclei extends with impressive success to data for deformed even-mass nuclei where more than one particle is behaving (approximately) independently. The current understanding of structure for essentially all strongly deformed nuclei appears to be very good for independent-particle motion.

Nuclear Data

An independent-particle motion view

David Jenkins and John L Wood

Appendix A

Ground-state spins and parities

These tabulations are based on assignments made in [1]. It is important to note that, following the basis of [1], we do not assign spin-parities where spectroscopic proof is absent. We note that ENSDF contains many spin-parity assignments to nuclear ground states based on systematics, even on theory. This is problematic because the assignments are often interpreted as 'data', which contradicts the title of the project. As such, these assignments are useless for building systematic views (when the values were assigned using systematics), or for testing the model that was used to make the assignment. We observe that this is a growing problem with studies of isotopes far from stability, especially where models are being tested, notably the shell model with configuration interactions.

The following spin-parity assignment differs from [1], with the reason specified: ^{133}Pr, $5/2^+ \rightarrow 3/2^+$, determined from E3 decay of an isomer based on conversion-electron spectroscopy and location using coincidences between conversion electrons and gamma rays [2]. This disagrees with an atomic beam magnetic resonance, ABMR spectroscopy measurement [3]. ABMR is considered the gold standard for a spin assignment based on counting the number of magnetic substates; but in this case background contamination resulted in a less reliable spin assignment [private communication from Curt Ekstrom to JLW]. See exercise A.1 (tables A.1–A.8).

A.1 Exercises

A-1 Figure A.1 shows the spectroscopic evidence for the isomerism observed in ^{133}Pr. If the spin-parity assignment in [1] is correct, how would the observed decays have to be modified? (Note that this is not a definitive spectroscopic proof of the spin-parity assignment, but it places doubt on the $5/2^+$ assignment in [1].

doi:10.1088/978-0-7503-5648-0ch7

Table A.1. Ground-state spin-parities, odd Z: $1 \leqslant Z \leqslant 27$, $0 \leqslant N \leqslant 50$.

Legend: a $(5/2^+, 3/2^+)$

Values are listed for each element in order of increasing N.

Element	Ground-state spin-parities (increasing N →)
Co	$7/2^-$ $7/2^-$ $7/2^-$ $7/2^-$ $7/2^-$ $(7/2)^-$ $(7/2^-)$ $(7/2^-)$ $(7/2^-)$ $(7/2^-)$
Mn	$5/2^-$ $5/2^-$ $7/2^-$ $5/2^-$ $5/2^-$ $5/2^-$ $5/2^-$ $5/2^-$ $(5/2^-)$
V	$7/2^-$ $3/2^-$ $7/2^-$ $7/2^-$ $7/2^-$
Sc	$7/2^-$ $7/2^-$ $7/2^-$ $7/2^-$ $7/2^-$ $(7/2)^-$ $(7/2^-)$ $(7/2)^-$
K	$3/2^+$ $3/2^+$ $3/2^+$ $3/2^+$ $3/2^+$ $3/2^+$ $1/2^+$ $1/2^+$ $3/2^+$ $3/2^+$
Cl	$(1/2^+)$ $3/2^+$ $3/2^+$ $3/2^+$ $3/2^+$ $3/2^+$ $(1/2^+)$ $(3/2^+)$ $(3/2^+)$
P	$1/2^+$ $1/2^+$ $1/2^+$ $1/2^+$ $1/2^+$ $(1/2^+)$ $(1/2^+)$
Al	$5/2^+$ $5/2^+$ $5/2^+$ $5/2^+$ $5/2^+$ $5/2^+$ a
Na	$(5/2^+)$ $3/2^+$ $3/2^+$ $5/2^+$ $5/2^+$ $3/2^+$ $3/2^+$ $(3/2^+)$
F	$5/2^+$ $1/2^+$ $5/2^+$ $5/2^+$ $(5/2^+)$
N	$1/2^-$ $1/2^-$ $1/2^-$ $(1/2^-)$
B	$(3/2^-)$ $3/2^-$ $3/2^-$ $3/2^-$
Li	$3/2^-$ $3/2^-$ $3/2^-$ $3/2^-$
H	$1/2^+$ $1/2^+$

N axis: 2, 8, 20, 28, 40, 50.

Table A.2. Ground-state spin-parities, odd Z: $29 \leqslant Z \leqslant 73$, $28 \leqslant N \leqslant 82$.

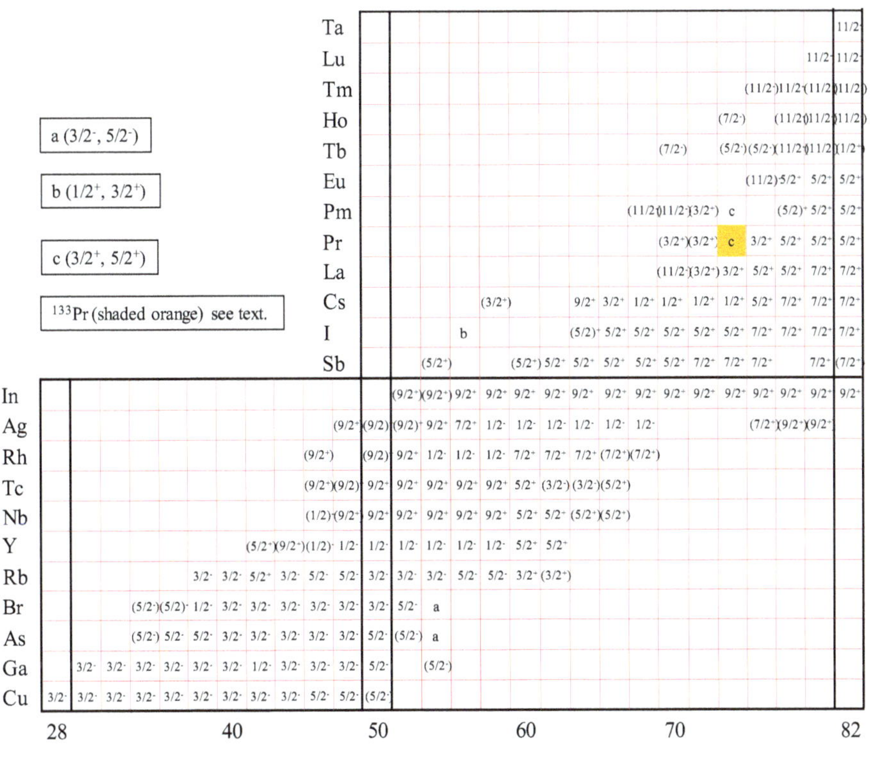

Legend: a $(3/2^-, 5/2^-)$; b $(1/2^+, 3/2^+)$; c $(3/2^+, 5/2^+)$; ^{133}Pr (shaded orange) see text.

Values are listed for each element in order of increasing N.

Element	Ground-state spin-parities (increasing N →)
Ta	$11/2^-$
Lu	$11/2^-$ $11/2^-$
Tm	$(11/2^-)$ $11/2^-$ $(11/2^-)$ $11/2^-$
Ho	$(7/2^-)$ $(11/2^-)$ $11/2^-$ $11/2^-$
Tb	$(7/2^-)$ $(5/2^-)$ $(5/2^-)$ $(11/2^-)$ $11/2^-$ $11/2^-$
Eu	$(11/2)$ $5/2^+$ $5/2^+$ $5/2^+$
Pm	$(11/2^-)$ $11/2^-$ $(3/2^+)$ c $(5/2)^+$ $5/2^+$ $5/2^+$
Pr	$(3/2^+)$ $(3/2^+)$ c $3/2^+$ $5/2^+$ $5/2^+$ $5/2^+$
La	$(11/2^-)$ $(3/2^+)$ $3/2^+$ $5/2^+$ $5/2^+$ $7/2^+$ $7/2^+$
Cs	$(3/2^+)$ $9/2^+$ $3/2^+$ $1/2^+$ $1/2^+$ $1/2^+$ $1/2^+$ $5/2^+$ $7/2^+$ $7/2^+$ $7/2^+$
I	b $(5/2)^+$ $5/2^+$ $5/2^+$ $5/2^+$ $5/2^+$ $5/2^+$ $7/2^+$ $7/2^+$ $7/2^+$ $7/2^+$
Sb	$(5/2^+)$ $(5/2^+)$ $5/2^+$ $5/2^+$ $5/2^+$ $5/2^+$ $5/2^+$ $7/2^+$ $7/2^+$ $7/2^+$ $7/2^+$ $(7/2^+)$
In	$(9/2^+)$ $(9/2^+)$ $9/2^+$ $9/2^+$ $9/2^+$ $9/2^+$ $9/2^+$ $9/2^+$ $9/2^+$ $9/2^+$ $9/2^+$ $9/2^+$ $9/2^+$ $9/2^+$ $9/2^+$ $9/2^+$ $9/2^+$
Ag	$(9/2^+)$ $(9/2)$ $(9/2)^+$ $9/2^+$ $7/2^+$ $1/2^-$ $1/2^-$ $1/2^-$ $1/2^-$ $1/2^-$ $1/2^-$ $(7/2^+)$ $(9/2^+)$ $(9/2^+)$
Rh	$(9/2^+)$ $(9/2)$ $9/2^+$ $1/2^-$ $1/2^-$ $1/2^-$ $7/2^+$ $7/2^+$ $7/2^+$ $(7/2^+)$ $(7/2^+)$
Tc	$(9/2^+)$ $(9/2)$ $9/2^+$ $9/2^+$ $9/2^+$ $9/2^+$ $9/2^+$ $5/2^+$ $(3/2^-)$ $(3/2^-)$ $(5/2^+)$
Nb	$(1/2)^-$ $(9/2^+)$ $9/2^+$ $9/2^+$ $9/2^+$ $9/2^+$ $9/2^+$ $5/2^+$ $5/2^+$ $(5/2^+)$ $(5/2^+)$
Y	$(5/2^+)$ $(9/2^+)$ $(1/2)^-$ $1/2^-$ $1/2^-$ $1/2^-$ $1/2^-$ $1/2^-$ $1/2^-$ $5/2^+$ $5/2^+$
Rb	$3/2^-$ $3/2^-$ $5/2^+$ $3/2^-$ $5/2^-$ $5/2^-$ $3/2^-$ $3/2^-$ $3/2^-$ $5/2^-$ $5/2^-$ $3/2^+$ $(3/2^+)$
Br	$(5/2^-)$ $(5/2)^-$ $1/2^-$ $3/2^-$ $3/2^-$ $3/2^-$ $3/2^-$ $3/2^-$ $3/2^-$ $5/2^-$ a
As	$(5/2^-)$ $5/2^-$ $5/2^-$ $3/2^-$ $3/2^-$ $3/2^-$ $3/2^-$ $3/2^-$ $5/2^-$ $(5/2^-)$ a
Ga	$3/2^-$ $3/2^-$ $3/2^-$ $3/2^-$ $3/2^-$ $3/2^-$ $1/2^-$ $3/2^-$ $3/2^-$ $3/2^-$ $5/2^-$ $(5/2^-)$
Cu	$3/2^-$ $3/2^-$ $3/2^-$ $3/2^-$ $3/2^-$ $3/2^-$ $3/2^-$ $3/2^-$ $3/2^-$ $5/2^-$ $5/2^-$ $(5/2^-)$

N axis: 28, 40, 50, 60, 70, 82.

Table A.3. Ground-state spin-parities, odd Z: $51 \leqslant Z \leqslant 89$, $82 \leqslant N \leqslant 126$.

Legend: **a** $= (1/2^+, 3/2^+)$

N = 82–97:

	82	83	84	85	86	87	88	89	90	91	92	93	94	95	96	97
Ac																
Fr																
At																
Bi																
Tl															(1/2+)	1/2+
Au											1/2+	1/2+	1/2+	1/2+	1/2+	3/2-
Ir									1/2+	(1/2+)	1/2+	a	(1/2+)	(5/2)-	5/2+	5/2-
Re					1/2+	1/2+	(1/2+)	9/2-	(9/2-)	(9/2-)	(5/2-)			5/2-	5/2+	5/2+
Ta	11/2-	1/2+	1/2+	(1/2+)	1/2+	a	(3/2+)	(5/2+)	(5/2+)	5/2-	7/2+	7/2+	7/2+	7/2+	7/2+	(7/2+)
Lu	11/2-	11/2-	(1/2+)	1/2+	1/2+	1/2+	1/2+	7/2+	7/2+	7/2+	7/2+	7/2+	7/2+	7/2+		
Tm	11/2-	(11/2-)	11/2-	1/2+	5/2+	7/2+	1/2+	1/2+	1/2+	1/2+	1/2+	(1/2+)	(1/2+)			
Ho	11/2-	11/2-	11/2-	5/2+	7/2-	7/2-	7/2-	7/2-	7/2-	7/2-	7/2-					
Tb	(1/2+)	1/2+	1/2+	5/2+	3/2+	3/2+	3/2+	3/2+	3/2+	(3/2+)	(3/2+)					
Eu	5/2+	5/2+	5/2+	5/2+	5/2+	5/2+	5/2+	5/2+								
Pm	5/2+	5/2+	7/2+	7/2+	5/2+	5/2-	(5/2-)	(5/2-)	(5/2-)	(5/2-)						
Pr	5/2+	5/2+	7/2+	3/2+	(5/2+)	(3/2-)										
La	7/2+	(7/2+)	(7/2)+	(5/2+)	(5/2+)	(3/2-)										
Cs	7/2+	7/2+	7/2+	3/2+	3/2+	(3/2+)										
I	7/2+															
Sb	(7/2+)	(7/2+)														

N = 98–113:

	98	99	100	101	102	103	104	105	106	107	108	109	110	111	112	113
Ac																
Fr																
At																
Bi																
Tl	1/2+	1/2+	1/2+	1/2+	1/2+	1/2+	1/2+	1/2+	1/2+	1/2+	1/2+	1/2+	1/2+	1/2+	1/2+	1/2+
Au	5/2-	5/2-	1/2+	1/2+	1/2+	3/2+	3/2+	3/2+	3/2+	3/2+	3/2+	3/2+	3/2+	3/2+	3/2+	3/2+
Ir	5/2-	5/2-	3/2+	3/2+	3/2+	3/2+	3/2+	3/2+	3/2+	3/2+	3/2+	3/2+	3/2+	3/2+	3/2+	3/2+
Re	5/2+	5/2+	5/2+	5/2+	(3/2+)											
Ta	(7/2+)															
Lu																
Tm																
Ho																
Tb																
Eu																
Pm																
Pr																
La																
Cs																
I																
Sb																

N = 114–128:

	114	115	116	117	118	119	120	121	122	123	124	125	126	127	128
Ac								9/2-	9/2-	9/2-	9/2-	9/2-	9/2-		
Fr							(9/2-)	9/2-	9/2-	9/2-	9/2-	9/2-	9/2-		
At			1/2+	1/2+	1/2+	9/2-	9/2-	9/2-	9/2-	9/2-	9/2-	9/2-	9/2-		
Bi	1/2+	(9/2-)	9/2-	9/2-	9/2-	9/2-	9/2-	9/2-	9/2-	9/2-	9/2-	9/2-	9/2-		
Tl	1/2+	1/2+	1/2+	1/2+	1/2+	1/2+	1/2+	1/2+	1/2+	1/2+	1/2+	1/2+	1/2+	1/2+	1/2+
Au	3/2+	3/2+	3/2+	3/2+	3/2+	3/2+	3/2+	3/2+	3/2+	3/2+	3/2+	3/2+			
Ir	3/2+	3/2+	3/2+	3/2+	3/2+	3/2+	3/2+	3/2+	3/2+						
Re															

Table A.4. Ground-state spin-parities, odd Z: $83 \leqslant Z \leqslant 109$, $126 \leqslant N \leqslant 158$.

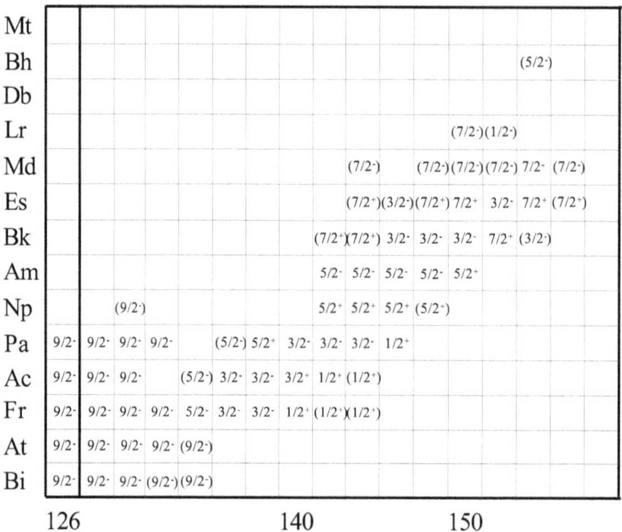

N = 126–141:

	126	127	128	129	130	131	132	133	134	135	136	137	138	139	140	141
Mt																
Bh																
Db																
Lr																
Md																
Es																
Bk															(7/2+)	(7/2+)
Am																
Np					(9/2-)											
Pa	9/2-	9/2-	9/2-	9/2-			(5/2-)	5/2+	3/2-	3/2-	3/2-	1/2+				
Ac	9/2-	9/2-	9/2-					(5/2-)	3/2-	3/2-	3/2-	1/2+	(1/2+)			
Fr	9/2-	9/2-	9/2-	9/2-				5/2-	3/2-	3/2-	1/2+	(1/2+)	(1/2+)			
At	9/2-	9/2-	9/2-	9/2-	(9/2-)											
Bi	9/2-	9/2-	9/2-	(9/2-)	(9/2-)											

N = 142–158:

	142	143	144	145	146	147	148	149	150	151	152	153	154	155	156	157	158
Mt																	
Bh						(5/2-)											
Db																	
Lr			(7/2-)	(1/2-)													
Md			(7/2-)				(7/2-)	(7/2-)	(7/2-)	7/2-	(7/2-)						
Es		(7/2+)	(3/2-)	(7/2+)	7/2+	3/2-	7/2+	(7/2+)									
Bk	3/2-	3/2-	3/2-	7/2+	(3/2-)												
Am	5/2-	5/2-	5/2-	5/2-	5/2+												
Np		5/2+	5/2+	5/2+	(5/2-)												
Pa																	
Ac																	
Fr																	
At																	
Bi																	

Table A.5. Ground-state spin-parities, odd N: $2 \leqslant Z \leqslant 26,\ 1 \leqslant N \leqslant 49$.

	1	3	5	7	9	11	13	15	17	19	21	23	25
Fe													$5/2^-$
Cr										$(3/2^+)$		$3/2^-$	$5/2^-$
Ti										$3/2^+$	$7/2^-$	$7/2^-$	$5/2^-$
Ca									$3/2^+$	$3/2^+$	$7/2^-$	$7/2^-$	$7/2^-$
Ar							$5/2^+$	$1/2^+$	$3/2^+$	$3/2^+$	$7/2^-$	$7/2^-$	$5/2^{(-)}$
S						$(5/2^+)$		$1/2^+$	$3/2^+$	$3/2^+$	$7/2^-$	$(7/2)^-$	
Si						$5/2^+$	$5/2^+$	$1/2^+$	$3/2^+$	$3/2^+$			
Mg					$5/2^+$	$3/2^+$	$5/2^+$	$1/2^+$	$3/2^+$	$1/2^+$	$3/2^-$	a	$(3/2^-)$
Ne			$(3/2^-)$	$1/2^-$	$1/2^+$	$3/2^+$	$5/2^+$	$1/2^+$	$(3/2^+)$	$(3/2^-)$	$(3/2^-)$		
O		$(3/2^-)$	$(3/2^-)$	$1/2^-$	$5/2^+$	$5/2^+$	$(5/2^+)$	$1/2^+$					
C		$3/2^-$	$3/2^-$	$1/2^-$	$1/2^+$	$3/2^+$	$1/2^+$						
Be		$3/2^-$	$3/2^-$	$1/2^+$	$(1/2^-)$	$(5/2^+)$							
He	$1/2^+$	$3/2^-$	$(3/2)^-$	$1/2^{(+)}$									

	27	29	31	33	35	37	39	41	43	45	47	49
Fe	$7/2^-$	$3/2^-$	$1/2^-$	$3/2^-$	$(3/2^-)$	$(5/2^-)$	$(1/2^-)$	$(1/2^-)$				
Cr	$7/2^-$	$3/2^-$	$3/2^-$	$(3/2^-)$	$(1/2^-)$	$(5/2^-)$						
Ti	$7/2^-$	$3/2^-$	$(3/2)^-$	$(1/2)^-$								
Ca	$7/2^-$	$3/2^-$	$3/2^-$									
Ar	b	$(3/2)^-$										

a $(3/2^-,\ 5/2^-)$

b $(5/2^-,\ 7/2^-)$

$$N$$

Table A.6. Ground-state spin-parities, odd N: $28 \leqslant Z \leqslant 72,\ 29 \leqslant N \leqslant 81$.

	29	31	33	35	37	39	41	43	45	47	49	51	53
Hf													
Yb													
Er													
Dy													
Gd													
Sm													
Nd													
Ce													
Ba													
Xe													
Te													
Sn												$(7/2^+)$	
Cd											$(9/2^+)$	$5/2^+$	$5/2^+$
Pd											$(9/2^+)$	$(5/2)^+$	$5/2^+$
Ru									$(9/2^+)$	$(9/2^+)$	$(9/2)$	$5/2^+$	$5/2^+$
Mo								$(1/2^+)$		$(9/2^+)$	$9/2^+$	$5/2^+$	$5/2^+$
Zr							$(3/2^-)$		$(7/2^+)$	$9/2^+$	$9/2^+$	$5/2^+$	$5/2^+$
Sr				$(5/2^-)$	$(3/2^-)$	$5/2^+$	$3/2^-$	$1/2^-$	$7/2^+$	$9/2^+$	$9/2^+$	$5/2^+$	$5/2^+$
Kr			$(5/2^-)$	$(5/2)^-$	$(3/2)^-$	$5/2^+$	$5/2^+$	$1/2^-$	$7/2^+$	$9/2^+$	$9/2^+$	$5/2^+$	$3/2^+$
Se				$1/2^-$	$(5/2^-)$	$9/2^+$	$5/2^+$	$1/2^-$	$7/2^+$	$1/2^-$	$9/2^+$	$(5/2)^+$	$(3/2^+)$
Ge		$3/2^-$	$3/2^-$	$1/2^-$	$5/2^-$	$1/2^-$	$9/2^+$	$1/2^-$	$7/2^+$			$(5/2^+)$	
Zn	$3/2^-$	$3/2^-$	$3/2^-$	$5/2^-$	$5/2^-$	$1/2^-$	$1/2^-$	$1/2^-$	$7/2^+$	$7/2^+$	$9/2^+$	a	
Ni	$3/2^-$	$3/2^-$	$3/2^-$	$1/2^-$	$5/2^-$	$1/2^-$	$(9/2^+)$	$(9/2^+)$	$(9/2^+)$				

	55	57	59	61	63	65	67	69	71	73	75	77	79	81
Hf														
Yb													$(1/2^+)$	$(1/2^+)$
Er												$(1/2^+)$	$(1/2^+)$	$(1/2^+)$
Dy										$(7/2^+)$	$(9/2^-)$	$(1/2^+)$	$(1/2^+)$	$(1/2^+)$
Gd									$(5/2^+)$			$(1/2^+)$	$1/2^+$	$1/2^+$
Sm									$(5/2^+)$	$(7/2^+)$	$(9/2^-)$	$1/2^+$	$1/2^+$	$3/2^+$
Nd									$(5/2^+)$	$(7/2^+)$	$9/2^-$	$1/2^+$	$3/2^+$	$3/2^+$
Ce						$(5/2)$	$(7/2^-)$	$(1/2^+)$	$(5/2^+)$	$7/2^+$	$1/2^+$	$1/2^{(+)}$	$3/2^+$	$3/2^+$
Ba				$(3/2^+)$	$(5/2^+)$	$5/2^+$	$5/2^+$	$1/2^+$	$1/2^+$	$1/2^+$	$1/2^+$	$1/2^+$	$3/2^+$	$3/2^+$
Xe				$(5/2^+)$	$5/2^+$	$5/2^+$	$5/2^+$	$1/2^+$	$1/2^+$	$1/2^+$	$1/2^+$	$3/2^+$	$3/2^+$	$3/2^+$
Te		$(5/2^+)$	$(5/2)^+$	$(7/2^+)$	$7/2^+$	$1/2^+$	$1/2^+$	$1/2^+$	$1/2^+$	$1/2^+$	$3/2^+$	$3/2^+$	$3/2^+$	
Sn	$(5/2^+)$	$(5/2^+)$	$5/2^+$	$7/2^+$	$1/2^+$	$3/2^+$	$1/2^+$	$1/2^+$	$3/2^+$	$11/2^-$	$11/2^-$	$11/2^-$	$3/2^+$	$3/2^+$
Cd	$5/2^+$	$5/2^+$	$5/2^+$	$5/2^+$	$1/2^+$	$1/2^+$	$1/2^+$	$1/2^+$	$1/2^+$	$3/2^+$	$3/2^+$	$3/2^+$	$3/2^+$	$11/2^-$
Pd	$5/2^+$	$5/2^+$	$5/2^+$	$5/2^+$	$5/2^+$	$(5/2^+)$	$(1/2)^+$	$(3/2^+)$						
Ru	$5/2^+$	$5/2^+$	$3/2^+$	$3/2^+$	$(5/2)^+$	$(5/2^+)$	$5/2^+$	$(1/2^+)$	$(1/2^+)$					
Mo	$5/2^+$	$1/2^+$	$1/2^+$	$3/2^+$	$(5/2^-)$	$(1/2^+)$	$(1/2^+)$							
Zr	$5/2^+$	$1/2^+$	$1/2^+$	$3/2^+$	$(5/2^-)$									
Sr	$5/2^+$	$1/2^+$	$1/2^+$	$3/2^+$	$(5/2^-)$									
Kr	$5/2^+$	$1/2^+$	$1/2^+$											
Se														
Ge														
Zn														
Ni														

a $(1/2^+,\ 5/2^+)$

$$N$$

Table A.7. Ground-state spin-parities, odd N: $50 \leqslant Z \leqslant 82$, $83 \leqslant N \leqslant 125$.

	83	85	87	89	91	93	95	97	99	101	103	105	107	109	111	113	115	117	119	121	123	125
Ra															$(3/2^-)$	$3/2^-$	$3/2^-$			$5/2^-$	$5/2^-$	$1/2^-$
Rn														$3/2^-$	$3/2^-$	$3/2^-$	$3/2^-$	$3/2^-$	$5/2^-$	$5/2^-$	$5/2^-$	$1/2^-$
Po											a	$(5/2^-)$	$3/2^-$	$3/2^-$	$3/2^-$	$(3/2^-)$	$3/2^-$	$3/2^-$	$5/2^-$	$5/2^-$	$5/2^-$	$1/2^-$
Pb							$(9/2^-)$	$(9/2^-)$	$3/2^-$	$3/2^-$	$3/2^-$	$3/2^-$	$3/2^-$			$3/2^-$	$3/2^-$	$3/2^-$	$5/2^-$	$5/2^-$	$5/2^-$	$1/2^-$
Hg						$(7/2^-)$	$(7/2^-)$	$7/2^-$	$7/2^-$	$1/2^-$	$1/2^-$	$1/2^-$	$3/2^{(-)}$	$3/2^-$	$3/2^{(-)}$	$3/2^{(-)}$	$1/2^-$	$1/2^-$	$1/2^-$	$3/2^-$	$5/2^-$	$1/2^-$
Pt					$(7/2^-)$	$7/2^-$	$(5/2^-)$	$(7/2^-)$	$5/2^-$	$1/2^-$	$1/2^-$	$1/2^-$	$9/2^+$	$3/2^-$	$3/2^-$	$3/2^-$	$1/2^-$	$1/2^-$	$1/2^-$	$3/2^-$	$(5/2^-)$	$(1/2^-)$
Os			$(7/2^-)$	$7/2^-$	$(7/2^-)$	$7/2^-$	$(5/2^-)$	$(5/2^-)$	$5/2^-$	$(5/2^-)$	$1/2^-$	$1/2^-$	$1/2^-$	$9/2^+$	$1/2^-$	$1/2^-$	$3/2^-$	$9/2^-$	$3/2^-$	$(3/2^-)$		
W	$(7/2^-)$			$7/2^-$	$(5/2^-)$	$(5/2^-)$		$5/2^-$	$5/2^-$	$(1/2^-)$	$1/2^-$	$7/2^-$	$9/2^+$	$1/2^-$	$3/2^-$	$3/2^-$						
Hf		$7/2^-$	$7/2^-$	$(7/2^-)$	$(5/2^-)$	$(5/2^-)$	$(5/2)^-$	$(5/2^-)$	$7/2^+$	$1/2^-$	$5/2^-$	$7/2^-$	$9/2^+$	$1/2^-$	$(3/2^-)$	$(9/2^-)$						
Yb	$7/2^-$	$7/2^-$	$7/2^-$	$5/2^-$	$3/2^-$	$3/2^-$	$5/2^-$	$5/2^-$	$7/2^+$	$1/2^-$	$5/2^-$	$7/2^-$	$9/2^+$	$(1/2^-)$								
Er	$(7/2^-)$	$(7/2^-)$	$7/2^-$	$3/2^-$	$3/2^-$	$3/2^-$	$5/2^-$	$5/2^-$	$7/2^+$	$1/2^-$	$5/2^-$	$(7/2^-)$										
Dy	$7/2^-$	$7/2^-$	$7/2^-$	$3/2^-$	$3/2^-$	$3/2^-$	$5/2^+$	$5/2^-$	$7/2^+$	$(1/2^-)$	$(5/2)^-$											
Gd	$7/2^-$	$7/2^-$	$7/2^-$	$3/2^-$	$3/2^-$	$3/2^-$	$3/2^-$	$5/2^-$	$7/2^+$													
Sm	$7/2^-$	$7/2^-$	$7/2^-$	$5/2^-$	$3/2^+$	$3/2^-$		$5/2^-$														
Nd	$7/2^-$	$7/2^-$	$5/2^-$	$5/2^-$	$3/2^+$	$(3/2)^-$	$(3/2^-)$															
Ce	$7/2^-$	$3/2^-$		$(5/2^-)$		$(3/2^-)$																
Ba	$7/2^-$	$3/2^-$	$5/2^-$	$5/2^-$	$5/2^-$																	
Xe	$7/2^-$	$3/2^-$	$5/2^-$	$5/2^-$																		
Te	$(7/2^-)$																					
Sn	$7/2^-$																					

a $(1/2^-, 5/2^-)$

N

Table A.8. Ground-state spin-parities, odd N: $80 \leqslant Z \leqslant 110$, $127 \leqslant N \leqslant 159$.

	127	129	131	133	135	137	139	141	143	145	147	149	151	153	155	157	159
Hs																	
Sg											$(11/2^-)$						
Rf										$(9/2^-)$	$(1/2^+)$						
No									$(7/2^+)$	$9/2^-$	$(1/2^+)$	$(3/2^+)$					
Fm								$(7/2^+)$	$7/2^+$	$9/2^-$	$1/2^-$	$7/2^+$	$9/2^+$				
Cf							$(1/2^+)$	$1/2^+$	$(7/2^+)$	$9/2^-$	$1/2^+$	$(7/2^+)$	$(7/2^+)$				
Cm									$1/2^+$	$5/2^+$	$7/2^+$	$9/2^-$	$1/2^+$	$(3/2^+)$			
Pu						$(3/2^+)$		$(5/2^+)$	$7/2^-$	$1/2^+$	$5/2^+$	$7/2^+$	$(9/2^-)$				
U					$(3/2^+)$	$3/2^+$		$5/2^+$	$7/2^-$	$1/2^+$	$5/2^+$	$7/2^+$	$9/2^-$				
Th				$(5/2)^+$	$3/2^+$	$(1/2^+)$	$5/2^+$	$5/2^+$	$1/2^+$								
Ra	$9/2^+$	$(9/2^+)$	$(7/2)^+$	$5/2^+$	$3/2^+$	$1/2^+$	$3/2^+$	$5/2^+$	$(5/2^+)$								
Rn		$9/2^+$	$9/2^+$	$5/2^+$	$7/2^+$	$7/2^+$	$7/2^-$	$(3/2^+)$	$(5/2^+)$								
Po	$9/2^+$	$9/2^+$	$9/2^+$	$(9/2^+)$													
Pb	$9/2^+$	$9/2^+$	$(9/2^+)$														
Hg	$9/2^+$																

N

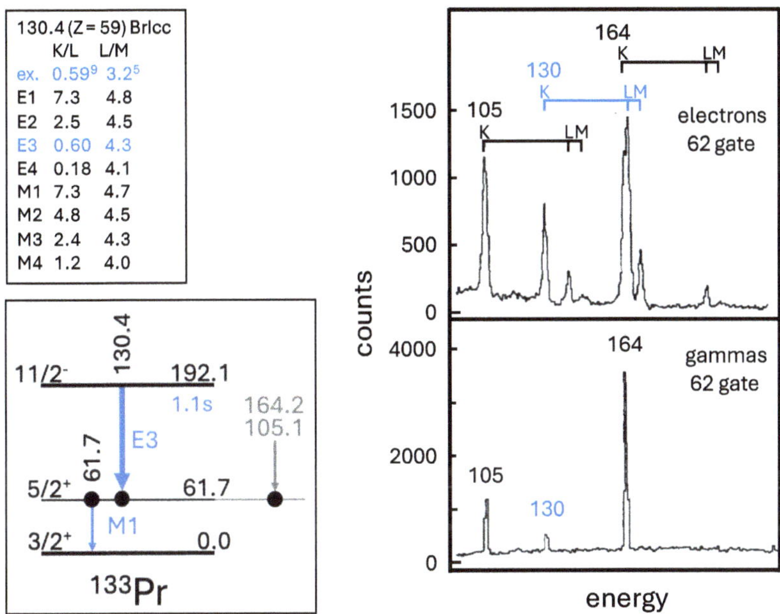

Figure A.1. Spectroscopic evidence for the $J^\pi = 11/2^-$ isomer in ^{133}Pr and indirectly, the basis for assigning $J^\pi = 3/2^+$ as the ground-state spin-parity. The spectroscopy is based on an E3 isomeric transition in prompt coincidence with a 62 keV transition. Conversion-coefficient ratios, computed using BrIcc, for a 130.4 keV transition in Pr ($Z = 59$) for all multipolarities of interest are tabulated. Note the unusual feature of an E3 K/L ratio less than unity. Reprinted from [2], copyright (1995) with the permission of Elsevier.

References

[1] Kondev F G *et al* 2021 The NUSBASE2020 evaluation of nuclear physics properties *Chin. Phys.* C **45** 030001

[2] Breitenbach J B, Wood J L, Jarrio M, Braga R A, Carter H K, Kormicki J and Semmes P B 1995 The decays of mass-separated ^{133m,g}Nd to ^{133}Pr *Nucl. Phys.* A **595** 481–98

[3] Ekström C, Ingelman S, Olsmats M, Wannberg B, Andersson G and Rosen A 1972 Nuclear spins of neutron-deficient Pr and Nd isotopes *Nucl. Phys.* A **196** 178–96

IOP Publishing

Nuclear Data
An independent-particle motion view
David Jenkins and John L Wood

Appendix B

Ground-state magnetic moments

These tabulations are based on magnetic moment values compiled in [1]. Because this compilation was published in 2005, where we are aware of new measurements, we have added to this compilation but without evaluation. We cite the sources of the added values (table B.1–B.8).

Table B.1. Ground-state magnetic moments in μ_N, odd Z: $1 \leqslant Z \leqslant 27$, $0 \leqslant N \leqslant 50$.

positive values
negative values
no sign

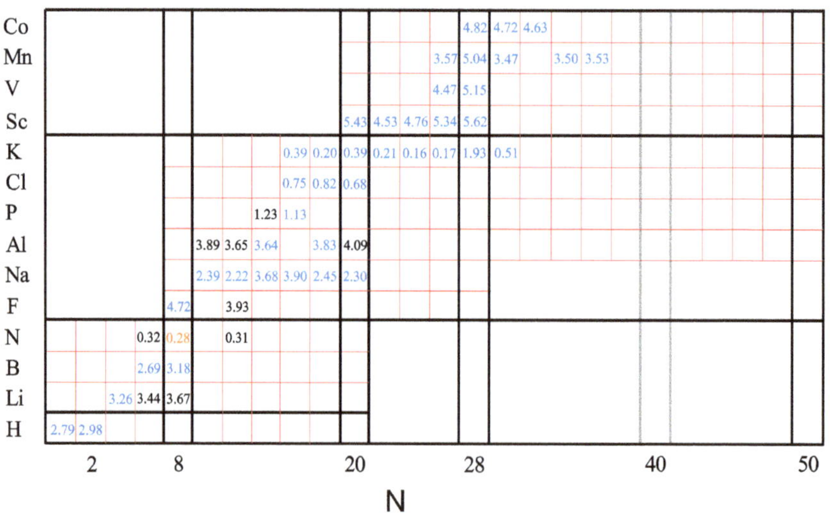

doi:10.1088/978-0-7503-5648-0ch8

Table B.2. Ground-state magnetic moments in μ_N, odd Z: $29 \leqslant Z \leqslant 73$, $28 \leqslant N \leqslant 82$.

positive values
negative values
no sign

Z \ N	28	30	32	34	36	38	40	42	44	46	48	50	52	54	56	58	60	62	64	66	68	70	72	74	76	78	80	82
Ta																												
Lu																												
Tm																												
Ho																												
Tb																												1.71
Eu																									6.1	3.49	3.67	3.99
Pm																												3.8
Pr																												4.28
La																										3.70	2.70	2.78
Cs																			5.46	0.77	1.38	1.41	1.46	1.49	3.54	2.58	2.73	2.84
I																			3.1	2.9	2.3	2.82	2.82	2.81	2.62	2.74	2.86	2.94
Sb																			3.46	3.43	3.45	3.36	2.55	2.59	2.70	2.82	2.89	3.07
In															5.68	5.59	5.54	5.50	5.53	5.54	5.52	5.52	5.50	5.49	5.50	5.53	5.60	6.31
Ag												6.13	5.81	5.63	4.47	**0.10**	*0.11*	*0.13*	*0.15*	**0.16**								
Rh																*0.09*	4.45											
Tc												6.32	5.94	5.80	5.68													
Nb											6.22	6.52	6.17	6.14	6.15	5.97												
Y											*0.19*	*0.14*	**0.16**	*0.14*	*0.16*	*0.12*	3.18	3.22										
Rb							0.65	3.36	2.06	1.42	1.35	2.75	2.38	2.18	1.41	1.33	1.84											
Br							0.76	0.97	2.11	2.27																		
As								1.44	1.29																			
Ga				1.77	1.85	2.02	2.56	0.21	1.84	2.02	1.05	1.75																
Cu	2.58	1.90	2.11	2.22	2.38	2.51	2.84	2.27	1.74	1.01	1.59																	

N

Table B.3. Ground-state magnetic moments in μ_N, odd Z: $51 \leqslant Z \leqslant 89$, $82 \leqslant N \leqslant 126$.

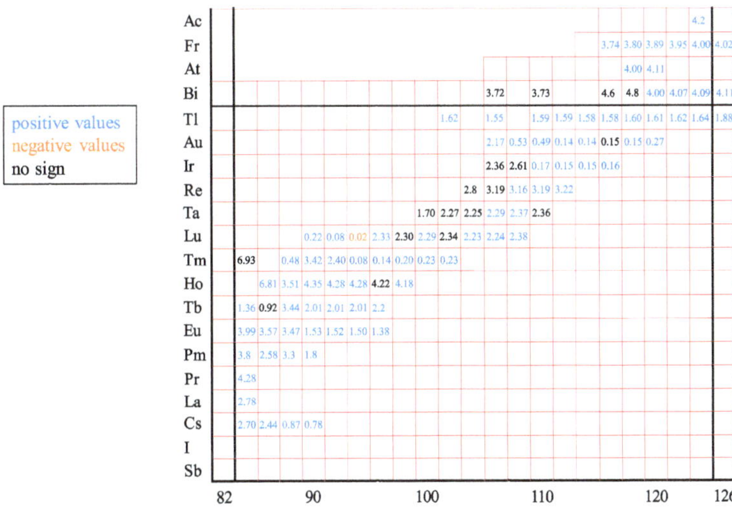

positive values
negative values
no sign

Z \ N	82	84	86	88	90	92	94	96	98	100	102	104	106	108	110	112	114	116	118	120	122	124	126
Ac																						4.2	
Fr																		3.74	3.80	3.89	3.95	4.00	4.02
At																			4.00	4.11			
Bi													**3.72**		**3.73**			**4.6**	**4.8**	4.00	4.07	4.09	4.11
Tl											1.62		1.55		1.59	1.59	1.58	1.58	1.60	1.61	1.62	1.64	1.88
Au													2.17	0.53	0.49	0.14	0.14	**0.15**	0.15	0.27			
Ir													**2.36**	**2.61**	0.17	0.15	0.15	0.16					
Re												**2.8**	**3.19**	3.16	3.19	3.22							
Ta										**1.70**	**2.27**	**2.25**	2.29	2.37	**2.36**								
Lu					0.22	0.08	*0.02*	2.33	**2.30**	2.29	**2.34**	2.23	2.24	2.38									
Tm		**6.93**		0.48	3.42	2.40	0.08	0.14	0.20	0.23	0.23												
Ho			6.81	3.51	4.35	4.28	4.28	**4.22**	4.18														
Tb		1.36	**0.92**	3.44	2.01	2.01	2.01	2.2															
Eu		3.99	3.57	3.47	1.53	1.52	1.50	1.38															
Pm		3.8	2.58	3.3	1.8																		
Pr		4.28																					
La		2.78																					
Cs		2.70	2.44	0.87	0.78																		
I																							
Sb																							

N

Table B.4. Ground-state magnetic moments in μ_N, odd Z: $83 \leqslant Z \leqslant 101$, $126 \leqslant N \leqslant 158$. The data for ^{253,255}Es are from [2].

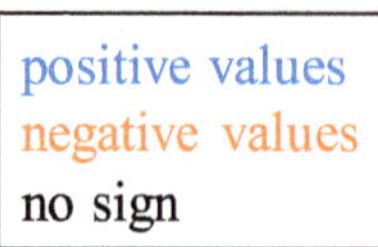

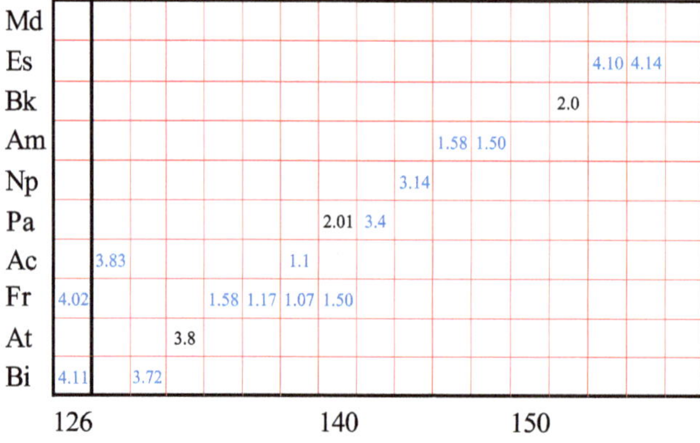

Table B.5. Ground-state magnetic moments in μ_N, odd N: $2 \leqslant Z \leqslant 26$, $1 \leqslant N \leqslant 49$.

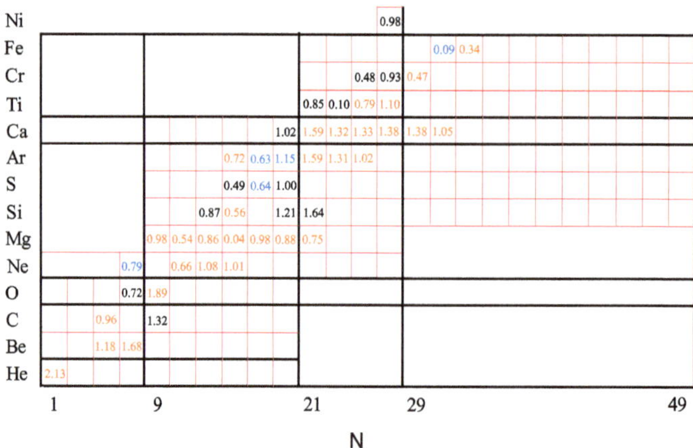

Table B.6. Ground-state magnetic moments in μ_N, odd N: $28 \leqslant Z \leqslant 72$, $29 \leqslant N \leqslant 81$.

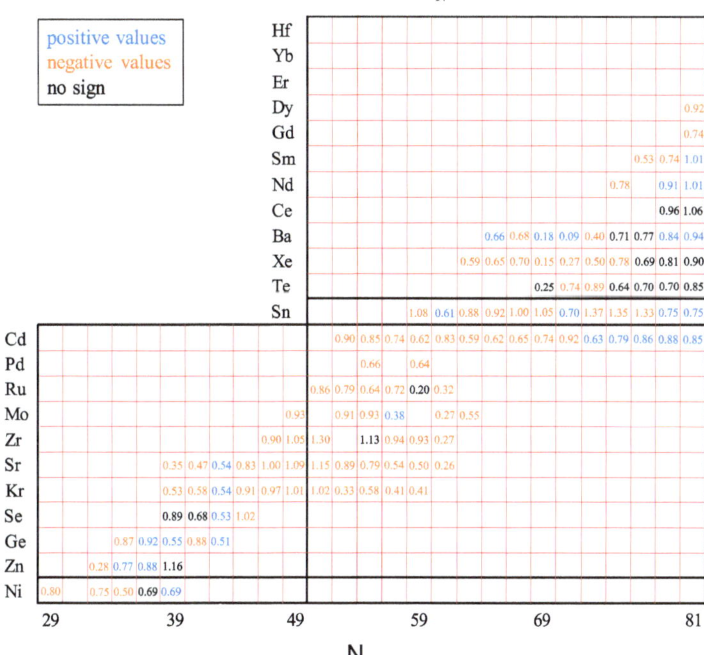

(Colour legend on the chart: blue = positive values, orange = negative values, black = no sign.)

Columns are neutron number N; rows are the element. Part A covers $N = 29$–53, Part B covers $N = 55$–81.

Part A ($N = 29$–53):

Z	29	31	33	35	37	39	41	43	45	47	49	51	53
Hf													
Yb													
Er													
Dy													
Gd													
Sm													
Nd													
Ce													
Ba													
Xe													
Te													
Sn													
Cd													0.90
Pd													
Ru												0.86	0.79
Mo											0.93		0.91
Zr										0.90	1.05	1.30	
Sr						0.35	0.47	0.54	0.83	1.00	1.09	1.15	0.89
Kr						0.53	0.58	0.54	0.91	0.97	1.01	1.02	0.33
Se						0.89	0.68	0.53	1.02				
Ge				0.87	0.92	0.55	0.88	0.51					
Zn			0.28	0.77	0.88	1.16							
Ni	0.80		0.75	0.50	0.69	0.69							

Part B ($N = 55$–81):

Z	55	57	59	61	63	65	67	69	71	73	75	77	79	81
Hf														
Yb														
Er														
Dy														0.92
Gd														0.74
Sm												0.53	0.74	1.01
Nd											0.78		0.91	1.01
Ce													0.96	1.06
Ba						0.66	0.68	0.18	0.09	0.40	0.71	0.77	0.84	0.94
Xe					0.59	0.65	0.70	0.15	0.27	0.50	0.78	0.69	0.81	0.90
Te								0.25	0.74	0.89	0.64	0.70	0.70	0.85
Sn			1.08	0.61	0.88	0.92	1.00	1.05	0.70	1.37	1.35	1.33	0.75	0.75
Cd	0.85	0.74	0.62	0.83	0.59	0.62	0.65	0.74	0.92	0.63	0.79	0.86	0.88	0.85
Pd	0.66		0.64											
Ru	0.64	0.72	0.20	0.32										
Mo	0.93	0.38		0.27	0.55									
Zr	1.13	0.94	0.93	0.27										
Sr	0.79	0.54	0.50	0.26										
Kr	0.58	0.41	0.41											
Se														
Ge														
Zn														
Ni														

Table B.7. Ground-state magnetic moments in μ_N, odd N: $50 \leqslant Z \leqslant 82$, $83 \leqslant N \leqslant 125$.

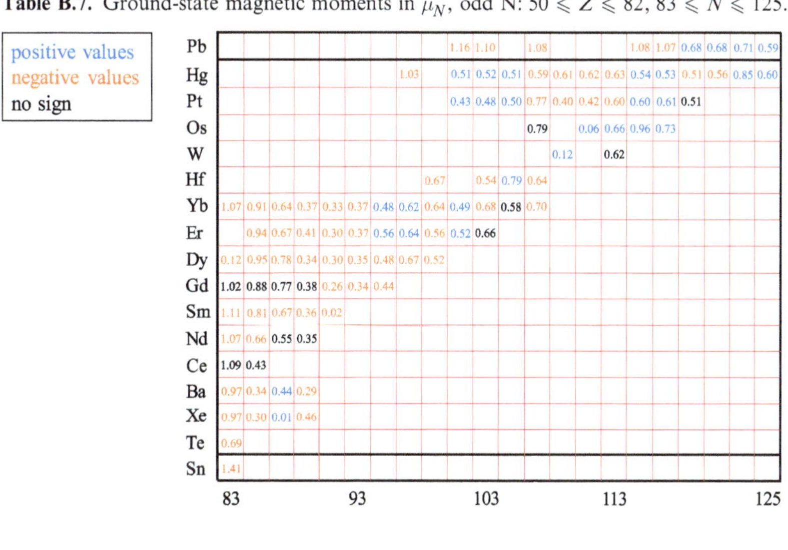

(Colour legend on the chart: blue = positive values, orange = negative values, black = no sign.)

Columns are neutron number N; rows are the element. Part A covers $N = 83$–103, Part B covers $N = 105$–125.

Part A ($N = 83$–103):

Z	83	85	87	89	91	93	95	97	99	101	103
Pb											1.16
Hg								1.03		0.51	0.52
Pt										0.43	0.48
Os											
W											
Hf									0.67		0.54
Yb	1.07	0.91	0.64	0.37	0.33	0.37	0.48	0.62	0.64	0.49	0.68
Er		0.94	0.67	0.41	0.30	0.37	0.56	0.64	0.56	0.52	0.66
Dy	0.12	0.95	0.78	0.34	0.30	0.35	0.48	0.67	0.52		
Gd	1.02	0.88	0.77	0.38	0.26	0.34	0.44				
Sm	1.11	0.81	0.67	0.36	0.02						
Nd	1.07	0.66	0.55	0.35							
Ce	1.09	0.43									
Ba	0.97	0.34	0.44	0.29							
Xe	0.97	0.30	0.01	0.46							
Te	0.69										
Sn	1.41										

Part B ($N = 105$–125):

Z	105	107	109	111	113	115	117	119	121	123	125
Pb	1.10			1.08		1.08	1.07	0.68	0.68	0.71	0.59
Hg	0.51	0.59	0.61	0.62	0.63	0.54	0.53	0.51	0.56	0.85	0.60
Pt	0.50	0.77	0.40	0.42	0.60	0.60	0.61	0.51			
Os	0.79			0.06	0.66	0.96	0.73				
W			0.12		0.62						
Hf	0.79	0.64									
Yb	0.58	0.70									
Er											
Dy											
Gd											
Sm											
Nd											
Ce											
Ba											
Xe											
Te											
Sn											

Table B.8. Ground-state magnetic moments in μ_N, odd N: $80 \leqslant Z \leqslant 106$, $127 \leqslant N \leqslant 157$. The data for ^{249,251,253}Cf are taken from [3] and the data for ^{253}No are from [4].

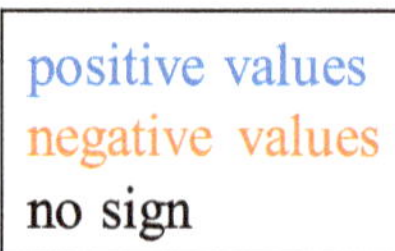

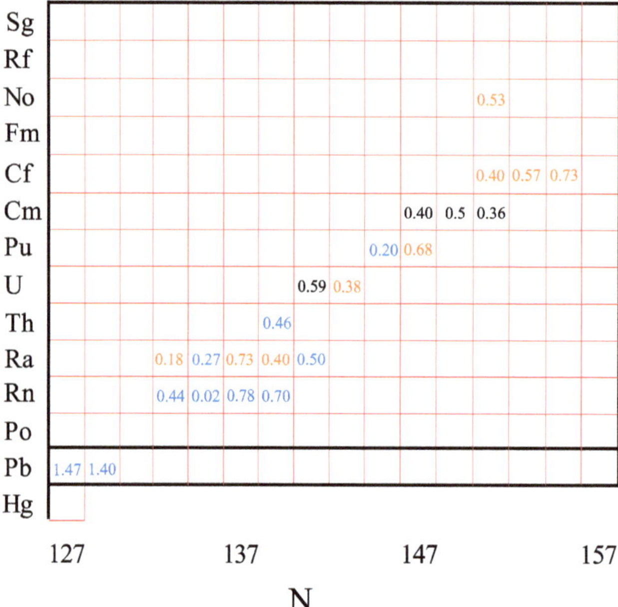

References

[1] Stone N J 2005 table of nuclear magnetic dipole and electric quadrupole moments *At. Data Nucl. Data tables* **90** 75–176

[2] Nothhelfer S *et al* 2022 Nuclear structure investigations of $^{253-255}$Es by laser spectroscopy *Phys. Rev.* C **105** L021302

[3] Weber F *et al* 2023 Nuclear moments and isotope shifts of the actinide isotopes $^{249-253}$Cf probed by laser spectroscopy *Phys. Rev.* C **107** 034313

[4] Raeder S *et al* 2018 Probing sizes and shapes of Nobelium isotopes by laser spectroscopy *Phys. Rev. Lett.* **120** 232503

IOP Publishing

Nuclear Data
An independent-particle motion view
David Jenkins and John L Wood

Appendix C

Mixing of Nilsson model quantum numbers

There are two primary sources of mixing for Nilsson model quantum numbers: $\Delta\Omega = \pm 1$ mixing which stems from the $\mathbf{I} \cdot \mathbf{j}$ term, cf equation (4.6); $\Delta N = \pm 2$ mixing which stems from the V_{def} term, cf equation (3.6). The occurrence of $\Delta\Omega = \pm 1$ mixing is singularly evident for $K = 1/2$ band energies because of a 'self-interaction' energy which results from the coupling of the $K = +1/2$ and $K = -1/2$ ($\Omega = +1/2$ and $\Omega = -1/2$) components of the symmetrized linear combination. This is already introduced in chapter 4.

There are further effects of $\Delta\Omega = \pm 1$ mixing on electromagnetic properties of Nilsson bands and on single-nucleon transfer reaction fingerprints, also introduced in chapter 5. Other occurrences of $\Delta\Omega = \pm 1$ mixing are identifiable in more narrowly defined structural situations. Notably, mixing between bands that differ in $K(=\Omega)$ quantum number by one unit, with identical parities. There is also an impact of $\Delta\Omega = \pm 1$ mixing on spectroscopic factors, C_{jl} coefficients, recall equation (4.9),

$$|\Omega\pi\rangle = \sum_j C_{jl} |Nl\,j\Omega\pi\rangle, \tag{C.1}$$

as observed in the need to modify one-nucleon transfer reaction fingerprints, cf figures 4.12(a) and (b). Some details of such occurrences are given shortly. Algebraic derivations involved in the theory of $\Delta\Omega = \pm 1$ mixing are presented in section C.1.

Occurrences of $\Delta N = \pm 2$ mixing are relatively rare and are most pronounced in spectroscopic factors for one-nucleon transfer reactions. This is most directly seen in fragmentation of one-nucleon transfer reaction strength involving specific Nilsson configurations. Other sources of mixing of Nilsson bands are observed. They are generally weak and appear to be the result of diverse effects without clear categories. These occurrences of mixing are beyond the scope of the present narrative.

doi:10.1088/978-0-7503-5648-0ch9　　　　　C-1　　　　　© IOP Publishing Ltd 2024. All rights, including for text and data mining (TDM), artificial intelligence (AI) training, and similar technologies, are reserved.

C.1 $\Delta\Omega = \pm 1$ mixing of Nilsson bands

The occurrence of $\Delta\Omega = \pm 1$ mixing stems from the $\mathbf{I} \cdot \mathbf{j}$ term in the rotational particle coupling, RPC model view of deformed nuclei. This emerges directly from the decomposition,

$$\mathbf{I} \cdot \mathbf{j} = 1/2(I_+ j_- + I_- j_+) + I_3 j_3, \tag{C.2}$$

where

$$I_{\pm} = I_1 \pm i I_2 \tag{C.3}$$

and

$$j_{\pm} = j_1 \pm i j_2. \tag{C.4}$$

The operators $I_{\pm}$ act on the rotational eigenstates, $|IMK\rangle$ as

$$I_{\pm}|IMK\rangle = \sqrt{(I \pm K)(I \mp K + 1)}\,\hbar|IMK \mp 1\rangle, \tag{C.5}$$

where note that I_+ has a lowering action on K and I_- has a raising action on K (this is derived in appendix A of [1]). The operators $j_{\pm}$ act on the independent-particle motion basis states, $|j\Omega\pi\rangle$ as

$$j_{\pm}|j\Omega\pi\rangle = \sqrt{(j \mp \Omega)(j \pm \Omega + 1)}\,\hbar|j\Omega \pm 1, \pi\rangle. \tag{C.6}$$

The operators $I_+ j_-$ and $I_- j_+$ act to couple rotational and independent-particle motions. From equations (4.8) and (4.9), viz.

$$|IMK; \Omega\pi\rangle = \frac{1}{\sqrt{2}}\{|IM, +K\rangle|+\Omega\pi\rangle + (-1)^{I+K}|IM, -K\rangle|-\Omega\pi\rangle\} \tag{C.7}$$

$$|\Omega\pi\rangle = \sum_{jl} C_{jl}|Nl\,j\Omega\pi\rangle, \tag{C.8}$$

with the abbreviation $|IMK\rangle = |IK\rangle$, two important actions of the RPC term are identified,

$$I_+ j_-|IK; \Omega\pi\rangle = I_+|IK\rangle j_-|\Omega\pi\rangle, \tag{C.9}$$

whence, from equation (C.5) the first action is

$$I_+|IK\rangle = \sqrt{(I + K)(I - K + 1)}\,\hbar|IMK, -1\rangle, \tag{C.10}$$

and from equation (C.6), with the abbreviation $|Njl\Omega\pi\rangle = |j\Omega\rangle$, the second action is

$$\begin{aligned}
j_-|\Omega\pi\rangle &= j_-\sum_{jl} C_{jl}|j\Omega\rangle \\
&= \sum_{jl} C_{jl} j_-|j\Omega\rangle \\
&= \sum_{jl} C_{jl}\sqrt{(j + \Omega)(j - \Omega + 1)}\,\hbar|j\Omega - 1\rangle.
\end{aligned} \tag{C.11}$$

Matrix elements of the RPC term then follow, here retaining the separation of the I_+ and j_- operators as factors,

$$\langle IK - 1|I_+|IK\rangle = \sqrt{(I + K)(I - K + 1)}\,\hbar \tag{C.12}$$

and

$$
\begin{aligned}
\langle \Omega - 1\pi|j_-|\Omega\pi\rangle &= \sum_{jl} C_{jl}^{\Omega-1} C_{jl}^{\Omega}\langle j\Omega - 1|j_-|j\Omega\rangle \\
&= \sum_{jl} C_{jl}^{\Omega-1} C_{jl}^{\Omega}(-1)^{j-1/2}\sqrt{(j + \Omega)(j - \Omega + 1)}\,\hbar,
\end{aligned}
\tag{C.13}
$$

Note that the operator j_- cannot change N, j, l or π, and the C_{jl} coefficients are real.

The important consequences of the RPC term can be classified into effects on $K = 1/2$ bands and mixing effects on bands differing in K by one unit. These effects are treated separately in the following two sections.

C.2 $K = 1/2$ bands and $\Delta\Omega = \pm 1$ mixing

The effect of $\Delta\Omega = \pm 1$ mixing on $K = 1/2$ bands is distinctive. The most evident effects are depicted in figures 4.7(a) and (b) and 4.8; the effect on energies is quantified in equation (4.10), viz.

$$E = A[I(I + 1) + a(-1)^{I+1/2}(I + 1/2)]. \tag{C.14}$$

The details follow directly from equation (C.12) with $K = \Omega = 1/2$, i.e.

$$
\begin{aligned}
\langle I, -1/2|I_+|I, +1/2\rangle &= \sqrt{(I + 1/2)(I - 1/2 + 1)}\,\hbar \\
&= (I + 1/2)\hbar,
\end{aligned}
\tag{C.15}
$$

and for $\Omega = +1/2(=K)$, equation (C.13) becomes

$$
\begin{aligned}
\langle -1/2, \pi|j_-|+1/2, \pi\rangle &= \sum_{jl} C_{jl}\, C_{jl}(-1)^{j-1/2}\sqrt{(j + 1/2)(j - 1/2 + 1)}\,\hbar \\
&= \sum_{jl} C_{jl}^2(j + 1/2)(-1)^{j-1/2}\hbar.
\end{aligned}
\tag{C.16}
$$

Thus, the $\mathbf{I} \cdot \mathbf{j}$ term yields the matrix element

$$\langle I, -1/2; -1/2, \pi|I_+j_-|I, +1/2;+1/2, \pi\rangle = (I + 1/2)\hbar^2\sum_{jl} C_{jl}^2(j + 1/2)(-1)^{j-1/2} \tag{C.17}$$

and this leads to an energy contribution from the RPC term, $-\mathbf{I} \cdot \mathbf{j}/\mathcal{J}$

$$E_{RPC} = -(-1)^{I+1/2}(I + 1/2)\sum_{jl} C_{jl}^2(j + 1/2)(-1)^{j-1/2}\hbar^2/2\mathcal{J}. \tag{C.18}$$

(The RPC term possesses a factor $I_3 j_3$, which yields a term proportional to K^2: this term is ignored herein as it contributes only a constant term to a rotational band.) Introducing the definition

$$a = -\sum_{jl} C_{jl}^2 (j + 1/2)(-1)^{j-1/2}, \qquad (C.19)$$

the rotational energy formula for a $K = 1/2$ band, equation (C.14) is obtained.

The effect of $\Delta\Omega = \pm 1$ mixing on the electromagnetic properties of $K = 1/2$ bands is more complex and lies beyond the scope of the present narrative.

C.3 Nilsson band $\Delta\Omega = \pm 1$ interband mixing

Mixing occurs between bands differing in K $(= \Omega)$ by ± 1. This is particularly important when such pairs of bands are close in energy. It is also important between Nilsson bands stemming from a high j value, between the multiplet members with $\Omega = 1/2, 3/2, 5/2, \ldots$ Details follow directly from equations (C.12) and (C.13). It is standard practice to combine these equations in the form

$$\langle I, K - 1; \Omega - 1, \pi | I_+ j_- | IK; \Omega\pi \rangle = \sqrt{(I + K)(I - K + 1)}\, \hbar \sum_{jl} C_{jl}^{\Omega-1} C_{jl}^{\Omega} \langle j, \Omega - 1 | j_- | j\Omega \rangle, \quad (C.20)$$

and to tabulate the matrix elements

$$\langle j\Omega - 1 | j_- | j\Omega \rangle = \langle jK - 1 | j_- | jK \rangle := A_{K, K-1}. \qquad (C.21)$$

The details of interband mixing applied to specific nuclei is a more specialized topic and we limit details to the results of specific example which is shown in figure C.1.

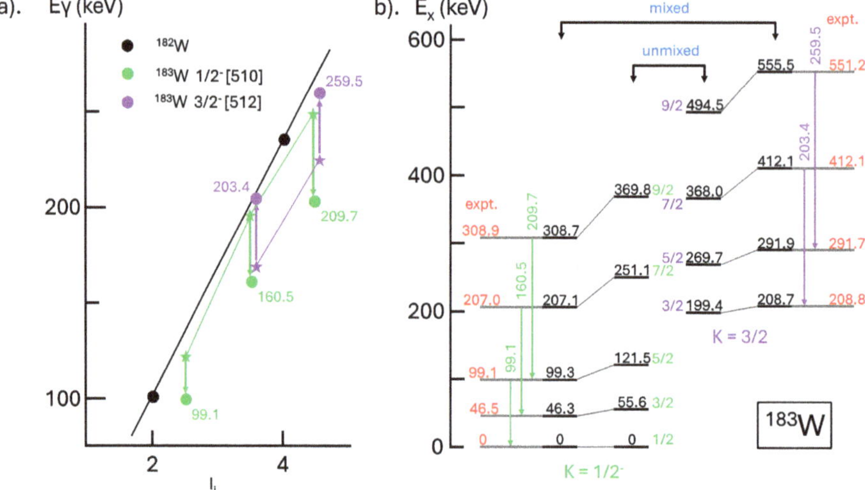

Figure C.1. An example of two-band mixing resulting from the rotation-particle coupling term. (a) Rotation-alignment plot for the 1/2⁻[510] and 3/2⁻[512] Nilsson bands in ¹⁸³W. (b) Details of the two-band mixing for the 1/2⁻[510] and 3/2⁻[512] Nilsson bands in ¹⁸³W. The theoretical energies are taken from [2]. Note: there is only one spin-1/2 state and so this remains unmixed and provides the energy-zero reference; thus, the energy shifts due to mixing can be identified, e.g. for the spin-3/2 states the unmixed configurations are 'repelled' by 9.3 keV. The $\Delta I = 2$ transition energies are colour coded: green for the 1/2⁻[510] band and violet for the 3/2⁻[512] band. Further, circles depict observed transition energies and stars depict the unmixed transition energies, e.g. 7/2⁻ → 3/2⁻ is $251.1 - 55.6 = 195.5$ keV in the [510] band. The resulting effect on E_γ values in (a) is shown using vertical arrows.

References

[1] Jenkins D G and Wood J L 2023 *Nuclear Data: A Collective Motion View* (Bristol: IOP Publishing)

[2] Stephens F S 1975 Coriolis effects and rotation alignment in nuclei *Rev. Mod. Phys.* **47** 43

www.ingramcontent.com/pod-product-compliance
Ingram Content Group UK Ltd.
Pitfield, Milton Keynes, MK11 3LW, UK
UKHW051938150726
7214IPUK00005B/42